Progress in Biological Control

Volume 19

Series editor
Heikki M.T. Hokkanen
Department of Applied Zoology, University of Helsinki, Helsinki, Finland

Biological control of pests, weeds, and plant and animal diseases utilising their natural antagonists is a well-established but rapidly evolving field of applied ecology. Despite its documented applications and systematic development efforts for longer than a century, biological control still remains a grossly underexploited method of pest management. Its formidable potential represents the best hope to providing lasting, environmentally sound, and socially acceptable control of most problem pests in agriculture, and of invasive alien organisms threatening global biodiversity. Based on the overwhelmingly positive features of biological control, it is the prime candidate in the search for reducing dependency on chemical pesticides. Public demand for finding solutions based on biological control is the main driving force in the rapid development in the various strategies of utilising natural enemies for controlling noxious organisms. This book series is intended to accelerate these developments through exploring the progress made within the various aspects of biological control, and via documenting these advances to the benefit of fellow scientists, students, public officials, and the public at large. Each of the books in this series is expected to provide a comprehensive, authoritative synthesis of the topic, likely to stand the test of time.

More information about this series at http://www.springer.com/series/6417

Daniel Carrillo • Gilberto José de Moraes
Jorge E. Peña
Editors

Prospects for Biological Control of Plant Feeding Mites and Other Harmful Organisms

Springer

Editors
Daniel Carrillo
Department of Entomology
and Nematology
Tropical Research and Education Center
University of Florida - IFAS
Homestead, FL, USA

Gilberto José de Moraes
Departamento de Entomologia e Acarologia
Escola Superior de Agricultura "Luiz de Queiroz"
Universidade de São Paulo
Piracicaba, SP, Brazil

Jorge E. Peña
Department of Entomology
and Nematology
Tropical Research and Education Center
University of Florida - IFAS
Homestead, FL, USA

Progress in Biological Control
ISBN 978-3-319-15041-3 ISBN 978-3-319-15042-0 (eBook)
DOI 10.1007/978-3-319-15042-0

Library of Congress Control Number: 2015937360

Springer Cham Heidelberg New York Dordrecht London

Printed on acid-free paper

Springer International Publishing AG Switzerland is part of Springer Science+Business Media (www.springer.com)

Obituary Maus Sabelis (1950–2015)

Maurice Sabelis passed away on January 7, 2015, after having been ill for 2 years.

He started his scientific career with a PhD in theoretical production ecology at Wageningen University in The Netherlands, where he was subsequently appointed assistant professor. After being employed as associate professor of Animal Ecology at the University of Leiden for a period of 4 years, he became full professor in Population Biology at the University of Amsterdam.

Maurice was a scientist in heart and soul, a Jack of all trades and a master of many. He was a theoretician, an experimenter and a modeller, and covered ecology, evolutionary biology, epidemiology, acarology, biological control and behavioural ecology. With others, he discovered how plants defend themselves against herbivores by recruiting predators. His research also resulted in the discovery of various natural enemies that are currently used in biological control of pests. In 2006, he was awarded the prize of Academy Professor of the Royal Netherlands Academy of Arts and Sciences.

A creative and innovative scientist, Maurice was a source of inspiration for many young scientists all over the world. Many of the dozens of PhD students that were supervised by him are now authorities in their areas of scientific expertise. His network of collaborators encompassed the globe. To be sure, they will all dearly miss his stimulating, critical and warm personality.

by Arne Janssen, Associate professor, University of Amsterdam

Series Preface

Biological control of pests, weeds, and plant and animal diseases utilising their natural antagonists is a well-established and rapidly evolving field of science. Despite its stunning successes worldwide and a steadily growing number of applications, biological control has remained grossly underexploited. Its untapped potential, however, represents the best hope to providing lasting, environmentally sound, and socially acceptable pest management. Such techniques are urgently needed for the control of an increasing number of problem pests affecting agriculture and forestry and to suppress invasive organisms which threaten natural habitats and global biodiversity.

Based on the positive features of biological control such as its target specificity and the lack of negative impacts on humans, it is the prime candidate in the search for reducing dependency on chemical pesticides. Replacement of chemical control by biological control—even partially as in many IPM programs—has important positive but so far neglected socioeconomic, humanitarian, environmental, and ethical implications. Change from chemical to biological control substantially contributes to the conservation of natural resources and results in a considerable reduction of environmental pollution. It eliminates human exposure to toxic pesticides, improves sustainability of production systems, and enhances biodiversity. Public demand for finding solutions based on biological control is the main driving force in the increasing utilization of natural enemies for controlling noxious organisms.

This book series is intended to accelerate these developments through exploring the progress made within the various aspects of biological control and via documenting these advances to the benefit of fellow scientists, students, public officials, policy-makers, and the public at large. Each of the books in this series is expected to provide a comprehensive, authoritative synthesis of the topic, likely to stand the test of time.

Helsinki, Finland — Heikki M.T. Hokkanen

Preface

This book is intended to facilitate future investigations on the potential implementation of biological control of harmful organisms using predatory mites. These beneficial arthropods represent an important resource for development of biological control programs, a subject of considerable interest for acarologists, entomologists, integrated pest management, and medical and veterinarian specialists who are actively searching for efficient and sustainable methods of pest control.

Applied biological control must be based on fundamental studies that elucidate the taxonomy, biology, mode of reproduction, behavior, prey range, and rearing techniques of the potential biological control agent. These studies are considered as the prelude for ecological field research designed to evaluate the relative importance of predatory mites with respect to their regulation of the target pest. The above mentioned knowledge must configure a solid basis to develop effective classical, augmentative, or conservation biological control programs. This book provides information on the levels of knowledge on several groups of mites within the Mesostigmata and Prostigmata, with regard to their current or potential use in biological control.

Mesostigmata or Gamasida predators are encountered in soil and litter, on aerial parts of plants, or in patchy habitats such as nests or galleries of insects, mammals, and birds, where they feed on small insects and nematodes or on phytophagous and mycophagous mites, showing a remarkable range of morphological and behavioral traits. Most of them are free-living predators, but some have established symbiotic relationships with other arthropods and disperse by phoresy. The first five chapters of this book refer to different groups of mesostigmatid predators. Chapter 1 provides a general description of the morphological and biological characteristics of species in this order and specifically mites in the superfamilies Rhodacaroidea and Parasitoidea, which are encountered in the soil or litter, or associated with insect galleries and mammal nests. Trends in efforts to understand the diversity, distribution, habitats, ecology, reproduction, and the practical use of species within these groups in biological control efforts are discussed. In addition, this covers other groups of the infraorder Gamasina that have received less attention than the former, including the hyporders Epicriiae, Arctacariae, Dermanyssiae and Parasitiae, and the suborders Trigynaspida and Sejida. Chapter 2 reviews the potential of individuals from the families Ascidae, Blattisociidae, and Melicharidae for control of damaging mites and small insects. Special emphasis is given to their association with other organisms, dispersion mechanisms, feeding habits, and general biology. Mites within the genera *Blattisocius* Keegan, *Lasioseius* Berlese, and *Proctolaelaps* Berlese, which have been the subject of more detailed studies, are discussed here. Chapter 3 addresses predators within the Laelapidae family with a special focus on free-living predators of the subfamily Hypoaspididae, which are successfully used in biological control of thrips (Thysanoptera: Thripidae), sciarid flies (Diptera: Sciaridae), nematodes (Rhabditida: Steinernematidae: Heterorhabditidae; Tylenchida: Tylenchulidae), and poultry mites (Acari: Dermanyssidae). The Macrochelidae family (Chap. 4) is composed of a diverse group of predators that live mostly in animal excrements. Representatives of these groups inhabit mostly patchy transient habitats and thus rely on other organisms for dispersal by phoresis. The interesting dispersal mechanisms displayed by macrochelids and the implications that they have on biocontrol programs are discussed in detail. These mites have great potential for control of different fly species (Diptera: Muscidae), thrips, and fungus gnats (Diptera: Sciaridae). The last group of mesostigmatid mites covered in this book is the family Phytoseiidae (Chap. 5). This group is referred to as the most important family of plant inhabiting predatory mites, which have been extensively used as biological control agents for management of mite and insect pests of crops grown in greenhouses and in open field conditions. This chapter provides statistics about the taxonomy, distribution, and habitats, together with a summary of the most recent investigations on specific species of phytoseiids.

Prostigmata predators are also found in diverse habitats, including soils and overlaying litter layers as well as aerial parts of plants. Chapter 6 provides a general description of the morphological and biological characteristics of the predatory species in the suborder, which also has a large number of phytophagous mites. Subsequently, Chap. 7 focuses on the generalist predators of the family Stigmaeidae, providing a review of the taxonomy, basic biology, and reproduction patterns observed in the group. The potential target pests, efficacy studies, seasonality, mass-rearing techniques, compatibility with other IPM practices, and the economic and commercial considerations of using stigmaeids in biological control are discussed. Chapter 8 deals with the Erythraeidae family, with a special reference to the potential use of the red velvet mite *Balaustium leanderi* (Heitlinger) as a biological control agent of spider mites, whiteflies, and thrips, pests of flower and vegetable crops in Colombia. The last family of prostigmatid predators covered in this book is the Cheyletidae (Chap. 9). This provides general information on the taxonomy and behavior of these predators, with special reference to *Cheyletus* species used for control of various arthropods that damage stored products.

In order to substantiate the chapters referring to larger taxonomical groups, Chaps. 10 and 11 document long-term investigations on the practical use of predatory mites targeting pest mites. The first example (Chap. 10) describes an intelligent approach referred to as "food web engineering" in which selected cover crops are used to enhance biological control of *Tetranychus urticae* (Tetranychidae) by resident phytoseiid mites in citrus orchards in Valencia, Spain. Chapter 11 documents multiple tactics used throughout decades of sustained efforts to biologically manage a number of spider mites on avocado in California, USA. This chapter discusses the use of several native and exotic phytoseiid predators and reviews the value of stigmaeid mites as predators. Lastly, Chap. 12 deals with pathogenic fungi that are highly specific natural enemies of pest mites and discusses their compatibility and complementary role with the use of predatory mites.

This book was possible because of the generous efforts by a group of proactive acarologists and biocontrol practitioners from around the globe in response to the growing public demand for a sustainable agriculture.

Homestead, FL, USA — Daniel Carrillo
Piracicaba, SP, Brazil — Gilberto José de Moraes
Homestead, FL, USA — Jorge E. Peña

Contents

Chapter 1
Mesostigmata as Biological Control Agents, with Emphasis on Rhodacaroidea and Parasitoidea

Raphael C. Castilho, Renan Venancio, and João Paulo Z. Narita

1.1 Introduction

The Mesostigmata is a large mite order consisting of about 11,500 species, corresponding to about 20 % of all known mite species (Beaulieu et al. 2011; Zhang 2013). These are found in different habitats, but are most abundant and diverse in the soil.

The life cycle of these mites most often includes the following developmental stages: egg, larva, protonymph, deutonymph and adults of both sexes. They usually move fast in comparison with most other mites, using the first pair of legs mainly as sensors, rather than as locomotory structures. Most often, adults reach up to 0.5 mm in length and their color usually varies between light yellow to dark brown. Their body cuticle is usually well sclerotized for most of its extent, at least in the adult stage. All post-larval stages usually have a pair of stigmata, most often laterad of the region between coxae III–IV (sometimes between coxae II–III). Each stigma is usually associated with an elongate peritreme.

Their gnathosoma usually bears chelate chelicerae and five-segmented palpi, each palpus bearing subterminal claws (apotele). The four pairs of legs of the post-larval stages bear a pair of claws and a flap-like structure (pulvillus) distally.

R.C. Castilho (✉)
Departamento de Fitossanidade, Faculdade de Ciências Agrárias e Veterinárias (FCAV), Universidade Estadual Paulista (UNESP), 14884-900 Jaboticabal, SP, Brazil
e-mail: rcastilho@outlook.com

R. Venancio • J.P.Z. Narita
Departamento de Entomologia e Acarologia, Escola Superior de Agricultura "Luiz de Queiroz", Universidade de São Paulo, 13418-900 Piracicaba, SP, Brazil
e-mail: renan453@yahoo.com.br; jpnarita@gmail.com

D. Carrillo et al. (eds.), *Prospects for Biological Control of Plant Feeding Mites and Other Harmful Organisms*, Progress in Biological Control 19,
DOI 10.1007/978-3-319-15042-0_1

Table 1.1 Higher groups of Mesostigmata (Based in Beaulieu et al. 2011)

Monogynaspida
Gamasina
Dermanyssiae
Rhodacaroidea
Veigaioidea
Eviphidoidea
Dermanyssoidea
Phytoseioidea
Ascoidea
Parasitiae
Parasitoidea
Epicriiiae
Epicrioidea
Arctacariae
Arctacaroidea
Uropodina
Uropodoidea
Microgynioidea
Thinozerconoidea
Diathrophalloidea
Trigynaspida
Sejida

Approximate numbers of taxa of Mesostigmata in each family and higher groups were given by Beaulieu et al. (2011). Table 1.1 shows the higher mesostigmatid groups according to the latter authors.

Different morphological characteristics are often associated with the occupation of different habitats and the possession of different habits, including the ability to use different items as food. It is presumed that predation represents the most primitive feeding habit of the Mesostigmata, and most of them retain this habit, feeding mainly on other mites, small insects, nematodes and Collembola. However, many are also able to feed on fungi, while others have evolved to feed on pollen, nectar or to parasitize different animal groups.

Many of the predatory Mesostigmata have been studied because of their potential to be used as biological control agents of agricultural pests or of parasites. Biological control with the use of predatory mites have long been used, but most recently the interest for this method of pest control has increased for different reasons, especially for the growing interest of people to consume agricultural items produced with a minimum or no pesticide use. Another reason is the progressively greater difficulty faced by growers to control pests with the use of chemicals, to which they became resistant.

Several mesostigmatid groups have been little studied, especially for not being abundant or for being difficult to handle. A few have received some attention, among which two superfamilies of the infraorder Gamasina, Rhodacaroidea (in the hyporder Dermanyssiae) and Parasitoidea (in the hyporder Parasitiae). These will be considered subsequently in this chapter.

1.2 Rhodacaroidea

According to Lindquist et al. (2009), this superfamily includes the families Digamasellidae, Halolaelapidae, Laelaptonyssidae, Ologamasidae, Rhodacaridae and Teranyssidae. These comprise free-living, cosmopolitan mites, generally reported in the literature as predators (Lindquist et al. 2009; Castilho et al. 2012a).

They are widespread in soil, litter, manure, decaying woods, galleries of bark beetles, associated with termites and small mammals (Hirschmann 1960; Lee 1970; Lindquist 1975; Shcherbak 1980; Krantz 2000; Halliday 2006; Lindquist et al. 2009; Castilho et al. 2012a).

1.2.1 Morphological Characterization

Lindquist et al. (2009) provided a revised classification of the Mesostigmata, with an expanded concept of the Rhodacaroidea, which was retained by Castilho et al. (2012a). Mites of this superfamily are mostly recognized by having almost invariably seta *st*4 on the sternal shield (thus, sternal shield with four pairs of setae) and by generally having a ventrianal shield with 1–9 pairs of preanal setae.

The most diverse families are Digamasellidae, Ologamasidae and Rhodacaridae. The Digamasellidae are distinguished by having palp apotele two-tined and dorsal shield usually divided; their dorsal and ventral shields are about homogeneously sclerotized and do not have punctate bands; mites of this family usually have densely sclerotized internal structures (scleronoduli) seen by transparency at the anterior half of dorsal shield. The Rhodacaridae are very similar to Digamasellidae, differing from them by having palp apotele three-tined and dorsal and ventral shields with desclerotized, puncate transversal bands. The Ologamasidae also have palp apotele three-tined, but their dorsal shield may not be divided, and most often they do not have scleronoduli. The two other families, Laelaptonyssidae and Teranyssidae, are very small and rarely found, while some authors still question whether the sixth family, Halolaelapidae, should or not be included in Rhodacaroidea.

1.2.2 Trends in the Efforts to Understand the Groups

Knowledge about the morphology and taxonomy of Rhodacaroidea increased greatly in the last 60 years. Several detailed studies about this superfamily were conducted in this period (eg. Hirschmann 1960; Ryke 1962; Evans 1963a; Lee 1970; Karg 1971; Lindquist 1975; Shcherbak 1980; Hirschmann and Wiśniewski 1982; Karg 1993; Castilho et al. 2012a).

The first characterization of Rhodacaroidea was done by Krantz (1978), but Lindquist et al. (2009) expanded the concept of the superfamily. A catalog of the family Rhodacaridae was recently published (Castilho et al. 2012a). The preparation of a catalog of the Ologamasidae is about to be concluded by R.C. Castilho,

E.S. Silva, B. Halliday and G.J. de Moraes. A total of 460 papers about the taxonomy of the Rhodacaroidea have been detected while preparing those catalogs and while starting the preparation of a catalog of the Digamasellidae.

The first two descriptions of species of this group [*Euryparasitus emarginatus* (Koch) and *Stylochirus haemisphaericus* (Koch), both Ologamasidae] were published by Koch (1839). Very few species were described since then until late 1940s. Most of the rhodacaroid species were described after 1950 (Fig. 1.1). The number of descriptions reached a maximum between 1970 and 1980, but it dropped considerably thereafter, especially after 2000. Nearly 54 % of digamasellid species were described by W. Hirschmann and J. Wiśniewski; 49 % of the ologamasid species by W. Karg and D.C. Lee, and 59 % of the rhodacarid species by W. Karg, G.C. Loots and H.W. Hurlbutt (Table 1.2).

1.2.3 *Diversity, Distribution and Habitats*

According to Beaulieu et al. (2011), the largest rhodacaroid family is Ologamasidae (45 genera containing 452 species), followed by Digamasellidae (13 genera containing 261 species), Rhodacaridae (15 genera containing 148 species), Laelaptonyssidae (6 species within a single genus) and Teranyssidae (one species). Since that publication, the number of described Ologamasidae and Rhodacaridae

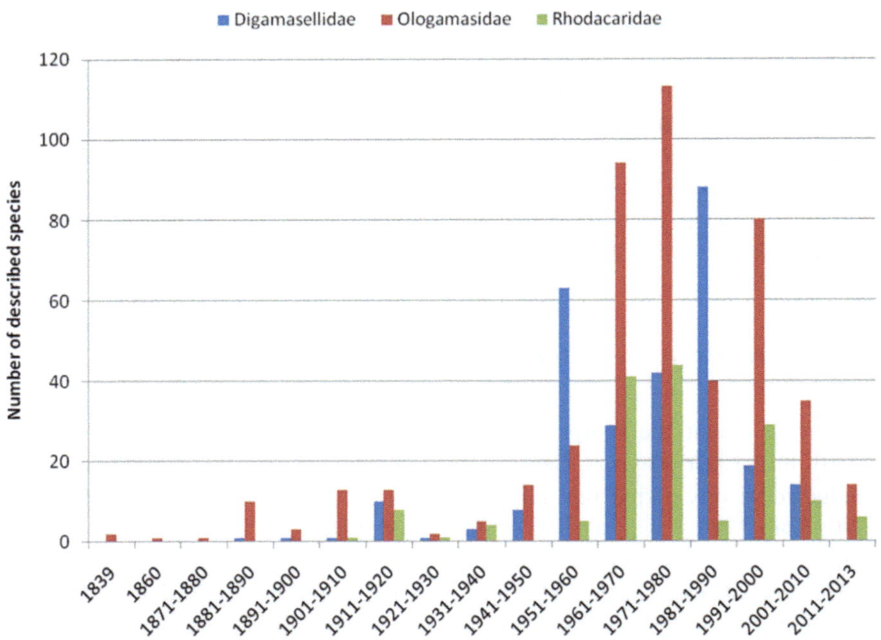

Fig. 1.1 Number of descriptions of species of the Rhodacaroidea since 1839

Table 1.2 Contribution of different authors to the description of rhodacaroid species

Digamasellidae		Ologamasidae	
Author	Number of described species	Author	Number of described species
Hirschmann, W.	141	Karg, W.	172
Wiśniewski, J.	80	Lee, D.C.	51
Shcherbak, G.I.	22	Womersley, H.	29
Karg, W.	12	Berlese, A.	20
Berlese, A.	11	Schorlemmer, A.	20
Others	<10	Others	<15
Rhodacaridae		**Laelaptonyssidae**	
Karg, W.	39	Halliday, R.B.	3
Loots, G.C.	30	Krantz, G.W.	2
Hurlbutt, H.W.	21	Others	1
Ryke, P.A.J.	13		
Shcherbak, G.I.	11	**Teranyssidae**	
Others	<10	Halliday, R.B.	1

species increased to 460 and 153, respectively (Castilho et al. 2012a, b; Marchenko 2013a, b; Santos et al. 2013; Rueda-Ramirez et al. 2013; Abo-Shnaf et al. 2013; Karg and Schorlemmer 2013). A list of the rhodacaroid genera was presented by Castilho et al. (2012a).

The most diverse genera of each family are: Ologamasidae – *Gamasellus* Berlese and *Gamasiphis* Berlese (about 16 and 15 % of the species of the family); Digamasellidae – *Dendrolaelaps* Halbert and *Multidendrolaelaps* Hirschmann (about 65 and 10 %); Rhodacaridae – *Afrodacarellus* Hurlbutt and *Rhodacarus* Oudemans (about 20 and 19 %) (Table 1.3).

Rhodacaroids have been described from a wide range of habitats, but mainly from soil (Table 1.3). Approximately 90 and 93 % of the ologamasid and rhodacarid species, respectively, were described from soil, litter, humus, mosses or dead organic matter on the soil surface; about 5 % of the ologamasid species were described in association with rodents or their nests; no rhodacarid was described from the latter habitats. However, only about 35 % of the digamasellid species were described from soil, litter, humus, mosses or dead organic matter on the soil surface, whereas about 26 % were described on Coleoptera or in Coleoptera galleries (mainly Cerambycidae and Curculionidae: Scolytinae). Approximately 88 % of the laelaptonyssid species as well as the single teranyssid species were described from Isoptera nests.

Most Rhodacaroidea species were described from the Palearctic region (~32 %), followed by the Neotropical (~22 %), Australasian (~19 %), Afrotropical (~14 %), Oriental (~8 %) and Nearctic regions (~4 %) (Fig. 1.2). However, there is a high variation between the numbers of species of each family described from each region. Most digamasellids were described from the Palearctic region (~64 %); most ologamasids were described from the Austalasian and Neotropical regions (~32 % each);

Table 1.3 Diversity of the families comprising the Rhodacaroidea and habitats from which species of each genus were described (numbers in parentheses indicate number of species described from the respective habitat)

Genus	Approx. nr. species	Habitats (number of species)
Digamasellidae		
Dendrolaelaps	179	Soil (68), wood (38), Scolytinae or Scolytinae galleries (29), tree bark (23), Formicidae nest (15), Cerambycidae (5), mammal nest (4), bee or bee nest (4), manure (4), Curculionidae (3), bird nest (3), plant (2), Dynastidae (1), Passalidae (1), Scarabaeidae (1), Lucanidae (1), Limoniidae (1), Cecidomyiidae (1); others or unspecified (14)
Multidendrolaelaps	30	Soil (7), wood (7), Cerambycidae (4), Scolytinae galleries (4), tree bark (3), fungus (3), bird nest (2), Formicidae nest (2), Molytinae gallery (1), Diptera (1)
Insectolaelaps	20	Scolytinae or Scolytinae galleries (11), wood (5), tree bark (3), Cerambycidae (2), soil (2), Cucujidae (1), Scarabaeidae (1), Formicidae nest (1), bird nest (1)
Dendrolaelaspis	19	Soil (12), tree bark (4), rodent nest (1), fungus (1), unspecified (1)
Longoseius	9	Wood (5), soil (3), Cerambycidae or Cerambycidae gallery (3), Elateridae or Elateridae gallery (3), Scolytinae gallery (1) Pirochroidae (1)
Dendroseius	5	Soil (2), tree bark (1), manure (1), Scarabaeidae (1)
Oligodentatus	4	Manure (2), rodent nest (1), dust (1)
Panteniphis	4	Soil (1), marshy meadow (1), dead leaves (1), unspecified (1)
Digamasellus	3	Soil (1), tree bark (1), manure (1)
Pontiolaelaps	3	Soil (1), algae (1), Crustacea (1)
Orientolaelaps	1	Rodent nest (1)
Laelaptonyssidae		
Starkovia	8	Isoptera or Isoptera nest (7), house fly colony (1)
Ologamasidae		
Gamasiphis	73	Soil (58), Formicidae nest (1), rodents (1), unspecified (13)
Gamasellus	71	Soil (67), rodents (1), unspecified (3)
Geogamasus	33	Soil (26), weeds (1), grass (1), manure (1), unspecified (4)
Gamasiphoides	22	Soil (18), weeds (2), Formicidae nest (1), unspecified (1)

Hydrogamasellus	21	Soil (16), tree bark (2), manure (1), unspecified (2)
Acugamasus	20	Soil (18), rodents nest (1), unspecified (1)
Ologamasus	20	Soil (11), tree bark (1), weeds (1), Formicidae nest (1), manure (1), algae (1), unspecified (4)
Athiasella	18	Soil (17), unspecified (1)
Neogamasellevans	14	Soil (11), algae (1), plant (1), unspecified (1)
Sessiluncus	14	Soil (10), cave (1), plant (1), unspecified (2)
Euryparasitus	13	Rodents or rodents nest (10), soil (1), mammal nest (1), unspecified (1)
Cyrtolaelaps	12	Rodent nest (8), manure (2), bat detritus (1), soil (1)
Rhodacaroides	10	Soil (9), beach sand (1)
Antennolaelaps	8	Soil (7), unspecified (1)
Euepicrius	8	Soil (8)
Stylochirus	8	Soil (3), marshy meadow (1), unspecified (4)
Caliphis	7	Soil (4), unspecified (3)
Cymiphis	7	Soil (6), unspecified (1)
Gamasellevans	7	Soil (7)
Rykellus	7	Soil (5), Formicidae nest (1), rocks (1)
Desectophis	5	Soil (5)
Gamaselliphis	5	Soil (4), unspecified (1)
Heterogamasus	5	Soil (4), manure (1)
Heydeniella	5	Soil (2), manure (1), rabbit nest (1), unspecified (1)
Periseius	5	Soil (4), algae (l)
Gamasellopsis	4	Soil (4)
Hydrogamasus	4	Soil (2), algae (1), Crustacea (1)
Laelaptiella	4	Soil (3), unspecified (1)
Onchogamasus	4	Soil (4)
Parasitiphis	4	Soil (4)

(continued)

Table 1.3 (continued)

Genus	Approx. nr. species	Habitats (number of species)
Acuphis	3	Soil (3)
Allogamasellus	2	Soil (2)
Evansellus	2	Soil (2)
Hiniphis	2	Soil (2)
Litogamasus	2	Soil (1), rocks (1)
Pachymasiphis	2	Soil (2)
Pilellus	2	Soil (1), mat (1)
Queensladolaelaps	2	Soil (2)
Gamasitus	1	Soil (1)
Laelogamasus	1	unspecified (1)
Notogamasellus	1	Soil (1)
Oriflamella	1	Soil (1)
Podonotogamasellus	1	Soil (1)
Pyriphis	1	Littoral soil (1)
Solugamasus	1	Soil (1)
Rhodacaridae		
Afrodacarellus	30	Soil (29), unspecified (1)
Rhodacarus	29	Soil (28), unspecified (1)
Afrogamasellus	24	Soil (21), unspecified (3)
Rhodacarellus	22	Soil (17), decomposed bark (3), flood plain (2)
Multidentorhodacarus	19	Soil (18), unspecified (1)
Protogamasellopsis	8	Soil (6), manure (1), yam (1)
Rhodacaropsis	6	Soil (6)
Poropodalius	5	Soil (5)

Pennarhodeus	4	Soil (4)
Paragamasellevans	2	Soil (2)
Binodacarus	1	Soil (1)
Interrhodeus	1	Soil (1)
Mediorhodacarus	1	Soil (1)
Minirhodacarellus	1	Soil (1)
Pararhodacarus	1	Soil (1)
Teranyssidae		
Teranyssus	1	Isoptera nest (1)

Soil: including litter, grasses, mosses, rotting trunks on the soil surface

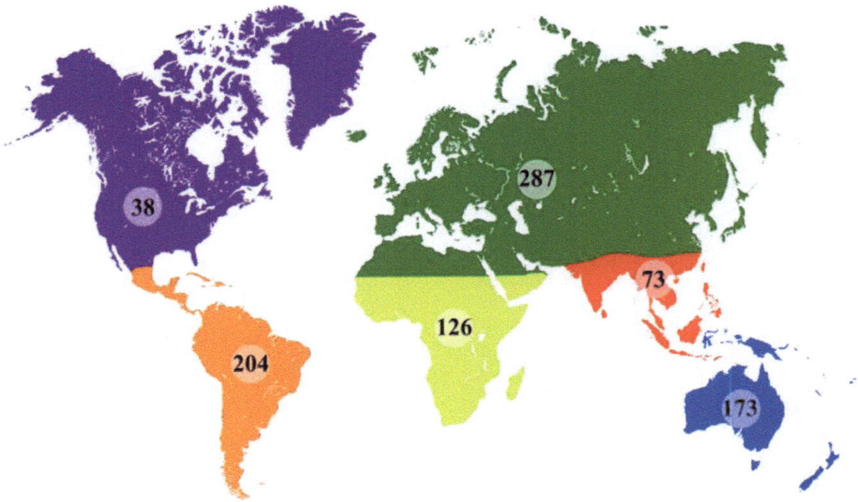

Fig. 1.2 Numbers of publications with descriptions of rhodacaroid species. Nearctic (*purple*), Neotropical (*orange*), Palearctic (*green*), Ethiopian (*yellow*), Oriental (*red*) and Australasian (*blue*) (Color figure online)

and most rhodacarids were described from the Afrotropical (~42 %) and Palearctic (~27 %) regions. Laelaptonyssids are only known from Australia (five species), USA and China (one species each), whereas the single teranyssid species was described from Australia.

1.2.4 Biological Information

The published information regarding biological aspects of Rhodacaroidea is relatively scarce. A discussion about available information is subsequently presented.

1.2.4.1 Digamasellidae

There are several reports about predation of digamasellids on nematodes. Kinn (1967) reported *Insectolaelaps quadrisetus* (Berlese) feeding on nematodes found abundantly in galleries of the bark beetle *Ips confusus* (LeConte) (Coleoptera: Scolytinae). Enda and Tamura (1977) observed *Dendrolaelaps fukikoae* Ishikawa and *Multidendrolaelaps unispinatus* (Ishikawa) feeding on pinewood nematodes in the pupal chambers and in the spiracles of pine sawyer beetles (*Monochamus* species, Coleoptera: Cerambycidae). The same authors reported that those mites completed their life cycle when reared on nematodes in the laboratory (Enda and Tamura 1980). Kinn (1980) suggested a mutualism between *Dendrolaelaps neodisetus* (Hurlbutt) and the bark beetle *Dendroctonus frontalis* Zimmermann (Coleoptera: Curculionidae: Scolytinae), because while phoretically dispersing on the beetle, *D. neodisetus* would attack nematodes infecting its carrier. Sardar and Murphy (1987) observed that *Dendroseius reticulatus* (Sheals) reproduced on an unidentified *Rhabditis* (Rhabditidae) nematode. Walter and Ikonen (1989) evaluated the ability of *Dendrolaelaps zwoelferi* Hirschmann, *Dendrolaelaps* nr. *strenzkei* and *Dendrolaelaps* nr. *procornutus*, to develop and reproduce on unidentified nematodes of the genus *Acrobeloides* (Rhabditidae), with reproductive periods of 20.0, 14.0 and 15.8 days, oviposition of 4.7, 3.9 and 4.2 eggs/day, and consumption of 3.8, 8.3 and 7.7 nematodes/day, respectively. They also observed *Dendrolaelaps* nr. *latior* to feed on the same nematode. Abou-El-Soud and Shoeib (2000) observed *Dendrolaelaps rasmii* Nasr and Mersal to feed on the root-knot nematode *Meloidogyne javanica* (Treub) (Heteroderidae).

There are also reports of predation of digamasellids on bark beetles. Kiełczewski and Bałazy (1966) apud Lindquist et al. (2009) mentioned some *Dendrolaelaps* species to prey upon eggs and first instar larvae of their bark beetle associates. Kinn (1967) reported *Insectolaelaps quadrisetus* (Berlese) feeding on first instar larvae of *I. confusus*. Moser (1975) observed six digamasellid species [*Dendrolaelaps neocornutus* (Hurlbutt), *Dendrolaelaps neodisetus* (Hurlbutt), *Dendrolaelaps rotoni* (Hurlbutt), *Dendrolaelaps varipunctatus* (Hurlbutt), *Longoseius cuniculus* Chant and *Multidendrolaelaps isodentatus* (Hurlbutt)] to prey on eggs and first instar larvae of *D. frontalis*. The author referred to *D. neocornutus*, *D. neodisetus* and *M. isodentatus* as possible natural biological control agents of this bark beetle.

Predation on Collembola has also been reported. Sheals (1956) observed predation on *Tullbergia krausbaueri* (Böerner) (Tullbergidae) and *Ceratophysella denticulata* (Bagnall) (Hipogastruridae) by *D. reticulatus*, whereas Sardar and Murphy (1987) reported predation of an unidentified *Tullbergia* species by the same predator.

Predation on mites seems uncommon. Sardar and Murphy (1987) reported predation of the Oribatida *Liebstadia similis* (Michael) (Liebstadiidae) by *D. reticulatus* and Abou-El-Soud and Shoeib (2000) observed *D. rasmii* to feed on *Acarus siro* Linnaeus (Acaridae).

Walter and Ikonen (1989) studied the biology of *D. zwoelferi*, *Dendrolaelaps* nr. *procornutus*, *Dendrolaelaps* nr. *stenzkei* and *Dendrolaelaps* nr. *latior*. The authors determined that development from egg to adult requires 8.3–9.0 days and that each female of the first three species laid totals of 54.3–94.4 eggs at 25 °C, when offered the nematode *Acrobeloides* sp. as food. They also determined that mating was required for the reproduction of those species.

1.2.4.2 Ologamasidae

Predation by ologamasid mites was first reported over a hundred years ago. Michael (1886) reported *Euryparasitus emarginatus* (Koch) (cited as *Gamasus terribilis*, its junior synonym) feeding on “wire-worms, beetle larvae, small worms and mole fleas”.

Gressit (1967) reported *Hydrogamasellus racovitzai* (Trouessart) to be commonly seen feeding on Collembola. Lee (1974) observed the feeding behavior of ten ologamasid species, dividing them in two groups, collembolan feeders and non-collembolan feeders. In the first group he mentioned *Athiasella dentata* (Womersley) feeding on *Isotoma* sp. (Isotomidae), *Acanthurella* sp (Entomobryidae), *Lepidosira* sp. (Entomobryidae), and on unidentified species of Onychuridae and Sminthuridae; and *Euepicrius filamentosus* Womersley and *Gamasellus concinnus* (Womersley) feeding on *Isotoma* sp. and *Lepidosira* sp. The author classified *Gamasellus tragardhi* (Womersley) as non-collembolan feeder, but cited feeding of this species on *Lepidosira* sp.

Using eletroctophoretic analysis of field populations, Lister et al. (1988) concluded that *H. racovitzai* preys primarily on collembolans, mainly *Cryptopygus antarcticus* Willem (Entomobryidae), in the maritime Antarctic zone. Beaulieu and Walter (2007) observed predation on *Folsomia candida* Willem (Isotomidae) by 20 ologamasid species of the genera *Acugamasus* Lee, *Antennolaelaps* Womersley, *Athiasella* Lee, *Caliphis* Lee, *Euepicrius* Womersley, *Gamasellus*, *Gamasiphis*, *Gamasiphoides* Womersley, *Geogamasus* Lee, *Laelaptiella* Womersley and *Queenslandolaelaps* Womersley. Predation of *Gamasiphis paulista* Castilho, Moraes and Narita on an unidentified Collembola was also reported by Castilho et al. (2010).

All of the collembolan feeders reported by Lee (1974) were also observed to prey on mites, as follows: *A. dentata* on *Eupodes* sp. (Eupodidae), *Geogamasus howardi* Lee (Ologamasidae), a species of Oribatida, and *Tyrophagus putrescentiae* (Schrank) (Acaridae); *Euepicrius filamentosus* Womersley on *Eupodes* sp., a species of Cunaxidae, and *T. putrescentiae*; and *Gamasellus concinnus* (Womersley) on *T. putrescentiae*. Similarly, all the non-collembolan feeders were observed by Lee (1974) to feed on mites, namely *Acugamasus semipunctatus* (Womersley), *Gamasiphis fornicatus* Lee and *G. howardi* on *T. putrescentiae*; *Gamasiphis saccus* Lee and *Geogamasus minimus* Lee on *Nanorchestes* sp. (Nanorchestidae); *G. tragardhi* on *Nanorchestes* sp., *T. putrescentiae* and a species cited as *Lionobella* sp.; and *Gamasiphoides propinquus* (Womersley) on *T. putrescentiae*. Castilho et al. (2010) reported predation of *G. paulista* on *T. putrescentiae*.

Nematodes have also been observed to be preyed upon by ologamasids. Lee (1974) reported *G. howardi* to feed on an unidentified *Rhabditis* species, while Muraoka and Ishibashi (1976) reported unidentified species of *Gamasellus*, *Gamasiphis*, *Heydeniella* Richters and *Hydrogamasus* Berlese to feed on *Cephalobus* sp. (Cephalobidae). In extensive laboratory studies, Beaulieu and Walter (2007) observed 18 ologamasid species preying upon Rhabditida nematodes; the ologamasids included species of the same genera reported by Beaulieu and Walter (2007) to feed on Collembola (listed previously), except for *Gamasiphoides*. However, different species of *Laelaptiella* had difficulty or did not feed on nematodes, and a species of *Athiasella* did not lay eggs when offered nematodes as prey.

Beaulieu and Walter (2007) reported predation by an *Antennolaelaps* species on the adult of an unidentified thrips, as well the consumption of the haemolymph of crushed caterpillars, oribatid mites and Collembola. They also reported the apparent consumption of eggs of the moth *Sitotroga* sp. (Gelechiidae) by species of *Athiasella* and *Gamasiphis*.

Within the ologamasids, the absence of males in *Gamasellus vibrissatus* Emberson from Canada (Hurlbutt 1979) led Norton et al. (1993) to suspect that this species reproduced by thelytoky.

1.2.4.3 Rhodacaridae

Nematodes have been reported many times to be preyed by rhodacarids. Predation on unidentified nematodes has been mentioned by Karg (1961) for *Rhodacarus roseus* Oudemans and *Rhodacarellus silesiacus* Willmann; Karg (1971) for *Rhodacarellus calcarulatus* Berlese; and Walter et al. (1988) for *Multidentorhodacarus denticulatus* (Berlese), *R. silesiacus* and *Rhodacarellus subterraneus* Willmann. Walter and Kaplan (1990) reported the association of *Afrodacarellus* sp., *Afrodacarellus citri* (Loots), *Protogamasellopsis* sp. and *M. denticulatus* with citrus roots in Florida, USA, suspecting that they could be feeding on nematodes. Sharma (1971) reported little effect of *R. roseus* as a control agent of the nematode *Tylenchorhynchus dubius* (Bütschli) (Belonolaimidae) in potted grasses. Predation was also reported by Van den Bund (1972) for *R. roseus* on *Rotylenchus* sp. (Haplolaimidae) and on an unidentified Tylenchidae, and by Sardar and Murphy (1987) for *R. roseus* and *R. silesiacus* on *Rhabditis* sp. and *Pelodera strongyloides* (Scheider) (Rhabditidae). Castilho et al. (2009) reported a high daily oviposition rate of *Protogamasellopsis posnaniensis* Wiśniewski and Hirschmann (6.3 eggs/female) (Fig. 1.3) when fed with *Protorhabditis* sp. (Rhabditidae).

Predation on unidentified Collembola has been mentioned by Karg (1971) for *R. calcarulatus*; by Sardar and Murphy (1987) for *R. roseus* and *R. silesiacus*; and by Walter et al. (1988) for *M. denticulatus*, *R. silesiacus* and *R. subterraneus*. Predation was also reported by Karg (1961) for *R. silesiacus* on *T. krausbaueri* and by Van den Bund (1972) for *R. roseus* on *Onychiurus armatus* (Tullberg) (Onychiuridae).

Fig. 1.3 Adult female of *Protogamasellopsis posnaniensis* Wiśniewski and Hirschmann

There are also reports of mites being preyed upon by rhodacarids. Unidentified mites were reported to be attacked by *R. roseus* (Van den Bund 1972), as well as by *M. denticulatus*, *R. silesiacus* and *R. subterraneus* (Walter et al. 1988). *Rhodacarus roseus* has also been reported by Karg (1961) to prey on *Tyrophagus* sp. and by Sardar and Murphy (1987) to prey on *T. putrescentiae* and on the Oribatida *L. similis*. Castilho et al. (2009) reported high daily oviposition rate of *P. posnaniensis* (7.6 eggs/female) fed with *T. putrescentiae*, but much lower oviposition (1.9 eggs/female) when it was fed with *Rhizoglyphus echinopus* (Fumouze and Robin) (also an Acaridae).

The possible effect of rhodacarids on insects has not been well studied. Castilho et al. (2009) evaluated the oviposition of *P. posnaniensis* on two insect species, reporting a low rate on the fungus gnat (Diptera: Sciaridae) *Bradysia matogrossensis* (Lane), but a high rate on the thrips *Frankliniella occidentalis* (Pergande) (Thripidae) (respectively 0.5 and 5.6 eggs/female/day), with the consumption of 1.8 larvae of *B. matogrossensis* or 4.3 pupae of *F. occidentalis* per day.

Walter and Ikonen (1989) studied the biology of two rhodacarid species, *R. silesiacus* and *M. denticulatus*. They report the developmental times of certain rhodacarids to be long (about a month at 23 °C) in comparison with other mesostigmatid groups. They determined that at 25 °C *R. silesiacus* developed from egg to adult in about 23.8 days, that each female laid a total of 67.2 eggs when offered the nematode *Acrobeloides* sp. as food, and that it reproduced through thelytokous parthenogenesis. Walter and Kaplan (1990) investigated the development of *Protogamasellopsis* sp. and *M. denticulatus* on the nematodes *Radopholus similis* (Cobb) (Pratylenchidae) and *Radopholus citropholus* Huettel, Kaplan and Dickson. They observed that at 25 °C *Protogamasellopsis* sp. and *M. denticulatus* required approximately 10.8 and 26.3 days to develop from egg to adult and that they laid totals of 204 and 48 eggs, respectively.

Thelytoky seems common in rhodacarids. This type of reproduction was reported for *R. silesiacus* (Walter and Ikonen 1989; Walter and Oliver 1989), *M. denticulatus* and an undetermined species of *Protogamasellopsis* Evans and Purvis (Walter and Kaplan 1990), and for *P. posnaniensis* (Castilho et al. 2009).

1.3 Parasitoidea

This superfamily contains a single family, Parasitidae, which is a large group of the suborder Gamasina (Hyatt 1980). Members of this family have been commonly reported as predators of small arthropods and nematodes, especially in the soil and in nests of insects, birds and mammals, as well as in insect galleries (Lindquist et al. 2009).

1.3.1 Morphological Characterization

The morphological characterization of Parasitidae was provided by several authors, including Micherdzinski (1969), Tseng (1995), Tichomirov (1977), Lindquist et al. (2009) and Kazemi et al. (2013).

Parasitidae females can be separated from other mesostigmatids by the following characteristics: dorsal shields divided or fused, in the latter case, with or without a pair of lateral incisions at the level of fusion. Sternal shield with three pairs of setae; genital shield triangular, flanked by a pair of large metasternal shields, each bearing a seta, abutting or (rarely) fused with a ventrianal shield. Palp apotele three-tined. Males with fused podonotal and opisthonotal shields; with or without a transverse line at the level of fusion. Genital opening at anterior margin of holoventral or sternogenital shield. Each chelicera with a specialized structure to transfer sperm to female genital opening, consisting of an elongate piece fused basally and distally with the movable digit (= spermatotreme). Typically with highly developed spurs and apophyses on legs II.

1.3.2 Diversity

The classification of this group is in need of revision. Different authors have adopted different taxonomic hierarchic positions for different subgroups as well as different definitions of genera, which are sometimes separated based on single characters, leading to incompatible groupings among researchers (Al-Atawi et al. 2002). As discussed by Lindquist et al. (2009), the family is classified in few genera by some authors and in considerably higher number of genera by others.

Parasitidae was revised by Micherdzinski (1969). Regional revisions were presented by Tichomirov (1977) for specimens from the former Soviet Union, Tseng (1995) for specimens from Taiwan, and Kazemi et al. (2013) for specimens from Iran. Juvara-Bals (1972) proposed a division of the family in two subfamilies, Parasitinae and Pergamasinae. These have been considered separate families by Woolley (1988). Beaulieu et al. (2011) considered the family to contain 426 species arranged in 35 genera.

1.3.3 Ecology

Species of Parasitinae predominate in transient habitats, such as dung, compost, tidal debris, or in nests of insects, birds and mammals, whereas species of Pergamasinae predominate in more stable habitats, particularly forest and grassland soils (Juvara-Bals 1972; Hyatt 1980). The potential usefulness of parasitids as biological indicators of soil conditions was discussed by Karg and Freier (1995).

The occupation of ephemeral and discontinuous habitats by mites usually requires their adaptation for transport between patches of the habitat on other organisms (= phoresy). Parasitids of the subfamily Parasitinae (but not of the subfamily Pergamasinae) ride on insects for dispersal (Evans 1992), as detailed in the following paragraph. They have been observed to arrive in animal dung earlier than other mite groups (Axtell 1962, 1963). Ignatowicz (1974) presumed that parasitids can inhibit the colonization by mites of the family Macrochelidae (Gamasina). His studies revealed that adults and deutonymphs of the parasitid mite *Parasitus coleoptratorum* (L.) feed on immatures and adult males of *Macrocheles glaber* (Müller) (Macrochelidae).

1.3.4 Dispersion

The occurrence of parasitid mites on insects led early observers to think that they were parasitizing the latter. The name of the type genus, *Parasitus*, was proposed by Latreille (1795) for *Acarus fucorum* De Geer, 1778. The name of the new genus was apparently selected to reflect the possible parasitism on the carrier insect (Lindquist et al. 2009). This species, now placed in *Parasitellus* Willmann, lives in nests of bumblebees (Hymenoptera: Apidae: *Bombus*), dispersing on these insects (Koulianos and Schwarz 1999).

Dispersal of both sexes of parasitids occurs when mites are in the deutonymphal stage (Norton et al. 1993). In some species, dispersal at this stage is obligatory. Deutonymphs of *Phorytocarpais* (= *Parasitus*) *fimetorum* (Berlese) do not feed before their phoretic dispersal (Binns 1982; Perotti and Braig 2009). Many authors have studied the preference of phoretic parasitines for specific carriers (Schaefer and Peckham 1968; Krantz and Mellot 1972; Schwarz 1996; Schwarz et al. 1996; Schwarz and Huck 1997). Huck et al. (1998) studied the dispersal behaviour of *P. fucorum* conducting choice experiments, concluding that deutonymphs preferred bumblebee queens to workers, but did not show significant difference between queens and males. This result corroborated by Stebbing (1965) and Richards and Richards (1976), who suggested the existence of a mechanism enabling deutonymphs to recognize queens. Possibly only queens produce a chemical cue that the mite uses to locate the carrier.

Deutonymphs of *Poecilochirus carabi* G. and R. Canestrini are known to identify their carrier beetles (*Nicrophorus* Fabricius; Silphidae) using a chemoreceptor located on tarsus I (Neumann 1943; Korn 1983). Brown and Wilson (1994) presented an overview of host specialization of *P. carabi*. It has also been demonstrated that within a breeding site, deutonymphs of *P. carabi* prefer riding beetles near the periphery (males in most cases), which are most likely to depart the site first (Schwarz and Müller 1992).

1.3.5 Biological Information

General information about the predatory aspects of parasitids was most recently summarized by Gerson et al. (2003) and Lindquist et al. (2009). Among insects, predation has been reported on beetles and flies. Specimens of a species referred to the genus *Parasitus* have been reported by Kinn (1971) to prey on larvae of bark beetles. In that same publication, the authors refer to a report by Walters and Campbell (1955) about the predation of bark beetles by unidentified *Parasitus* species. Those reports could actually refer to species of *Schizothetus* Athias-Henriot (Athias-Henriot 1982; Lindquist et al. 2009). Predation on different fly larvae (Fig. 1.4) is reported for several parasitids, including *Parasitus bituberosus* Karg (Karg 1972; Al-Amidi and Downes 1990), *Parasitus coleoptratorum* (L.) (Axtell 1986), *Parasitus gregarius* Ito (Ito 1977) and *Poecilochirus monospinosus* Wise, Henessey and Axtell (Axtell 1986; Axtell and Rutz 1986; Wise et al. 1988).

Parasitid species are also reported to prey on nematodes. Available information include an unidentified *Eugamasus* (probably *Schizothetus* Athias-Henriot, according to Lindquist et al. 2009) species (Kinn 1971), *P. bituberosus* (Karg 1972; Al-Amidi and Downes 1990), and *P. gregarius* (Ito 1977).

Ito (1977) observed that adults of *P. gregarius* preferred feeding on house fly larvae over eggs, and deutonymphs preferred nematodes over house fly eggs and larvae. Axtell (1986) and Axtell and Rutz (1986) mentioned that the deutonymphs and adults of *P. monospinosus* (referred by the authors as an undescribed species of *Poecilochirus*, according to Wise et al. 1988) fed on house fly eggs and first instar larvae, with a preference for the latter.

Al-Amidi and Downes (1990) studied the prey preference of deutonymphs of *P. bituberosus*. First instar larvae of *Heteropeza pygmaea* (Diptera: Cecidomyiidae) was preferred over eggs of the fungus gnat *Lycoriella solani* Winnertz (Diptera: Sciaridae) or collembolans.

Koulianos and Schwarz (1999) provided basic information on life-history traits and dietary requirements of *P. fucorum*, collected in association with the bumble bee *Bombus terrestris* (L.). They observed that development from egg to protonymph was completed in about 9.5 days, and that field collected deutonymphs molted to adults in 1–7 days. Fecundity was about 102 eggs. Larvae were reported to feed on dead mealworms (*Tenebrio molitor* L., Coleoptera: Tenebrionidae) and eggs of unspecified mites, while deutonymphs and females were reported to feed on pollen.

Fig. 1.4 Different stages of *Parasitus* sp. feeding on a larva of Diptera

Predation of parasitids on pest mites has been more rarely reported. Lesna et al. (1995) reported *P. fimetorum*, *Phorytocarpais* (= *Parasitus*) *hyalinus* (Willmann) and *Parasitus primitivus* Heim and Oudemans to develop and oviposit when offered the astigmatid mites *Rhizoglyphus robini* Claparède (Acaridae) and *T. putrescentiae* as prey.

1.3.6 Reproduction

Both sexes of parasitid mites are diploid (Sokolov 1934). However, the sex-determination is unknown (Norton et al. 1993). *Phorytocarpais hyalinus* is suspected to reproduce by thelytokous parthenogenesis (Wrensch and Johnston 1983).

Yasui (1997) studied the mating behaviour of *P. fimetorum*, concluding that multiple mating in female increased fecundity. Koulianos and Schwarz (1999) observed that *P. fucorum* females mated immediately after molting to adult, as in other parasitids associated with nidiculous insects (Rapp 1959; Korn 1982; Schwarz and Walzl 1996).

1.3.7 Practical Use in Biological Control Programs

Some parasitid species have been studied as potential biological control agents (Lesna et al. 1995; Gerson et al. 2003).

Al-Amidi and Downes (1990) studied the ability of *P. bituberosus* to control the fungus gnat *H. pygmaea* in commercial production of the mushroom, *Agaricus biosporus* (Lange) Pilate. They reported that the predator could delay or reduce the peak of *H. pygmaea*, thus reducing yield loss. Al-Amidi et al. (1991) evaluated the performance of *P. bituberosus* against the fungus gnat *L. solani* in commercial mushroom houses. The number of *L. solani* from bags inoculated with the predator was reduced three-fold compared with bags without predators, indicating the potential use of this predator in integrated pest management.

Promising results were obtained by Lesna et al. (1995) in a small scale experiment simulating lily bulb propagation to test the ability of *P. fimetorum* to control *R. robini* using peat (but not vermiculite) as substrate. Over a period of 4 weeks, prey density was 150 lower in the units where the predator was released.

The commercial use of parasitids as biological control agents requires the development of adequate mass rearing procedures. The first information about rearing parasitid mites under laboratory conditions was published by Pereira and Castro (1946), who used "brick chambers", earlier proposed by Autuori (1942) for the study of insects. The predators were fed with house fly eggs. Rapp (1959) cited the crucial importance of the presence of manure in the rearing chamber for the development of deutonymphs of *P. coleoptratorum*. Wise et al. (1988) observed that survivorship and molting of deutonymphs of *P. monospinosus* to adults were greatly improved in the presence of fresh chicken manure.

We have maintained a colony of an unidentified *Pergamasus* sp. in the laboratory in containers filled with moistened vermiculite, offering all stages of the mite *Aleuroglyphus ovatus* (Troupeau) (Astigmatina: Acaridae) as food (unpublished).

1.4 Other Groups of the Infraorder Gamasina

Other groups of the infraorder Gamasina have received limited attention, especially in relation to their potential as biological control agents. A brief discussion is subsequently presented about those groups.

Information about the constitution of each group subsequently presented was taken from Beaulieu et al. (2011), updated with the data of Walter (2013) for the suborder Sejida and Mašán and Halliday (2014) for the Pachylaelapidae.

1.4.1 Hyporder Epicriiae

This hyporder comprises three superfamilies: Epicrioidea, with Epicriidae; Heatherelloidea, with Heatherellidae; and Zerconoidea, with Coprozerconidae and Zerconidae. These are edaphic organisms.

Altogether, the hyporder comprises almost 450 species divided in 42 genera. The morphology of their chelicera and the reduced tritosternum suggest that epicriid species feed on fungi (Lindquist et al. 2009). Heatherellid species are known to feed on nematodes under laboratory conditions (Walter 1997). Species of Zerconidae, the largest family of this group, with about 390 species of 36 genera, are thought to be oligophagous predators, while the only species of Coprozerconidae was collected from rat feces in Kentucky, USA.

1.4.2 Hyporder Arctacariae

This hyporder comprises a single superfamily, Arctacaroidea, which includes a single family, Arctacaridae, containing six species arranged in two genera. They have well sclerotized chelicera. Their feeding habits are unknown (Lindquist et al. 2009).

1.4.3 Hyporder Dermanyssiae

This large hyporder is composed of six superfamilies. The superfamily Veigaioidea comprises a single family, the Veigaiidae, with 95 species of four genera. Members of this family are predators in litter habitats, feeding on Collembola, Protura, Pauropoda and other mites (Lindquist et al. 2009). *Veigaia pusilla* (Berlese) was observed not to feed on any of 11 species of the following genera of nematodes in a laboratory experiment: *Acrobeloides*, *Aporcelaimus*, *Chiloplacus*, *Chrysonema*, *Deladenus*, *Longidorus*, *Mesodorylaimus*, *Pelodera*, *Rhabditis* and *Steinernema* (Walter and Ikonen 1989).

The Eviphidoidea comprises five families: Eviphididae, Leptolaelapidae, Macrochelidae, Pachylaelapidae and Parholaspididae. Eviphididae comprises 108 species arranged in 19 genera, some of which inhabit stable substrates and are non-phoretic; others inhabit ephemeral habitats and for that reason are phoretic on insects as deutonymphs (Mašán and Halliday 2010). They are predators of nematodes and other members of the microfauna, including insect eggs and larvae. The Leptolaelapidae comprises 48 species of 12 genera. Two species were reported to feed on dead arthropods, and one of these (*Hunteracarus* sp.) was observed to feed on nematodes (Beaulieu and Walter 2007). Lindquist et al. (2009) mentioned that the diversity among the genera of this family suggests that they represent a polyphiletic group. The Macrochelidae will be treated in Chap. 4 of this book.

A recent review of the Pachylaelapidae was recently published (Mašán and Halliday 2014). Most of the information here provided was obtained from that publication. This family comprises 226 species of 16 genera. A new characterization of the family and its genera was given by Mašán and Halliday (2014). The pachylaelapids inhabit soil, litter and nests of birds, mammals and social insects, feeding on a range of microinvertebrates (Lindquist et al. 2009). Many species are associated with coprophilous insects (especially scarabaeid beetles) and some seem mirmecophilous. Their potential as biological control agents has not been adequately investigated. Gomaa et al. (1989) reported the postlarval instars of *Pachylaelaps aegypticus* Hafez and Nasr to feed voraciously on house fly larvae in laboratory colonies. The Parholaspididae comprises 96 species of 12 genera. They are edaphic predators and none of its species is known to be phoretic (Lindquist et al. 2009). The biology of *Parholaspus ochraeus* (Ishikawa) was studied by Ishikawa (1980), who observed it to feed on nematodes and microarthropods.

The Dermanyssoidea comprises 17 families, most of which are vertebrate parasites. The latter will not be discussed in this book. One of the families, Varroidae contains major parasites of honeybee. Among the remaining families, the most important is Laelapidae, many of which are also parasitic, mainly of vertebrates, but also of arthropods, and many are free-living mites Laelapidae mites will be treated in Chap. 3 of this book.

The Phytoseioidea comprises four families. Blattisociidae and Phytoseiidae will be treated in other chapters. The Podocinidae comprises 25 species arranged in two genera. They inhabit litter, humus, grain field debris, mammal and insects nests (Lindquist et al. 2009). Podocinids have been reported to prey on Collembola and other mites. The Otopheidomenidae comprises 28 species arranged in ten genera. All species of this family are associated with insects. The shape of their chelicera (with reduced fixed digit and pruning-saw shaped movable digit) and the absent or reduced tritosternum suggest them to be parasite of arthropods.

The Ascoidea comprises the families Ameroseiidae, Ascidae and Melicharidae (Lindquist et al. 2009). The latter two families will be discussed in Chap. 2. About

150 species of eight genera are placed in Ameroseiidae. A ninth genus, *Sinoseius*, was proposed by Bai et al. (1995), however this genus has close similarities with a previous described genus of Ameroseiidae, *Ameroseius*, thus it does not allow a clear distinction between the genera. Morphologically, adult females of this family are characterized having entire dorsal shield, seta *J5* absent, sternal shield usually with two pairs of seta (rarely with one or three pairs), palp apotele bifurcate or trifurcate, corniculi undivided, bifurcate or trifurcate; *pilus dentilis* of the chelicera absent and replaced by a lobe.

Selected publications regarding the taxonomy of this group in different regions are: America – Elsen and Whitaker (1985); Narita et al. (2013b), Asia – Evans (1963b), Australia – Halliday (1997); Africa – Elsen (1972, 1973); Europe – Westerboer and Bernhard (1963), Bregetova (1977) and Karg (1993). A few keys for the identification of ameroseiid species have been published: *Afrocypholaelaps* Elsen – Ho et al. (2010); *Ameroseiella* Bregetova – Bregetova (1977), *Ameroseius* Berlese – Bregetova (1977), Narita et al. (2013b); *Epicriopsis* Berlese – Bregetova (1977); *Neocypholaelaps* Vitzthum – de Moraes and Narita (2010), Narita et al. (2011, 2013a).

These mites can be found in very different habitats, including soil, litter and composts (Westerboer and Bernhard 1963), plants (Evans 1963b; Scoble and Clarke 2006), associated with insects (Elsen 1972, 1973; Seeman and Walter 1995), birds (Noske 1993; Halliday 1997), rodents and their nests (Elsen and Whitaker 1985; Villegas-Guzmán et al. 2004), fungi (Mašán 1998; Salmane and Brumelis 2010), and in stored food (Hughes 1976). The phoretic association between species of *Afrocypholaelaps* and *Neocypholaelaps* with bees is well documented (Ramanan and Ghai 1984; Seeman and Walter 1995; Fan and Jiang 2014).

The ameroseiids have not been demonstrated to be predators. However, published information suggests that they could play a role as biological control agents of plant pathogenic fungi. *Ameroseius kosi* (El-Badry, Nasr and Hafez) has been reported to reduce the incidence of damping-off of green beans caused by *Rhizoctonia solani* J.G. Kühn and *Sclerotium bataticola* Taubenh (Moustafa and El-Hady 2006). A few other studies indicate *Ameroseius* species to have a potential role in the dissemination of fungi, carrying spores on the body or ingesting fungi structures without digesting the spores, which are able to originate new fungal colonies after released with their feces (Flechtmann 1985; Roets et al. 2011). Further studies about *Ameroseius* biology and their relation with fungi are necessary to understand the importance as control agents or vectors of pathogens.

1.4.4 Infraorder Uropodina

This infraorder comprises a cohort within the suborder Monogynaspida. Beaulieu et al. (2011) considered Uropodina to comprise four superfamilies, containing a total of 2,224 species of 35 families. These mites are commonly found in soil and litter, on decaying trunks or associated with insects, in tropical and temperate regions of the world.

These mites are characterized by usually having dorsal and ventral shields coalesced into a turtle-like armor; genital shield usually ovoid, without setae; idiosoma of some groups with ventrolateral depressions to fit the legs when contracted; gnathosoma of some species being withdrawn within an anteroventral recess of the idiosoma when not in use; hypostomal setae longitudinally aligned (Lindquist et al. 2009); and usually elongate chelicerae (up to more than half of the length of the body), terminating in reduced digits.

Uropodines disperse phoretically as deutonymphs, mainly on Coleoptera (Wiśniewski and Hirschmann 1993; Bajerlein and Witaliński 2014). These attach to the carrier through a specialized structure known as anal pedicel, produced by a secretion from pedicellar glands (Bajerlein and Witaliński 2012). While part of the species was described as adults, many were described as deutonymphs, collected from insects, turning species identification difficult (Wiśniewski and Hirschmann 1993; Kontschán et al. 2012).

Some uropodines have been mentioned as predators, while others feed on decaying organic matter and fungi (Gerson et al. 2003; Lindquist et al. 2009). *Nenteria hypotrichus* (El-Borolossy and El-Banhawy) (Uropodidae) was able to develop when feeding on nematodes and acarid mites (El-Banhawy et al. 1997). *Urobovella marginata* (Koch) [= *Fuscuropoda vegetans* (De Geer)] (Uropodidae) was reported to play a role as predator of immature flies (Gerson et al. 2003). Jalil and Rodriguez (1970) reported higher oviposition by that predator on nematodes than on house fly eggs. Another study has shown that *U. marginata* can also feed on slugs (Raut 1996 *apud* Gerson et al. 2003). *Fuscuropoda agitans* (Banks) (= *Uropoda agitans*), known as earthworm mite, can be a problem in earthworm beds, mainly by competing for food with the earthworms. It has also been reported to attack earth worm cocoons (Stone and Ogles 1953; Tappan 1959).

Jalil and Rodriguez (1970) reported that only fertilized females of *U. marginata* were able to oviposit. The virtual absence of males has led Norton et al. (1993) to suspect that several uropodine species reproduce by thelytokous parthenogenesis. This was proven for a few species. However, at least one species was shown to reproduce also sexually.

1.4.5 Suborder Trigynaspida

This suborder comprises two infraorders, the Cercomegistina, which has one superfamily containing about 25 species of six families; and Anthennophorina, which contains seven superfamilies with about 320 species of 21 families (Beaulieu et al. 2011).

Trigynaspids are morphologically characterized by having sternal region with four pairs of setae and three pairs of lyrifissures; genital oviporus covered by two to four shields (a mesogynial shield and latigynial shields); one to three excrescences on the movable cheliceral digit; and leg IV with an intercalary sclerite between the basitarsus and telotarsus bearing two setae (*av4* and *pv4*). The superfamily Celaenopsoidea is the most diverse group of Trigynaspida (Kethley 1977; Lindquist et al. 2009).

These mites are mostly found in tropical regions of the world, including the Neotropics, Africa, Southern Asia and Oceania. Most of them are associated with insects, myriapods or reptiles (Funk 1980; Lindquist et al. 2009). Members of the Antennophorina group can be phoretic in the adult stage and predatory in the deutonymphal stage. *Antennophorus grandis* (Berlese) (Antennophorina: Antennophoridae) is known to act as a kleptoparasite of ants (Franks et al. 1991). Some species have been reported from bark, insect galleries, soil and litter, and associated with fungi (Datta 1985; Seeman and Walter 1997; Salmane and Brumelis 2010).

Most studies show trigynaspids to be predators. Among the Celaenopsoidea (Antennophorina), Seeman and Walter (1997) observed *Funkotriplogynium iagobadius* Seeman and Walter to be aggressive, feeding on fly larvae, Collembola, nematodes and other mites. A similar habit was observed for *Funkotriplogynium irapora* (Flechtmann), which also feeds on fly larvae, nematodes, beetle eggs and larvae (*Diabrotica* sp.), the acarid mite *R. echinopus*, and the dry bulb mite *Aceria tulipae* (Keifer) (Prostigmata: Eriophyidae) (our unpublished data). Euzerconidae species have been reported to feed on Collembola and beetle larvae (Hunter and Davis 1965). In the Megisthanoidea (Antennophorina), *Megisthanus floridanus* Banks (Megisthanidae) was reported to prey on other mites, Collembola and nematodes (Butler and Hunter 1968). Walter (2000) observed the Cercomegistoidea (Cercomegistina) *Saltiseius hunteri* Walter to prey on nematodes.

1.4.6 Suborder Sejida

This suborder comprises two superfamilies, the Sejoidea and the Heterozerconoidea (Klompen et al. 2007; Beaulieu et al. 2011). The first superfamily contains three families: Sejidae, Ichthyostomatogasteridae and Uropodellidae (Walter 2013). The most diverse of those families is Sejidae, with five genera and 46 species, which are litter inhabitants, feeding on small arthropods and nematodes (Lindquist et al. 2009). The uropodellids are fluid feeders, but their biology is unknown, while ichthyostomatogasterids can feed on fungi and mites. The Heterozerconoidea comprises two families, the Discozerconidae and Heterozerconidae (Beaulieu et al. 2011). The Discozerconidae comprises only three species collected from centipedes (Seeman and Baker 2013) and Heterozerconidae comprises 13 species; the adults of most of them are found on millipedes, while their immatures are free-living predators, feeding on mites and small arthropods (Klompen et al. 2013).

1.5 Final Remarks

As a whole, Mesostigmata have not been sufficiently studied, especially those living in the soil, where they are most numerous.

Knowledge about the diversity of soil mites is still very small in most of the world. Diversity studies should be supported, to determine new candidates to be evaluated as new alternatives for practical use. The preparation of catalogs and reviews of each group is highly desirable as these may facilitate future biodiversity studies. The availability of a catalog of the family Rhodacaridae and the revision of the family Pachylaelapidae are good starting points. Similar works should be conducted with other mite groups in an effort to enhance the practical use of these predators, for the control of pests and other undesirable organisms (as fly species).

Knowledge of the biology and ecology of edaphic mesostigmata is still more limited than knowledge about their taxonomy. The number of biological and ecological studies on the described mesostigmatid species (let aside the number of undescribed species), is very limited. Available studies most often are restricted to the evaluation of the ability of particular species to feed on pre-determined food sources, without proceeding to experimental setups simulating greenhouse or field conditions.

The limited information available indicates that some species of Rhodacaroidea are promising biological control agents, mainly of nematodes, given their ability to feed on these organisms and their body structure, allowing them to reach habitats where edaphic nematodes of several crops are found, as suggested by Stirling (2014). Their relatively tiny body and divided dorsal shield allow them sufficient maneuvering ability to penetrate deeper in the soil than other mite groups.

Still for rhodacaroids and other Mesostigmata, it seems possible to make use of these organisms using a different biological control strategy, that is, conservation. This was elegantly discussed by Stirling (2014), and involves the use of plant residues to cover the soil surface (mulching), in an attempt to turn the soil in agroecosystems more similar to areas with natural vegetation, where those mites are usually abundant. This would probably allow better control of pest organisms that spend part of their life in the soil, as fungus gnat larvae, thrips pre-pupae and pupae, as well as nematodes in the top soil layer.

There are some reports about the predatory habit of trigynaspids mites. Most of these reports are based on field observations but few research works were conducted up to now. The family Triplogyniidae seems to have an interesting potential as biocontrol agents, considering observations by Seeman and Walter (1997), and our observations on the species *Funkotriplogynium irapora*. These mites have a generalist feeding behavior, big size and apparently a high reproductive capacity.

Encouraging examples warranting further studies include the results presented by Al-Amidi and Downes (1990) and Al-Amidi et al. (1991) with the parasitid *P. bituberosus* used against the flies *H. pygmaea* and *L. solani*, Lesna et al. (1995) with the parasitid *P. fimetorum* against the mite *R. robini*, and Castilho et al. (2009) with the rhodacarid *P. posnaniensis* for the control of the thrips *F. occidentalis*. It is possible to envision large scale production of some of these mites for periodic releases on selected crops. So far, no species of these groups have been produced commercially for practical use.

Acknowledgement To G.J. de Moraes and C.H.W. Flechtmann (Escola Superior de Agricultura "Luiz de Queiroz", Universidade de São Paulo) for evaluating a previous version of this manuscript and making very important suggestions.

References

Abo-Shnaf RIA, Castilho RC, de Moraes GJ (2013) Two new species of Rhodacaridae (Acari: Mesostigmata) from Egypt and a key to the species of the family from the Mediterranean region. Zootaxa 3718:28–38

Abou-El-Soud AB, Shoeib AA (2000) Root-knot nematode *Meloidogyne javanica* and acarid mite, *Acarus siro* L. predation by digamasellid mite, *Dendrolaelaps rasmii* Nasr and Mersal. Al-Azhar J Agri Res 31:45–50

Al-Amidi AHK, Downes MJ (1990) *Parasitus bituberosus* (Acari: Parasitidae), a possible agent for biological control of *Heteropeza pygmaea* (Diptera: Cecidomyiidae) in mushroom compost. Exp Appl Acarol 8:13–25

Al-Amidi AHK, Dunne R, Downes MJ (1991) *Parasitus bituberosus* Karg 1972 an agent for control of *Lycoriella solani* (Diptera: Sciaridae) in mushroom crops. Exp Appl Acarol 11:159–166

Al-Atawi F, Klompen H, Moser JC (2002) Redescription of *Schizosthetus lyriformis* (McGraw and Farrier, 1969) (Parasitiformes: Parasitidae), with revision of the genus. Int J Acarol 28:341–360

Athias-Henriot C (1982) *Schizosthetus* n.g. (type *Eugamasus lyriformis* McGr. and Farr 1969) avec deux espèces nouvelles (Parasitiformes: Parasitidae). Acarologia 23:207–214

Autuori M (1942) Contribuição para o conhecimento da sauva (*Atta* spp. – Hymenoptera - Formicidae). II – O sauveiro inicial (*Atta sexdens rubropilosa* Forel, 1908). Arq Inst Biol 13:67–86

Axtell RC (1962) Macrochelidae (Acarina: Mesostigmata) inhabiting manure and their effect on house fly production. Dissertation, Cornell University, Ithaca

Axtell RC (1963) Effect of Macrochelidae (Acarina: Mesostigmata) on house fly production from dairy cattle manure. J Econ Entomol 56:317–321

Axtell RC (1986) Fly management in poultry production: cultural, biological and chemical. Poult Sci 65:657–667

Axtell RC, Rutz DA (1986) Role of parasites and predators as biological fly control agents in poultry production facilities. Misc Publ 61:88–100

Bai XL, Gu YM, Fang L (1995) A new genus and species of Ameroseiidae. Acta Zootaxonomica Sinica 20:435–438

Bajerlein D, Witaliński W (2012) Anatomy and fine structure of pedicellar glands in phoretic deutonymphs of uropodid mites (Acari: Mesostigmata). Arthropod Structure Dev 41:245–257

Bajerlein D, Witaliński W (2014) Localization and density of phoretic deutonymphs of the mite *Uropoda orbicularis* (Parasitiformes: Mesostigmata) on *Aphodius* beetles (Aphodiidae) affect pedicel length. Naturwissenschaften 101:265–272

Beaulieu F, Walter DE (2007) Predation in perched and forest floor soils: observations on Australian mesostigmatic mites. Acarologia 47:43–54

Beaulieu F, Dowling APG, Klompen H, de Moraes GJ, Walter DE (2011) Superorder Parasitiformes Reuter, 1909. In: Zhang Z-Q (ed) Animal biodiversity: an outline of higher-level classification and survey of taxonomic richness. Zootaxa 3148:123–128

Binns ES (1982) Phoresy as migration – some functional aspects of phoresy in mites. Biol Rev 57:571–620

Bregetova NG (1977) Family Ameroseiidae (Berlese, 1919) Evans, 1961. In: Gilyarov MS, Bregetova NG (eds) Key to the soil inhabiting mites. Mesostigmata. Akademia Nauka, Leningrad, pp 148–169

Brown JM, Wilson DS (1994) *Poecilochirus carabi*: behavioral and life-history adaptations to different hosts and the consequences of geographical shifts in host communities. In: Houck

MA (ed) Mites: ecological and evolutionary analyses of life-history patterns. Chapman and Hall, Inc., New York, p 357
Butler L, Hunter PE (1968) Redescription of *Megisthanus floridanus* with observations on its biology (Acarina: Megisthanidae). Fla Entomol 51:187–197
Castilho RC, de Moraes GJ, Silva ES, Silva LO (2009) Predation potential and biology of *Protogamasellopsis posnaniensis* Wiśniewski and Hirschmann (Acari: Rhodacaridae). Biol Control 48:164–167
Castilho RC, Iwanicki N, Narita JPZ, de Moraes GJ (2010) Predation potential of *Gamasiphis paulista* Castilho, Moraes and Narita (Acari: Ologamasidae). Abstract book 13th international congress of Acarology, Brazil, p 47
Castilho RC, de Moraes GJ, Halliday B (2012a) Catalogue of the mite family Rhodacaridae Oudemans, with notes on the classification of the Rhodacaroidea (Acari: Mesostigmata). Zootaxa 3471:1–69
Castilho RC, Narita JPZ, de Moraes GJ (2012b) Three new species of *Gamasiphis* (Acari: Mesostigmata: Ologamasidae) from Brazil, with complementary information about *Gamasiphis plenosetosus* Karg and a key to the world species of the genus. J Nat Hist 46:1969–1998
Datta AK (1985) A new genus and species of the family Triplogyniidae (Acari: Mesostigmata) from Assam, India. Indian J Acarol 9:48–56
de Moraes GJ, Narita JPZ (2010) Description of a new species of *Neocypholaelaps* (Acari: Ameroseiidae) from Brazil, with a key to the world species. Zootaxa 2554:37–44
El-Banhawy EM, El-Borolossy MA, El-Sawaf BM, Afia I (1997) Biological aspects and feeding behavior of the predaceous soil mite *Nenteria hypotrichus* (Uropodina: Uropodidae). Acarologia 38:357–360
Elsen P (1972) Sept nouvelles espèces de *Neocypholaelaps* Vitzthum, 1941 (Mesostigmata: Ameroseiidae) phoretiques d'Hymenoptères africains. Rev Zool Bot Afr 86:17–30
Elsen P (1973) Nouvelles espèces d'*Ameroseius* Berlese 1903 (Acari: Mesostigmata: Ameroseiidae) phoretiques d'Hymenoptères et de Coleoptères. Rev Zool Bot Afr 87:729–757
Elsen P, Whitaker JO Jr (1985) *Sertitympanum*, a new genus of Ameroseiidae (Acarina, Mesostigmata) taken from rodents in United States: Including descriptions of three new species in the genus. Acarologia 26:117–122
Enda N, Tamura H (1977) Nematophagous mites found on adults of the Japanese pine sawyer. Shinrin Boecki 26:188–190
Enda N, Tamura H (1980) Nematode feeding mites found on adults of the Japanese pine sawyer. Shinrin Boecki 29:17–20
Evans GO (1963a) Observations on the chaetotaxy of the legs in the free–living Gamasina (Acari: Mesostigmata). Bull Nat Hist Mus Zool 10:277–303
Evans GO (1963b) The Genus *Neocypholaelaps* Vitzthum (Acari: Mesostigmata). Ann Mag Nat Hist 6:209–230
Evans GO (1992) Principles of acarology. CAB International, Wallingford, p 563
Fan QH, Jiang F (2014) The tiny flower fairies *Neocypholaelaps indica* Evans, 1963 (Acari: Ameroseiidae). Syst Appl Acarol 19:248–249
Flechtmann CAH (1985) On the biology of *Ameroseius dendrovagans* (Acari: Mesostigmata: Ameroseiidae). Rev Bras Zool 2:397–399
Franks NR, Healey KJ, Byrom L (1991) Studies on the relationship between the ant ectoparasite *Antennophorus grandis* (Acarina: Antennophoridae) and its hosts *Lasius flavus* (Hymenoptera: Formicidae). J Zool 225:59–70
Funk RC (1980) Generic revision of the family Euzerconidae (Mesostigmata: Celaenopsoidea) with description of seven new genera. Int J Acarol 6:313–349
Gerson U, Smiley RL, Ochoa R (2003) Mites (Acari) for pest control. Blackwell Science, Oxford, p 539
Gomaa EA, Abow-Awad BA, Nasr AK, Abou-Elela MM (1989) Life-history studies and feeding behaviour of the predatory mite, *Pachylaelaps aegypticus*, with description of immature stages (Acari: Pachylaelapidae). Insect Sci Appl 10:691–698

Gressit JL (1967) Notes on arthropod populations in the Antarctic Peninsula – South Shetland Islands – South Orkney Islands area. In: Gressit JL (ed) Entomol Antarctica. Antarctic Res Ser, American Geophysical Union, Washington, 10:373–395

Halliday RB (1997) Revision of the Australian Ameroseiidae. Invertebr Taxon 10:179–201

Halliday RB (2006) New taxa of mites associated with Australian termites (Acari: Mesostigmata). Int J Acarol 32:27–38

Hirschmann W (1960) Gangsystematik der Parasitiformes Teil 3. Die Gattung *Dendrolaelaps* Halbert, 1915. Acarologie Schriftenreihe für Vergleichende Milbenkunde 3:1–27

Hirschmann W, Wiśniewski J (1982) Weltweite revision der Gattungen *Dendrolaelaps* Halbert 1915 und *Longoseius* Chant 1961 (Parasitiformes). Acarologie. Schriftenreihe für Vergleichende Milbenkunde 29(1), 1–190, 29(2), 1–48 + Plates I–XIV + Plates 1–94

Ho CC, Ma LM, Wang SC, Severinghaus LL (2010) A new species and five new records of Ameroseiidae from Taiwan. Formosan Entomol 30:87–97

Huck K, Schwarz HH, Schmid-Hempel P (1998) Host choice in the phoretic mite *Parasitellus fucorum* (Mesostigmata: Parasitidae): which bumblebee caste is the best? Oecologia 115:385–390

Hughes AM (1976) The mites of stored food and houses, 2nd edn, Technical Bulletin 9. Ministry of Agriculture, Fisheries and Food. Her Majesty's Stationery Office, London, p 400

Hunter PE, Davis R (1965) Mites associated with the passalus beetle III. Life stages and observations on the biology of *Euzercon latus* (Banks) (Acarina: Euzerconidae). Acarologia 7:30–42

Hurlbutt HW (1979) *Gamasellus* (Acarina: Mesostigmata) from east African mountains. Acarologia 20:173–187

Hyatt KH (1980) Mites of the subfamily Parasitinae (Mesostigmata: Parasitidae) in the British Isles. Bull British Mus (Nat Hist) Zool 38:237–378

Ignatowicz S (1974) Inter and intraspecific competition in mites: *Parasitus coleoptratorum* (L.) and *Macrocheles glaber* (Müller). In: Proceedings of the 4th international congress of Acarology, Saalfelden am Steinernen Meer, 1974, pp 623–625

Ishikawa K (1980) Taxonomic and ecological studies in the family Parholaspidae (Acari: Mesostigmata) from Japan (Part 4). Rep Res Matsuyama Shinonome Jr College 11:183–190

Ito Y (1977) Predatory activity of mesostigmatid mites (Acarina: Mesostigmata) for house fly eggs and larvae under feeding of nematodes. Jpn J Sanit Zool 28:167–173

Jalil M, Rodriguez JG (1970) Biology of and odor perception by *Fuscuropoda vegetans* (Acarina: Uropodidae), a predator of the housefly. Ann Entomol Soc America 63:935–938

Juvara-Bals I (1972) *Mixogamasus* un nouveau genre de Parasitidae (Acariens anactinotriches) de Roumanie. Acarologia 14:3–114

Karg W (1961) Oekologische Untersuchungen von edaphischen Gamasiden (Acarina: Parasitiformes) 2 Teil. Pedobiologia 1:77–98

Karg W (1971) Acari (Acarina), Milben. Unterordnung Anactinochaeta (Parasitiformes). Die freilebenden Gamasina (Gamasides), Raubmilben. Die Tierwelt Deutschlands 59:1–475

Karg W (1972) Zur Kenntnis der Gattung *Parasitus* Latreille, 1795 (Acarina: Parasitiformes) aus Komposterden und Gurkenkulturen. Dtsche Entomol Z 19:55–63

Karg W (1993) Acari (Acarina), Milben Parasitiformes (Anactinochaeta) Cohors Gamasina Leach, Raubmilben. Die Tierwelt Deutschlands. 59. Teil, p 523

Karg W, Freier B (1995) Parasitiforme Raubmilben als Indikatoren für den ökologischen Zustand von Ökosystemen. Mitt Biol Bunden Land und Forstwirtschaft Berlin Dahlem 308:1–96

Karg W, Schorlemmer A (2013) Origin of five unique mite-genera in geological periods compared to other group of Gamasina (Acarina: Parasitiformes) and description of two new species of *Rykellus* Lee and *Oloopticus* Karg. Zoosystematics Evol 89:193–207

Kazemi S, Arjomandi E, Ahangaran Y (2013) A review of the Iranian Parasitidae (Acari: Mesostigmata). Persian J Acarol 2:159–180

Kethley JB (1977) A review of the higher categories of Trigynaspida (Acari: Parasitiformes). Int J Acarol 3:129–149

Kiełczewski B, Bałazy S (1966) Zagadnienie drapieznictwa roztoczy (Acarina) na jajach korników (Scolytidae: Coleoptera). Ekologia Polska, ser. B, 12:161–163 [in polish]
Kinn DN (1967) Notes on the life cycle and habitats of *Digamasellus quadrisetus* (Mesostigmata: Digamasellidae). Ann Entomol Soc America 60:862–865
Kinn DN (1971) The life cycle and behaviour of *Cercoleipus coleonotus* (Acarina: Mesostigmata) including a survey of phoretic mite associates of California Scolytidae. University of California Publications in Entomology. 65. University of California Press, Berkeley, p 62
Kinn DN (1980) Mutualism between *Dendrolaelaps neodisetus* and *Dendroctonus frontalis*. Environ Entomol 9:756–758
Klompen H, Lekvelshvili M, Black WC (2007) Phylogeny of parasitiform mites (Acari) based on rRNA. Mol Phylogenet Evol 43:936–951
Klompen H, Amin M, Gerdeman BS (2013) A revision of the genus *Afroheterozercon* (Acari: Heterozerconidae). Zootaxa 3626:301–325
Koch CL (1839) Deutschlands Crustaceen, Myriapoden und Arachniden. Ein Beitrag zur deutschen Fauna. Herausgegeben von Dr. Herrich–Schäffer. Friedrich Pustet, Regensburg 24, p 24
Kontschán J, Tambe JT, Riolo P (2012) *Urobovella phoenicicola* sp. n., a new Uropodina mite (Acari: Mesostigmata) associated with the African palm weevil (*Rhynchophorus phoenicis* Fabricius, 1801) from Cameroon. African Invertebr 53:593–600
Korn W (1982) Zur Fortpflanzung von *Poecilochirus carabi* G. and R. Canestrini 1882 (syn. *P. neerophori* Vitzt.) und *P. austroasiaticus* Vitzthum 1930 (Gamasina: Eugamasidae). Spixiana (Muench) 5:261–288
Korn W (1983) Zur Vergesellschaftung der Gamasidenarten *Poecilochirus carabi* G. and R. Canestrini 1882 (=*P. necrophori* Vitzthum 1930), *P. austroasiaticus* Vitzthum 1930 und *P. subterraneus* Müller 1859 mit Aaskäfern aus der Familie der Silphidae. Spixiana (Muench) 6:251–279
Koulianos S, Schwarz HH (1999) Reproduction, development and diet of *Parasitellus fucorum* (Mesostigmata: Parasitidae), a mite associated with bumblebees (Hymenoptera: Apidae). J Zool Soc London 248:267–276
Krantz GW (1978) A manual of acarology, 2nd edn. Oregon State University Bookstores, Corvallis, p 509
Krantz GW (2000) Two new species of the genus *Laelaptonyssus* Womersley from North America and Australia, with observations supporting the reinstatement to family level of the subfamily Laelaptonyssinae *sensu* Lee, 1970 (Acari: Mesostigmata: Rhodacaroidea). Acarologia 41:25–38
Krantz GW, Mellot JC (1972) Studies on phoretic specificity in *Macrocheles mycotrupetes* and *M. peltotrupetes* Krantz and Mellot, associates of geotrupine Scarabaeidae. Acarologia 14:317–344
Latreille PA (1795) Observations sur la variété des organes de la bouche des Tiques, et distribuition méthodique des insectes de cette familie. Magasin encyclopédique, ou Jornal des Sciences, des Lettres et des Arts, Redigé par Millin, Noel et Warens, Paris 4:15–19
Lee DC (1970) Rhodacaridae (Acari: Mesostigmata); classification, external morphology and distribution of genera. Rec South Aust Mus 16:1–219
Lee DC (1974) Rhodacaridae (Acari: Mesostigmata) from near Adelaide, Australia. III. – Behaviour and development. Acarologia 16:21–44
Lesna I, Sabelis MW, Bolland HR, Conijn CGM (1995) Candidate natural enemies for control of *Rhizoglyphus robini* Claparède (Acari: Astigmata) in lily bulbs: exploration in the Weld and preselection in the laboratory. Exp Appl Acarol 19:655–669
Lindquist EE (1975) *Digamasellus* Berlese, 1905, and *Dendrolaelaps* Halbert, 1915, with descriptions of new taxa of Digamasellidae (Acarina: Mesostigmata). Canadian Entomol 107:1–43
Lindquist EE, Kantz GW, Walter DE (2009) Order Mesostigmata. In: Krantz GW, Walter DE (eds) A manual of acarology, 3rd edn. Texas Tech University Press, Lubbock, pp 124–232
Lister A, Block W, Usher MB (1988) Predation in an Antarctic micro-arthropod community. J Anim Ecol 57:957–970

Marchenko I (2013a) A new species of *Gamasiphis* Berlese (Acari: Ologamasidae) from North Asia, with a key to the Eurasian species. Zootaxa 3626:381–390
Marchenko I (2013b) A new species of *Gamasiphis* Berlese (Acari: Ologamasidae) from Russia (Sakhalin and Kuril Islands) with a key to the Asian species. Zootaxa 3741:172–180
Mašán P (1998) *Ameroseius fungicolis* sp. n. and *A. callosus* sp. n., two new ameroseiid species (Acarina: Mesostigmata) associated with wood-destroying fungi. Biologia (Bratislava) 53:645–649
Mašán P, Halliday B (2010) Review of the European genera of Eviphididae (Acari: Mesostigmata) and the species occurring in Slovakia. Zootaxa 2585:1–122
Mašán P, Halliday B (2014) Review of the family Pachylaelapidae (Acari: Mesostigmata). Zootaxa 3776:1–66
Michael AD (1886) Observations upon a species of *Gamasus* supposed to be unrecorded. J Quekett Microscopical Club 2:260–268
Micherdzinski W (1969) Die Familie Parasitidae Oudemans 1901 (Acarina, Mesostigmata. Zaklad Zoologii Systematycznej). Polskiej Akademii Nauk, Cracowm, p 690
Moser JC (1975) Mite predators of the southern pine beetle. Ann Entomol Soc America 68:1113–1116
Moustafa MA, El-Hady MM (2006) Biocontrol of some sclerotia forming fungi attacking green beans plant using two fungivorous mites and *Trichoderma harzianum*. J Agric Sci, Mansoura Univ 31:3239–3245
Muraoka M, Ishibashi N (1976) Nematode-feeding mites and their feeding behaviour. Appl Entomol Zool 11:1–7
Narita JPZ, de Moraes GJ, Fernando LCP (2011) Two species of *Neocypholaelaps* from Sri Lanka (Acari: Ameroseiidae), with description of a new species. Zootaxa 2741:59–65
Narita JPZ, Pédelabat M, de Moraes GJ (2013a) A new species of *Neocypholaelaps* Vitzthum (Acari: Ameroseiidae), with notes on the cheliceral lobes and ventral pore-like structures of mites of this family. Zootaxa 3666:1–15
Narita JPZ, Bernardi LFO, Ferreira RL, de Moraes GJ (2013b) A new species of *Ameroseius* Berlese from Brazil, redescriptions of *Ameroseius plumosus* (Oudemans) and *Ameroseius plumigera* (Oudemans) (Acari: Mesostigmata: Ameroseiidae) based on the examination of type material. J Nat Hist 47:2311–2326
Neumann KW (1943) Die Lebensgeschichte der Kaefermilbe *Poecilochirus necrophori* Vitzth. nebst Beschreibung aller Entwicklungsstufen. Zoologischer Anzeiger 142:1–21
Norton RA, Kethley JB, Johnston DE, O'Connor BM (1993) Phylogenetic perspectives on genetic systems and reproductive modes of mites. In: Wrensch DL, Ebbert MA (eds) Evolution and diversity of sex ratio in insects and mites. Chapman and Hall Publishers, New York, pp 8–99
Noske RA (1993) *Bruguiera hainesii*: another bird-pollinated mangrove? Biotropica 25:481–483
Pereira C, Castro MP (1946) Cultura de ácaros e de alguns outros aracnídeos predadores. Livro de Homenagem a R.F. d'Almeida 33:285–287
Perotti MA, Braig HR (2009) Phoretic mites associated with animal and human decomposition. Exp Appl Acarol 48:85–124
Ramanan RV, Ghai S (1984) Observations on the mite *Neocypholaelaps indica* Evans and its relationship with the honey bee *Apis cerana indica* Fabricius and the flowering of *Eucalyptus* trees. Entomon 9:291–292
Rapp A (1959) Zur Biologie und Ethologie der Käfermilbe *Parasitus coleoptratorum* L. 1758. Ein Beitrag zum Phoresie Problem. Zool Jahrb, Abt Syst. Öekol Geog Tiere 86:303–366
Raut SK (1996) Factors determining the effectiveness of the mites *Fuscuropoda marginate* in the control of the slug pests *Laevicaulis alte*. In: Henderson IF (ed) Slug and snail pests in agriculture. British Crop Protection Council, Farnham, pp 247–254
Richards LA, Richards KW (1976) Parasitid mites associated with bumblebees in Alberta, Canada (Acarina: Parasitidae; Hymenoptera: Apidae). II. Biology. The Univ Kansas Sci Bull 51:1–18
Roets F, Wingfield MJ, Wingfield BD, Dreyer LL (2011) Mites are the most common vectors of the fungus *Gondwanamyces proteae* in *Protea* infructescences. Fungal Biol 115:343–350

Rueda-Ramirez D, Castilho RC, de Moraes GJ (2013) Mites of the superfamily Rhodacaroidea (Acari: Mesostigmata) from Colombia, with a key for the world species of *Desectophis* Karg (Ologamasidae). Zootaxa 3734:521–535

Ryke PAJ (1962) The subfamily Rhodacarinae with notes on a new subfamily Ologamasinae (Acarina: Rhodacaridae). Entomol Ber (Amst) 22:155–162

Salmane I, Brumelis G (2010) Species list and habitat preference of Mesostigmata mites (Acari: Parasitiformes) in Lativia. Acarologia 50:373–394

Santos JC, Castilho RC, Silva ES, de Moraes GJ (2013) A new species of *Hydrogamasellus* (Acari: Mesostigmata: Ologamasidae) from Brazil, with a key to the world species of the genus. Zootaxa 3718:81–88

Sardar MA, Murphy PH (1987) Feeding tests of grassland soil-inhabiting gamasine predators. Acarologia 28:117–121

Schaefer CW, Peckham DB (1968) Host preference studies on a mite infesting the cockroach *Gromphadorhina portentosa*. Ann Entomol Soc America 61:1475–1478

Schwarz HH (1996) Host range and behavioral preferences in German sibling species of the *Poecilochirus carabi* complex (Acari: Mesostigmata: Parasitidae). Int J Acarol 22:135–140

Schwarz HH, Huck K (1997) Phoretic mites use flowers to transfer between foraging bumblebees. Insectes Sociaux 44:303–310

Schwarz HH, Müller JK (1992) The dispersal behaviour of the phoretic mite *Poecilochirus carabi* (Mesostigmata: Parasitidae) – adaptation to the breeding biology of its carrier *Necrophorus vespilloides* (Coleoptera: Silphidae). Oecologia 89:487–493

Schwarz HH, Walzl MG (1996) Pairing, oviposition and development in two sibling species of phoretic mites (Acari: Mesostigmata: Parasitidae: *Poecilochirus* spp.) associated with burying bettles (Coleoptera: Silphidae: *Nicrophorus* spp.). J Nat Hist 30:1337–1348

Schwarz HH, Huck K, Schmid-Hempel P (1996) Prevalence and host preferences of mesostigmatic mites (Acari: Anactinochaeta) phoretic on Swiss bumble bees (Hymenoptera: Apidae). J Kansas Entomol Soc 69:35–42

Scoble J, Clarke MF (2006) Nectar availability and flower choice by eastern spinebills foraging on mountain correa. Anim Behav 72:1387–1394

Seeman OD, Baker M (2013) A new genus and species of Discozerconidae (Acari: Mesostigmata) from carabid beetles (Coleoptera: Carabidae) in New Zealand. Zootaxa 3750:130–142

Seeman OD, Walter DE (1995) Life history of *Afrocypholaelaps africana* (Evans) (Acari: Ameroseiidae), a mite inhabiting mangrove flowers and phoretic on honeybees. Aust J Entomol 34:45–50

Seeman OD, Walter DE (1997) A new species of Triplogyniidae (Mesostigmata: Celaenopsoidea) from Australian rainforests. Int J Acarol 23:49–59

Sharma RD (1971) Studies on the plant parasitic nematode *Tylenchorhynchus dubius*. Mededelingen Van de Faculteit Landbouwwetenschappen 71:1–154

Shcherbak GI (1980) Kleschi Semeĭstva Rhodacaridae Palearktiki. Naukova Dumka, Kiev, p 216

Sheals JG (1956) Notes on a collection of soil Acari. Entomol Monthly Mag 92:99–103

Sokolov II (1934) Untersuchungen über die Spermatogenese bei den Arachniden. V. Uber die Spermatogenese der Parasitidae. Z Zellforsch 21:42–109

Stebbing PD (1965) A study of the mite *Parasitus bomborum*. Essex Naturalist 3:284–287

Stirling GR (2014) Biological control of plant-parasitic nematodes: soil ecosystem management in sustainable agriculture. In: Stirling GR (ed) Biological control of plant-parasitic nematodes: soil ecosystem management in sustainable agriculture. Biological Crop Protection Pty Ltd, Brisbane, pp 157–192

Stone PC, Ogles GD (1953) *Uropoda agitans*, a mite pest in commercial fishworm beds. J Econ Entomol 46:711

Tappan WB (1959) Mite control in redworm beds. Fla Entomol 42:31–34

Tichomirov SI (1977) Parasitidae. In: Ghilyarov MS, Bregetova NG (eds) A key to the soil inhabiting mites. Mesostigmata. Zoological Institute of the Academy of Sciences, USSR, Leningrad, pp 55–107 (in Russian)

Tseng YH (1995) A taxonomical study of free-living gamasine mite family Parasitidae Oudemans (Acari: Mesostigmata) from Taiwan. J Taiwan Mus 48:11–81

Van den Bund CF (1972) Some observations on predatory action of mites on nematodes. Zeszyty Problemowe Postepow Nauk Rolniczych 129:103–110

Villegas-Guzmán GA, Montiel-Parra G, Vargas M, Polaco O (2004) Ameroséidos asociados a nidos de *Neotoma mexicana* Baird, 1855 (Rodentia: Muridae) y descripción de una nueva especie del género *Sertitympanum* Elsen y Whitaker, 1985. Acta Zool Mex 20:27–36

Walter DE (1997) Heatherellidae – a new family of Mesostigmata (Acari: Parasitiformes) based on two species form rainforest litter in Australia. Int J Acarol 23:167–175

Walter DE (2000) A jumping mesostigmatan mite, *Saltiseius hunteri* n. g., n. sp. (Acari: Trigynaspida: Saltiseiidae, n. fam.) from Australia. Int J Acarol 26:25–31

Walter DE (2013) A new genus and family of sejine mites (Acari: Parasitiformes: Mesostigmata, Sejoidea) based on new species from Lord Howe Island and Brazil, and a redescription of *Sejus americanus* (Banks, 1902). Zootaxa 3691:301–323

Walter DE, Ikonen EK (1989) Species, guilds, and functional groups: taxonomy and behavior in nematophagous arthropods. J Nematol 21:315–327

Walter DE, Kaplan DT (1990) A guild of thelytokous mites (Acari: Mesostigmata) associated with citrus roots in Florida. Environ Entomol 19:1338–1343

Walter DE, Oliver JH Jr (1989) *Geolaelaps oreithyiae* n. sp. (Acari: Laelapidae), a thelytokous predator of arthropods and nematodes, and a discussion of clonal reproduction in the Mesostigmata. Acarologia 30:291–303

Walter DE, Hunt HW, Elliott ET (1988) Guilds or functional groups? An analysis of predatory arthropods from a shortgrass steppe soil. Pedobiologia 31:247–260

Walters J, Campbell DK (1955) Mites of agents of natural control of Douglas-fir beetle. Canada Department of Forestry and Rural Development. Bi-Monthly Research Notes 11:3–4

Westerboer I, Bernhard F (1963) Die Familie Phytoseiidae Berlese 1916. In: Stammer H (ed) Beiträge zur Systematik und Ökologie mitteleuropaischer Acarina. Band II, Mesostigmata I. Akademische Verlagsgesellschaft Geest & Portig K.-G., Leipzig, pp 451–791

Wise GU, Kennessey MK, Axtell RC (1988) A new species of manure-inhabiting mite in the genus *Poecilochirus* (Acari: Mesostigmata: Parasitidae) predaceous on house fly eggs and larvae. Ann Entomol Soc America 81:209–224

Wiśniewski J, Hirschmann W (1993) Katalog der Ganggattungen, Untergattungen, Gruppen und Arten der Uropodiden der Erde (Taxonomie, Literatur, Grösse, Verbreitung, Vorkommen). Acarologie Schriftenreihe für Vergleichende Milbenkunde 40:1–220

Woolley TA (1988) Acarology: mites and human welfare. Wiley, New York, p 484

Wrensch DL, Johnston DE (1983) Sexual dimorphism in deutonymphs of mites of the family Parasitidae (Acari: Mesostigmata). Ann Entomol Soc America 76:473–474

Yasui Y (1997) Sperm competition and the significance of female multiple mating in the predatory mite *Parasitus fimetorum*. Exp Appl Acarol 21:651–664

Zhang Z-Q (2013) Phylum Arthropoda. Zootaxa 3703:17–26

Chapter 2
Potential of Ascidae, Blattisociidae and Melicharidae (Acari: Mesostigmata) as Biological Control Agents of Pest Organisms

Gilberto José de Moraes, Renan Venancio, Victor L.V. dos Santos, and Adilson D. Paschoal

2.1 Introduction

Interest in the use of natural enemies for the control of harmful organisms has increased considerably due to the growing interest for the consumption of goods produced without the application of pesticides, as well as because of problems related to pest resistance to pesticides. The interest of private companies in the commercial production of biological control agents has been instrumental to allow their practical use by growers.

Predatory mites are among the natural enemies considered for use in agricultural areas, mainly for the control of damaging mites and insects. Interest in the practical use of these mites has grown considerably during the last 50 years. Mites of the family Phytoseiidae (Mesostigmata) are considered the main predators of phytophagous mites and of some small insects on the aerial plant parts (McMurtry et al. 2013). Several species of this family have been widely used in agriculture.

Predatory mites of other families have been mostly used for the control of undesirable edaphic phytophagous mites and insects. Species presently produced commercially for that purpose belong to the families Macrochelidae and Laelapidae (both Mesostigmata), as observed in the websites of different companies in many countries.

In addition to the predatory mites already in use, it is conceivable that many others wait to be discovered for practical use. These new options could consist of species of those same or of other families, including species already described that still have not been adequately studied. Three of the prospective families correspond to a taxonomic group until recently referred to as Ascidae.

G.J. de Moraes (✉) • R. Venancio • V.L.V. dos Santos • A.D. Paschoal
Departamento de Entomologia e Acarologia, Escola Superior de Agricultura "Luiz de Queiroz", Universidade de São Paulo, 13418-900 Piracicaba, SP, Brazil
e-mail: moraesg@usp.br

D. Carrillo et al. (eds.), *Prospects for Biological Control of Plant Feeding Mites and Other Harmful Organisms*, Progress in Biological Control 19,
DOI 10.1007/978-3-319-15042-0_2

The term Ascidae stems from Ascinae, a subfamily proposed by Voigts and Oudemans (1905) within what they called "family Acarides" to contain only the genus *Asca* von Heyden. Baker and Wharton (1952) referred to mites of that subfamily as "Ascaidae Oudemans, 1905" [sic], which they considered to contain several genera, most of which are placed today in other families. Lindquist and Evans (1965) used the term Ascidae for the first time, including a total of 22 genera in it. The most recent taxonomic appraisal of the genera of this group was presented by Halliday et al. (1998). That concept of Ascidae remained valid until recently, when Lindquist et al. (2009) divided its members into three families, Ascidae (*sensu stricto*), Blattisociidae Garman and Melicharidae Hirschmann. For the sake of simplicity, mites of those families are referred to in this publication as the "ascid complex", and mites of the Ascidae *sensu stricto* are referred simply as Ascidae or ascid.

The main characteristic used by Lindquist et al. (2009) to separate Ascidae, Blattisociidae and Melicharidae was the constitution of their reproductive system (in accordance with what had been suggested by Athias-Henriot 1968), cheliceral elements and the extent of sclerotization of the venter of the idiosoma. Both ascids and melicharids have the so called "laelapid-type" while the blattisociids have the "phytoseiid-type" sperm access systems, often referred to as spermatheca, which contains the sperm reservoir and the associated ducts. Ascids and blattisociids have setiform *pilus dentilis* on the fixed cheliceral digit, whereas melicharids usually have a lobe instead of a setiform *pilus dentilis*. Most melicharids have reduced anal shields, whereas in many ascids and blattisociids the anal shield is fused with the ventral shield, constituting ventrianal shields of different sizes. Additionally, in ascid females the third pair of ventral lyrifissures (*iv3*) is usually located on the posterolateral corners of the sternal shield and the fourth pair of sternal setae (*st4*) is located on the unsclerotized cuticle behind the sternal shield, whereas in blattisociid and melicharid females *iv3* is usually off the sternal shield, and associated with *st4* on a pair of metasternal plates.

A large proportion of the melicharids (all *Rhinoseius* Baker and Yunker, *Spadiseius* Lindquist and Moraza, *Tropicoseius* Baker and Yunker, *Xanthippe* Naskrecki and Colwell as well as many *Proctolaelaps* Berlese), but also many ascids (mainly some *Asca*) and blattisociids (mainly some *Lasioseius* Berlese) evolved to live on plants. These melicharids feed mostly on nectar and/or pollen, a habit less common for mites of other groups of the ascid complex, but also found in some phytoseiids, mainly the *Amblyseius* Berlese and *Euseius* Wainstein (McMurtry et al. 2013).

Within the ascid complex, blattisociids are the mites morphologically most similar to phytoseiids. In addition to the shared type of sperm access system, the anatomical similarities between these groups include the low number of idiosomal setae in the blattisociid *Aceodromus* Muma and in the *phytoseioides* group of *Lasioseius*. Conversely, ascids and melicharids are most similar to Ameroseiidae. The latter also contains some genera consisting of plant inhabiting species, especially *Afrocypholaelaps* Elsen, *Hattena* Domrow and *Neocypholaelaps* Vitzthum, although the species of the other genera (actually, most species of this family) are found mainly in other habitats, especially in litter. No predaceous ameroseiid is known; instead, most ameroseiids are considered fungivorous, whereas some of the plant inhabiting ameroseiids have been shown to feed on pollen or nectar, as summarized by Narita et al. (2013).

Species of the ascid complex are most readily separated from phytoseiids by the usually larger number of dorsal shield setae (at least 20 pairs), the usual presence of seta *J1*, and the consistent presence of three or more pairs of marginal (*r-R*) setae. Species of the ascid complex and the phytoseiids are readily distinguished from the ameroseiids by possessing seta *J5* and by containing five dorsal setae on femur II (only four in ameroseiids).

2.2 Trends in the Efforts to Understand Members of the Ascid Complex

Ruf (1996) compared characteristics of development and reproduction of soil inhabiting mesostigmatid mites, including species of the ascid complex. As he then realized, possible conclusions about the role of those mites were limited, because of the scarcity of published information. Even within the most common groups, very little was known about the biology and the potential of the involved species. Unfortunately, this picture has not changed much since then.

Taxonomic studies of the ascid complex are much less numerous than those of the phytoseiids, at least in part because the species of the former group are less frequent on plants. For practical reasons, most emphasis has been placed on the study of plant inhabiting mites than on mites inhabiting the soil, where a large proportion of the ascid complex members is found.

A catalog of the species belonging to the ascid complex, under preparation by G.J. Moraes, E.P.J. Britto, J.L. Mineiro and R. B. Halliday, lists about 600 papers related to the taxonomy of this group, whereas Demite et al. (2012)'s database to the phytoseiid taxonomy lists more than 1,700 papers.

The number of studies about the biology, ecology and possible practical use of the ascid complex members has also been relatively small. In a literature review for the preparation of this chapter, about 170 papers dealing with these aspects were detected. There is no recent evaluation in this respect for the phytoseiids, but this number is certainly much higher.

Yet, the number of papers dealing with the taxonomy of the ascid complex, as well as with their biology and ecology, has increased sharply in the last decades (Fig. 2.1). Most of the publications about the taxonomy of the group refer to species from the Palearctic and Neotropical regions (Fig. 2.2), whereas those about other aspects refer to species from the Palearctic and Nearctic regions (Fig. 2.3). A single previous attempt to evaluate the zoogeographic distribution of these mites refers to an abstract (Bhattacharyya 2001) that relates the species found in India with species found elsewhere. The largest numbers of papers were published in decreasing order by W. Karg, E.E. Lindquist and L.M. Ma on taxonomy, and by E.E. Lindquist and D.E. Walter on biology/ecology (Table 2.1).

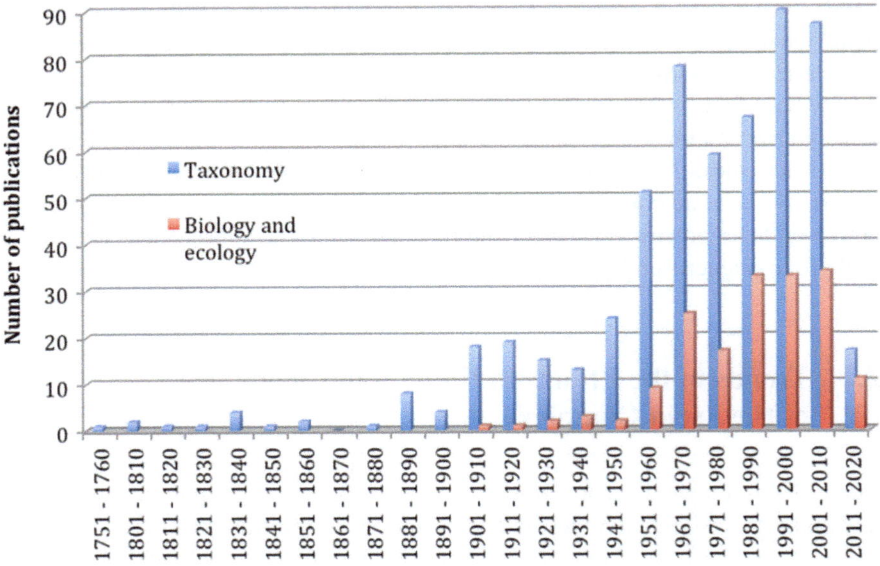

Fig. 2.1 Number of publications on "biology and ecology" and "taxonomy" of species of the ascid complex since 1751

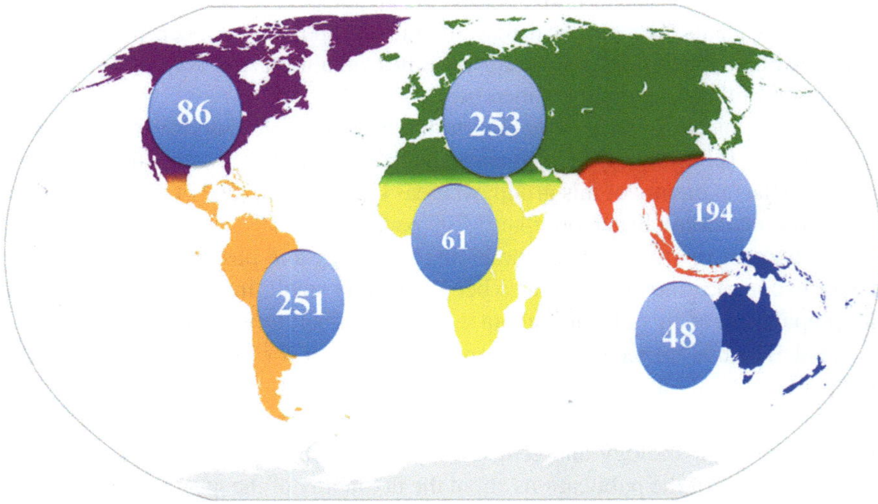

Fig. 2.2 Number of publications on taxonomy of species of the ascid complex. Nearctic (*purple*), Neotropical (*orange*), Palearctic (*green*), Ethiopian (*yellow*), Oriental (*red*) and Australasian (*blue*) (Color figure online)

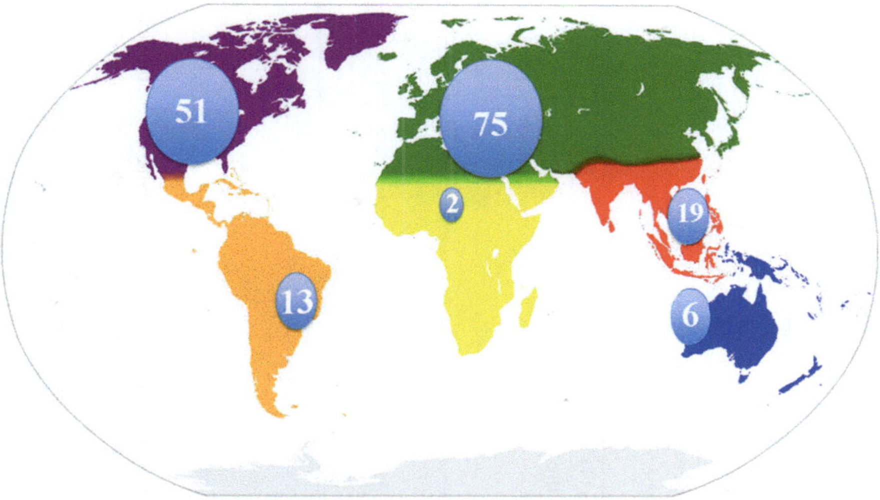

Fig. 2.3 Number of publications on biology and ecology of species of the ascid complex. Nearctic (*purple*), Neotropical (*orange*), Palearctic (*green*), Ethiopian (*yellow*), Oriental (*red*) and Australasian (*blue*) (Color figure online)

Table 2.1 Contribution of different authors to the knowledge of different aspects of the ascid complex

Taxonomy		Biology and ecology	
Author	Number of papers	Author	Number of papers
Karg, W.	41	Lindquist, E.E.	18
Lindquist, E.E.	32	Walter, D.E.	11
Ma, L.M.	25	Moraes, G.J.	7
Bhattacharyya, S.K.	24	Afifi, A.M.	5
Oudemans, A.C.	22	Colwell,	5
Berlese, A.	20	Gondim Jr, M.G.C.	5
Gwiazdowicz, D.J.	20	Karg, W.	5
Evans, G.O.	16	Mowafi, M.H.	5
Walter, D.E.	15	Nawar, M.S.	5
Fain, A.	12	Nielsen, P.S.	5
Others	<12	Others	<5

2.3 Diversity of the Ascid Complex and Association with Other Organisms

Since the most recent comprehensive appraisal of the world genera of this group (Halliday et al. 1998), few other genera have been described. In total, 41 genera containing slightly over 900 species (according to the work in preparation by G.J. Moraes, E.P.J. Britto, J.L. Mineiro and R.B. Halliday) are considered to belong to this complex (Table 2.2). The largest diversity is found in Ascidae,

Table 2.2 Diversity of the families comprising the ascid complex and habitats from which species of each genus were described (numbers in parentheses indicate numbers of species described from the respective habitat)

Genus	Approx. nr. species	Habitats (number of species)
Ascidae		
Asca	146	Soil (53), plants (50), mammals and their nests (4), tree bark (4), bird nests (3), spider nests (1), insects (1), lichen (2), unspecified (9)
Antennoseius	60	Soil (30), Carabidae (18), rodent nests (4), plants (5), unspecified (3)
Arctoseius	42	Soil (34), bark beetle galleries (1), mammal nest (1), unspecified (5)
Protogamasellus	27	Soil (23), insect nest (3), unspecified (2)
Gamasellodes	23	Soil (17), plants (4), insect nest (1), bark beetle galleries (1)
Leioseius	19	Soil (12), Crustacea (2), plants (2), tree bark (2), unspecified (1)
Zerconopsis	15	Soil (9), Diptera (1), fungi (1), unspecified (4)
Iphidozercon	10	Soil (6), fungi (1), unspecified (3)
Anephiasca	4	Soil (4)
Neojordensia	4	Soil (3), uspecified (1)
Anystipalpus	4	Carabidae (3), Dermaptera (1)
Xenoseius	3	Soil (2), Cerambycidae (1)
Iphidonopsis	2	Soil (2)
Arctoseiodes	1	Soil (1)
Athiashenriotis	1	Soil (1)
Ectoantennoseius	1	Plants (1)
Maxinia	1	Grasses (1)
Blattisociidae		
Lasioseius	191	Soil (86), plants (40), rodents or rodent nests (14), tree bark (9), insects (5), bark beetle galleries (5), bird and bird nests (5), Diplopoda (3), fungi (2), bulbs (2), other mammals or their nests (2), lab mite colonies (1), banana (2), unspecified (34)
Cheiroseius	117	Soil (85), plants (16), Tipulidae (3), rodent nests (1), bird nests (1), horse/cow dung (2), unspecified (11)
Platyseius	20	Soil (12), plants (3), unspecified (4)

Hoploseius	10	Fungus (7), Diptera (2), soil (1)
Aceodromus	2	Orthoptera (1), plant (1)
Fungiseius	2	Fungi (2)
Adhaerenseius	1	Flowers (1)
Arrhenoseius	1	Soil (1)
Cheiroseiulus	1	Plants (1)
Krantzoseius	1	Carabidae (1)
Opilioseius	1	Fungi (1)
Zercoseius	1	Soil (1)
Melicharidae		
Proctolaelaps	133	Soil (27), insects (32), plants (22), mammals (9), bark beetles and their galleries (5), hummingbirds (9), rotting fruits (4), other birds (3), stored products (2), bird and mammal nests (4), seaweed (1), tree bark (2), fungi (1), unspecified (11)
Tropicoseius	38	Plant (21), hummingbirds (17)
Rhinoseius	11	Hummingbirds (6), plants (2), flowers (3)
Mucroseius	8	Cerambycidae (8)
Melichares	3	Fruits (2), soil (1)
Orthadenella	2	Soil (2)
Spadiseius	2	Flowers (2)
Xanthippe	2	Flowers (2)
Mycolaelaps	1	Fungi (1)
Orolaelaps	1	Soil (1)
Proctogastrolaelaps	1	Bark beetle (1)

Soil: including litter, grasses, mosses, rotting trunks on the soil surface; Plant: unspecified plant part followed by Blattisociidae and Melicharidae (about 366, 365 and 204 species in 13, 17 and 11 genera, respectively).

Despite the relatively large number of species within each family, most of them belong to few genera. Only eight genera of Ascidae, five of Blattisociidae and three of Melicharidae have 10 or more species; most numerous genera are *Asca* in Ascidae (about 126 species), *Lasioseius* in Blattisociidae (about 191 species) and *Proctolaelaps* in Melicharidae (about 132 species) (Table 2.2).

Species of the ascid complex have been described from a wide range of habitats (Table 2.2). Altogether, about 53 % (452 out of 851 whose habitats were mentioned in the original description) of these species were described from soil, grasses, mosses or dead organic matter on the soil surface. However, the proportions of species from these substrates were much higher for ascids (67 % – 211 out of 311) and blattisociids (60 % – 200 out of 338) than for melicharids (20 % – 41 out of 202). The presence of these mites in the soil may interfere with nutrient cycling (Moore and Walter 1988), especially for their fungivorous behavior or for their predatory habit on other organisms involved in that process.

Among the ascids, about half of the species not described from soil were described from plant parts, whereas the remaining was mostly described in association with small animals (mostly invertebrates), which could serve as their transporters. About 45 % of the species of the *Asca* species were described from plant parts, most of them from tropical regions with high rainfall in the neotropics (De Leon 1967) or in Australia (Walter et al. 1993). Yet, *Asca foliata* Womersley has been reported from plants of a dry forest in Australia (Walter et al. 1993).

Similarly to the ascids, about half of the blattisociids not described from soil were described from plant parts, the remaining being associated mostly with small animals, especially rodents, insects and birds. Species of *Blattisocius* Keegan are particularly outstanding for their association with stored products. The association of species of this genus with insects might correspond in most cases to the transportation of the former along the fragmented habitats in which they live. Also outstanding is the almost exclusive association between *Hoploseius* Berlese with bracket fungi. The curved row of teeth at the apex of the fixed cheliceral digit of these mites apparently allows the latter to function as a rasping tool for feeding on fungal tissues (Lindquist 1970; Moraza and Lindquist 2011). The elongate idiosoma of *Hoploseius tenuis* Lindquist seems to adapt this species to live within pores of bracket fungi (Lindquist 1965; Moraza and Lindquist 2011). *Hoploseius* species have been mentioned to be morphologically adapted for phoretic dispersal on adult drosophilid flies (Lindquist 1963).

Most of the plant associate blattisociids belongs to *Lasioseius*; about 27 % of the species of this genus were described from plant parts from all over the world. Several *Lasioseius* species are known to inhabit flowers or bark beetle (Curculionidae: Scolytinae) galleries. Within the Blattisociidae, the single species of *Adhaerenseius* Loots and Theron is known only from flowers.

Among the melicharids, about 32 % and 27 % of the species not described from soil were described, respectively, from plant parts or from nectar feeding organisms (especially hummingbirds and hymenopteran insects). The vast majority of the species described in association with plants were reported to inhabit or assumed to inhabit flowers. This is particularly the case with the *Tropicoseius* and *Rhinoseius* species. Species described from birds and nectar feeding insects could also be somehow associated with plants. Similar to what was reported for *Hoploseius* within the blattisociids, is the outstanding association between the melicharids *Fungiseius* Moraza and Lindquist and *Mycolaelaps* Lindquist species with bracket fungi. The elongate idiosoma of *Mycolaelaps* species and *Fungiseius armatus* Moraza and Lindquist also seems to adapt them to live within pores of those fungi (Lindquist 1995; Moraza and Lindquist 2011); that adaptation represents a case of convergence of these mites with *H. tenuis*, which occupies the same niche (Lindquist 1995). Species of *Mucroseius* Lindquist are outstanding for their association with beetles of the genus *Monochamus* Megerle (Cerambycidae), a feared invasive pest group. The assumption that adult beetles were just serving as transporters to the mites is supported by the fact that only adults of the latter have been found on adult beetles (Lindquist 1970). This does not preclude the possible consumption by those mites of eggs of their carriers or of other mites living in the same habitat (Lindquist 1970), or perhaps of nematodes or fungi. However, several cases of predation or apparent parasitism have also been observed, as discussed in the following section.

2.4 Biological Information

Experimental data about life cycle and predation potential has been provided for species of thirteen genera of the ascid complex. However, by far most detailed information is available for species of *Arctoseius* Thor, *Gamasellodes* Athias-Henriot, *Protogamasellus* Karg, *Blattisocius*, *Lasioseius* and *Proctolaelaps*. This pattern is in part positively related to the diversity of the genera of the ascid complex, but the paucity of studies for *Asca* and *Cheiroseius* Berlese is striking, given their large species diversity. This seems particularly intriguing in relation to *Asca*, given their common occurrence on plants.

The main emphasis in this review lies on a discussion about aspects most related to the possible use of members of the ascid complex as biological control agents, namely on their association with other organisms (including their capacity to disperse between parts of the preferred habitats), feeding habits and general biology.

Functional aspects of phoresy in mites were discussed by Binns (1982), who included examples of the ascid complex. Phoresy is usually necessary for mite species specialized on living in patchy habitats that become periodically inadequate for their survival. In a detailed study about predator-prey relation, Flanders and Badgley (1963) related the ability of *Blattisocius tarsalis* Berlese to control *Anagasta kuehniella* (Zeller) in a controlled laboratory study to its phoretic behavior on the latter.

Mite feeding habits are obviously related to sensorial and physiological characteristics, as well as to the shape of the chelicerae, the main external structure involved in feeding. There is an extensive literature about the prey finding process by phytoseiids (Sznajder et al. 2010). Aspects of the sensory structures and of the physiology of members of the ascid complex have not been adequately treated in the literature. Karg (1983) reported the fact that despite not having eyes, predatory Gamasina (mesostigmatid group to which the ascid complex belongs) are sensitive to light. He also highlighted the importance of the first pair of legs in those mites as sensorial structures, being longer in more quickly moving species. The author pointed out the concentration of long and short club-shaped setae on the distal part of tarsus I, the former tactile and the latter chemo-sensitive structures, suggesting that the palps are used to scan the supporting substrate. He then described the feeding process in Gamasina.

Karg (1983) related the shape of chelicerae of Gamasina with their preference for nematophagy. According to that study, species with predominantly nematophagous feeding habit have stout chelicerae provided with short digits and multidentate fixed digit. In a detailed study by Walter and Ikonen (1989), shape of chelicerae was again related to preferences for different food items, but their conclusion did not totally agree with what was proposed by Karg (1983). Thus, relating cheliceral shape with food preference is not an easy task, most probably because cheliceral shape is just one of the factors involved in food acceptance (Esteca et al. 2014).

Most interesting results of experimental studies and reported ecological observations are subsequently presented for species representing genera of the ascid complex.

2.5 Ascidae

2.5.1 Antennoseius

Few studies have been dedicated to species of *Antennoseius* Berlese. Experimental data were reported only by Lindquist and Walter (1989) for *Antennoseius janus* Lindquist and Walter and by Halliday et al. (1998) for *Antennoseius lobochelus* Halliday, Walter and Lindquist. The former was the first Mesostigmata for which female dimorphism was experimentally demonstrated. One of the morphs is smooth (dispersing) and the other is granular (non-dispersing). The authors observed that the dispersing form was generated when mite density increased in the rearing units, but not necessarily because of food shortage. Several species of this genus have been reported on beetles of the family Carabidae, probably in phoretic associations. The fact that only adults of these ascids were reported in the respective original descriptions supports the assumption that the beetles were only serving as transporters.

The biology of *A. janus* was studied by Lindquist and Walter (1989). This species seems to be polyphagous. It was reported to complete immature development on nematodes, mites and Collembola. Non-dispersing females of this species begin oviposition within 3 days of mating, while dispersing females only initiated oviposition many weeks after mating took place. Mating has been mentioned to be required for oviposition of *A. janus* and *A. lobochelus* (Lindquist and Walter 1989; Walter and Lindquist 1995; Halliday et al. 1998). Thus, Lindquist and Walter (1989) stated that *A. janus* probably has a diplo-diploid sex-determining mechanism, without ruling out the possibility of pseudo-arrhenotoky.

2.5.2 Arctoseius

As the name of this genus suggests (a "seius" from the Arctic), these mites are typically found in cold regions of the world. They are the dominant mesostigmatid mites in the Alaskan tundra in terms of abundance and diversity (Lindquist 1961; Binns 1972), although some species have been found in areas not subjected to extremely low temperatures, such as *Arctoseius bilinear* Nasr, described from Giza, Egypt. Species of this genus have also been found in high places (above 4,000 m) in Ecuador (Karg 1998).

Karg (1961) apud Binns (1972) reported *Arctoseius cetratus* (Sellnick) not to feed on nematodes, while Binns (1972) reported the opposite. A few years later, Karg (1983) mentioned the shape of *Arctoseius* chelicerae to be appropriate for nematophagy. Laboratory investigations confirmed the ability of *A. cetratus* (Binns 1974; Sardar and Murphy 1987; Walter and Ikonen 1989) and *Arctoseius semiscissus* (Berlese) (Rudzinska 1998) to develop and reproduce on nematode prey.

Feeding on drosophilid larvae was reported by O'Farrell and Butler (1948) for *Arctoseius butleri* (Hughes). Binns (1972) reported *A. cetratus* to prey on eggs and first instar larvae of *Drosophila melanogaster* Meigen (Drosophilidae) and of *Lycoriella auripila* (Winnertz) (Sciaridae), as well as on small larvae [sic] of *Heteropeza pygmaea* Winnertz (Cecidomyiidae). Karg (1961) apud Binns (1972) reported *A. cetratus* to feed on collembolans but not on sciarid larvae. Laboratory investigations confirmed the ability of *A. cetratus* (Binns 1974; Walter and Ikonen 1989) and *A. semiscissus* (Rudzinska 1998) to develop and reproduce on sciarid larvae, and of the former species to feed on collembolans (Sardar and Murphy 1987).

Weis-Fogh (1948) apud Sardar and Murphy (1987) stated that *A. cetratus* did not feed on astigmatid mites (Astigmatina), and this was also observed by Sardar and Murphy (1987) when exposing *Acarus siro* L. and *Tyrophagus putrescentiae* (Schrank) to the same predator species. Karg (1961) apud Binns (1972) reported the same ascid species to feed on Oribatida; observations later confirmed by laboratory investigations (Sardar and Murphy 1987). Predation on another mite family, Tarsonemidae, has also been reported; Binns (1972) observed the predation of *Tarsonemus myceliophagus* Hussey by *A. cetratus*.

Phoresy seems to be common among mites of this genus. It has been reported for *A. cetratus* on *L. auripila* (Binns 1972), for *Arctoseius* sp. on *Trichocera* sp. (Trichoceridae) (Makarova 1995) and for *A. semiscissus* on unspecified sciarids (Dmoch 1995).

The life cycle of only two *Arctoseius* species has been studied. Mating has been reported as required for oviposition of *A. cetratus*, *Arctoseius confusus* Lindquist and *Arctoseius idiodactylus* Lindquist (Walter and Lindquist 1995).

2.5.3 Asca

Not much has been published about the association of these mites with other arthropods, but *Asca macromela* Walter, Halliday and Lindquist was suspected by Walter et al. (1993) to be a predator of *Brevipalpus* Donnadieu species (Tenuipalpidae), with which *A. macromela* was associated on *Diploglottis diphyllostegia* (F. Muell.) (Sapindaceae) about 26 m above ground level, near Atherton, Australia, where rainfall averages over 1,200 mm yearly.

Asca species have seemingly not been reported phoretically associated with insects, although a few species have been described from birds, mammals or from their nests (Table 2.2). Apparently, no study on the life cycle of *Asca* species has been conducted. However, laboratory investigations have shown that mating is a pre-requisite for oviposition of *Asca brachychaeta* Hurlbutt, *A. macromela*, *Asca mindi* Walter, Halliday and Lindquist, *Asca nesoica* Athias-Henriot and *Asca piloja* Hurlbutt (Walter and Lindquist 1995).

2.5.4 Gamasellodes

Gamasellodes vermivorax Walter was observed by Walter (1987) to feed on several nematode groups, namely species of *Acrobeloides* (Cobb), *Panagrolaimus* Fuchs, *Aphelenchoides* Fisher, as well as on *Aphelenchus avenae* Bastian and the entomopathogenic nematode *Neoaplectana carpocapsae* Weiser. *Gamasellodes adriannae* Walter was reported as a voracious predator of nematodes (Walter and Lindquist 2000; Walter 2003; Beaulieu and Weeks 2007). The almost exclusive findings of species of this genus in the soil and their small size suggest that nematodes could be one of their food sources. *Gamasellodes rectiventris* Lindquist was one of the few species of this genus not described from soil. The finding of this species in association with bark beetles (Lindquist 1971) might be related to the high abundance of nematodes in their galleries (Lindquist 1969). Predation on nematodes was confirmed experimentally under laboratory conditions for *G. adrianae* (Walter 2003), *Gamasellodes bicolor* (Berlese) (Walter and Ikonen 1989), *Gamasellodes claudiae* Walter (Walter and Ikonen 1989) and *G. vermivorax* (Walter 1987). These studies confirmed adequate development and reproduction of these mites on nematodes.

Walter (1987) observed *G. vermivorax* to be able to feed on the mobile stages (but not on eggs) of the collembolans *Isotoma notabilis* Schäffer and *Tullbergia granulata* Mills. He observed that predator to be able to feed on mites of very different groups (including predatory mites), namely of the families Alicorhagiidae and Tydeidae, as well as on *Asca nesoica* Athias-Henriot (Ascidae), *Cosmolaelaps vacua* (Michael) (Laelapidae), *Pilogalumna* sp. (Oribatida) and *T. putrescentiae*. However, *G. vermivorax* was unable to complete immature development on *T. putrescentiae*, whereas it completed immature development but did not oviposit when offered the mites *Opiella nova* Oudemans (Oribatida), *Alicorhagidia fragilis* Berlese (Alicorhagiidae) and *Eupodes* sp. (Eupodidae) as food sources.

Phoresy does not seem common for mites of this genus. A single species of *Gamasellodes* (Table 2.2) was described from a rodent nest [*Gamasellodes shealsi* Costa, junior synonym of *G. bicolor*]. Information on life cycle has been reported for four *Gamasellodes* species. Larval feeding was reported by Walter (1987) for *G. vermivorax*. Reproduction of *Gamasellodes* species seems to involve mostly arrhenotoky, as reported for *G. adrianae* (Walter 2003), *G. bicolor* (Walter and Lindquist 1995), *G. claudiae* (Walter and Ikonen 1989; Walter 2003) and *G. vermivorax* (Walter 1987). However, a population of *G. bicolor* was reported to reproduce by thelytoky (Walter and Lindquist 1995).

2.5.5 Protogamasellus

Protogamasellus mica (Athias-Henriot) was shown to develop to adulthood on the nematode *Acrobeloides* sp. by Walter and Lindquist (1989), while *Protogamasellus minutus* Nasr and *Protogamasellus primitivus similis* Genis, Loots and Ryke were reported to develop and reproduce on the nematodes *Acrobeloides* sp. and *Rhabditis* sp. (Afifi et al. 1986) and on an unspecified nematode species (Nawar and Nasr 1989), respectively.

Protogamasellus minutus was also reported by Afifi et al. (1986) to develop and reproduce on the collembolan *Lepidocyrtinus incertus* Handschin, as well as on the astigmatid mites *Rhizoglyphus robini* Claparède (Acaridae) and *T. putrescentiae*. Development and reproduction of *P. primitivus similis* on the same mite prey was observed by Nawar and Nasr (1989).

Mites of this genus have also been shown to be able to use fungi as food sources. *Protogamasellus mica* developed and reproduced on *Gibberella fujikurai* (Sawada) Wollenw (mentioned as *Fusarium moniliforme*) (Ibrahim et al. 1989), whereas *P. primitivus similis* was reported to develop and reproduce on *Aspergilus flavus* Link and *Haematonectria haematococca* (Berk and Broome) Samuels and Rossman (mentioned as *Fusarium solani*), but not on *Fusarium oxysporum* Schlecht, *Penicillium digitatum* (Pers.) Sacc. or *Rhizoctonia solani* Kühn (Nawar and Nasr 1989).

No species of this genus has been described in phoretic association with any animal group. The life cycle of four species has been investigated, by Afifi et al. (1986) (*P. minutus*), Walter and Lindquist (1989) (*P. mica*), Nawar and Nasr (1989) (*P. primitivus similis*) and *P. denticus* (Saber et al. 2007). Larval feeding was not observed for *Protogamasellus denticus* Nasr (Saber et al. 2007), *Protogamasellus dioscorus* Manson (Afifi 1989) and *P. minutus* (Afifi et al. 1986), but it was observed for *P. primitivus similis* (Nawar and Nasr 1989).

The main type of reproduction in this genus seems to be by thelytoky, which has been reported for *P. denticus* (Saber et al. 2007), *Protogamasellus hibernicus* Evans (Walter and Lindquist 1995), *P. mica* (Walter and Kaplan 1990; Nawar and Nasr 1989; Walter and Lindquist 1995) and six undescribed species (Walter and Lindquist 1995). However, arrhenotoky has been reported for a population of *P. mica* (Ibrahim et al. 1989) and for *P. minutus* (Afifi et al. 1986), while mating has been shown as a pre-requisite for oviposition of *P. primitivus similis* (Nawar and Nasr 1989).

Feeding habits of other ascid groups have been hardly studied. Nematophagy was reported for *Iphidozercon corticalis* Evans (Ascidae) by Halliday et al. (1998).

2.6 Blattisociidae

2.6.1 Aceodromus

Of the only two species described in this genus, *Aceodromus asternalis* Lindquist and Chant 1964 was described from specimens collected on *Caulopsis cuspidata* (Scudder) (Tettigoniidae), possibly as a phoront. Nothing else is known about that mite species. The other species, *Aceodromus convolvuli* Muma, has been reported several times on plants in tropical America. Nothing is known about its biology, but Cruz et al. (2014) showed that *A. convolvuli* occurred in considerably higher levels than phytoseiids on samples consisting of a combination of the leaves of the 11 most abundant and frequent plant species growing spontaneously in that region, especially on leaves of *Helicteres guazumifolia* Kunth (Malvaceae).

2.6.2 Blattisocius

This is one of the three most extensively studied genera of the ascid complex. Table 2.3 shows a list of studies on the life cycle and prey consumption of mites of this genus. The most extensively studied species have been often reported from storage facilities, associated with insects and mites. *Blattisocius* Keegan species have quite diverse cheliceral structures; while some have fixed and movable digits of similar length, others have the fixed digit much shorter than the movable digit (Haines 1979).

One of the most common species, *B. tarsalis*, has short fixed digit. It seems to feed mainly on insects, having been reported in association with the lepidopterans *Sitotroga cerealella* Oliver (Oudemans 1929), *A. kuehniella* (Flanders and Badgley 1963; Çobanoglu et al. 2007), *Plodia interpunctella* (Hübner) (Darst and King 1969; Riudavets et al. 2002a) and *Ephestia* (*Cadra*) *cautella* (Walker) (Graham 1970). Actual predation by *B. tarsalis* was observed on *A. kuehniella* by Burkhardt (1920) and Hase (1928), who misidentified the mite as *Sejus muricatus* Oudemans, according to Hase (1933) (who mentioned it as *Typhlodromus tineivorus* Koch) as well as by Flanders and Badgley (1963), Treat (1975) and Riudavets et al. (2002a). Predation by *B. tarsalis* was also reported on *Tineola bisselliella* (Hummel) and *Ephestia sericarium* (Scott) (Nesbitt 1951). In addition, the mite has been shown to develop and reproduce adequately on *A. kuehniella*, *E. cautella* and *P. interpunctella* (Flanders and Badgley 1963; Darst and King 1969; Haines 1981; Hafez et al. 1988; Nielsen 1999a, b, 2001, 2003; Riudavets et al. 2002b; Çobanoglu et al. 2007).

Table 2.3 Studies on the life cycle and prey consumption by *Blattisocius* species

Lasioseius species	Food source	Reference
Blattisocius dentriticus	*Tyrophagus putrescentiae*	Rivard (1960, 1962a, b)
Blattisocius keegani	*Cryptolestes turcicus*, *Amyelois transitella*, *Anagasta kuehniella*	Barker (1967)
B. keegani	*A. transitella*, *A. kuehniella*	Thomas et al. (2011)
Blattisocius tarsalis	*Plodia interpunctella*	Darst and King (1969)
B. tarsalis	*Ephestia cautella*, *Tribolium castaneum*	Haines (1981)
B. tarsalis	*T. castaneum*, *Orizophilus surinamensis*	Hafez et al. (1988)
B. tarsalis	*A. kuehniella*	Flanders and Badgley (1963), Nielsen (1999a, b, 2001, 2003)
B. tarsalis	*T. putrescentiae*, *Anagasta kuehniella*, *Plodia interpunctella*, *Acanthoscelides obtectus*, *Cryptolestes ferrugineus*, *Lasioderma serricorne*, *Rhyzopertha Dominica*, *Sitophilus oryzae*, *T. castaneum*, *Liposcelis bostrychophila*	Riudavets et al. (2002b)
B. tarsalis	*A. kuehniella*	Riudavets and Quero (2004)
B. tarsalis	*Acarus siro*	Thind and Ford (2006)
B. tarsalis	*A. kuehniella*	Çobanoglu et al. (2007)

Blattisocius tarsalis has also been mentioned to feed on several coleopteran species. Hase (1933) reported it (as *Blattisocius tineivorus* Oudemans) to prey on eggs of *Anthrenus verbasci* (L.), *Anthrenus scrophulariae* L. and *Anthrenus museorum* (L.), while Nesbitt (1951) reported it to prey on *Attagenus piceus* Olivier, *Tribolium castaneum* (Herbst) and *Tribolium confusum* (J. du Val). It was also reported to feed on *T. castaneum* by Graham (1970) and on *Lasioderma serricorne* (Fabricius), *Cryptolestes ferrugineus* (Stephens), *T. castaneum* and *Acanthoscelides obtectus* (Say) by Riudavets et al. (2002a). Further, it has been shown to develop and reproduce adequately on *L. serricorne*, *Oryzophilus surinamensis* L. and *T. castaneum* (Haines 1981; Hafez et al. 1988; Riudavets et al. 2002b). However, it was observed not to feed on eggs of *Sitophilus oryzae* (L.) or *Rhyzopertha dominica* (Fabricius) (Riudavets et al. 2002b). Riudavets and Quero (2003) showed the preference of this mite for eggs of *P. interpunctella* over eggs of *A. kuehniella* and *T. confusum*.

Feeding on other insect groups has occasionally been mentioned. Hase (1933) observed predation by *B. tarsalis* on eggs of the hemipteran *Cimex* sp. and Riudavets et al. (2002a), on the first nymphal instar of the psocopteran *Liposcelis bostrychophila* (Badonnel). However, its ability to develop and reproduce on these potential prey was not evaluated.

Despite the common occurrence of pest astigmatid mites in storage, little has been published about the relation between those organisms and *B. tarsalis*. Riudavets et al. (2002a) reported predation on *T. putrescentiae* eggs by this species, while Thind and Ford (2006) reported good performance of *B. tarsalis* on eggs but poor performance on the mobile stages of *A. siro*.

Another very common species, *Blattisocius keegani* Fox, whose fixed digit is much shorter than the movable digit, has also been reported to prey on lepidopteran species. Stein (1960) reported it to prey on an unspecified species of this group, whereas Trivedi et al. (1994) mentioned it to feed on eggs of *Phthorimaea operculella* (Zeller). It has been shown to develop and reproduce on *Amyelois transitella* (Walker) and *A. kuehniella* (Thomas et al. 2011). However, although attacking a relatively large number of eggs of *A. kuehniella* each day, very low oviposition was reported by Esteca et al. (2014) for that predator on the same prey.

Coleoptera have also been reported to be attacked by *B. keegani*. Pan (1985) suspected this species to prey on eggs of *Necrobia rufipes* (DeGeer) and *Necrobia ruficollis* (Fabricius). It has also been reported to feed on eggs of *Diaprepes abbreviatus* (L.) (Beavers et al. 1972), *Cryptolestes turcicus* (Grouvelle), *C. ferrugineus*, *O. surinamensis*, *T. castaneum*, *T. confusum* and *Trogoderma parabile* Beal (Barker 1967; Hughes 1976), and to develop and reproduce on *C. turcicus* (Barker 1967). However, oviposition was very low on *L. serricorne* and *Zabrotes subfasciatus* (Bohemann), despite the relatively high number of eggs of *L. serricorne* attacked each day (Esteca et al. 2014).

Blattisocius keegani has been reported to prey on astigmatid mites. Barker (1967) and Hughes (1976) reported it to feed on eggs and occasionally on the mobile stages of *Glycyphagus domesticus* (De Geer). However, it fed on eggs of *Aeroglyphus robustus* (Banks) only when starving, and did not feed on any developmental stage of *T. putrescentiae*. Conversely, Rezk (2000) reported to be able to maintain a colony of *B. keegani* on a mixture of *Aleuroglyphus ovatus* (Troupeau), *Dermatophagoides farinae* Hughes and *T. putrescentiae*, whereas Taha et al. (2007) reported adequate development and reproduction of this predator on larvae of *T. putrescentiae*.

Given the common occurrence of *B. keegani* in association with pest mites in storage, Athanassiou and Palyvos (2006) evaluated the effect of treatments of diatomaceous earth on this species. They concluded that *B. keegani* was susceptible to the two commercial formulations evaluated.

A possible case of parasitism by a species of this group on a noctuid moth was described by Treat (1966, 1969). The author reported *Blattisocius patagiorum* Treat, which also has fixed digit much shorter than movable digit, as a possible facultative parasite of noctuid moths (Treat 1966), as suggested by the fact that it was able to reproduce and develop on its carrier insects.

Unlike the previous *Blattisocius* species, *Blattisocius dentriticus* (Berlese) has cheliceral digits of similar lengths. Pan (1985) reported this species to feed on *T. putrescentiae*. Rivard (1960, 1962a, b) observed adequate development and reproduction of this species on the same prey. Development was faster and oviposition higher on larvae than on eggs of that prey, with progressively better survival, development and fecundity with increasing densities of prey larvae (Rivard 1960). Feeding by *B. dentriticus* on nymphs, adults and (possibly) eggs of the psocopteran *Liposcelis entomophila* (Enderlein) was shown by Mashaya (2002). In that same study, the author suggested the use of selective pesticides as a conservation measure to promote the control exerted by the predator.

Blattisocius mali (Oudemans), whose cheliceral digits are also of similar length, was reported by Hughes (1976) to be a possible predator of the astigmatid mites *A. siro*, *Carpoglyphus lactis* (L.) and *T. putrescentiae*. In a recent study conducted by Esteca et al. (2014), *Blattisocius everti* Britto, Lopes and Moraes, of digits of similar lengths, had better performance than *B. keegani* on an undescribed *Thyreophagus* Rondani species (Astigmatina).

Phoresy has been repeatedly reported for *Blattisocius* species from storage facilities. This is congruent with the need of those mites to move between parts of their patchy habitat (food storage sites). *Blattisocius tarsalis* (mentioned as *T. tineivorus*) was long reported to be commonly found on moths of *A. kuehniella* (Hase 1933). Treat (1966, 1975) and Hughes (1976) revised previous publications dealing with the *Blattisocius* species reported in association with moths, mentioning reports of *B. tarsalis*, *B. dentriticus*, *B. keegani* and *B. patagiorum* on species of a wide range of genera, all of which were microlepidopterans. Six *Blattisocius* species were described from specimens collected on insects, while one was described from specimens from rodents and one from bird nests (Table 2.2).

The studies conducted so far seem to suggest that *Blattisocius* species with reduced movable digit are better adapted to feeding on insect eggs and less adapted to feeding on mites, whereas the opposite seems to occur with *Blattisocius* species with digits of similar lengths.

Mating has been mentioned to be a pre-requisite for oviposition of *B. dentriticus* (Rivard 1960) and *B. tarsallis* (Darst and King 1969). Larval feeding has been reported as not required by *B. dentriticus* (Rivard 1960, 1962b), *B. keegani* (Barker 1967) and *B. tarsalis* (Darst and King 1969; Hafez et al. 1988), but feeding by larvae of *B. keegani* was reported by Taha et al. (2007).

The genetic system of those mites has not been adequately studied. Norton et al. (1993) considered that the chromosome counts of *B. patagiorum* by Treat (1966) suggested a haplo-diploid genetic system.

2.6.3 Cheiroseius *and* Platyseius

These are closely related genera, but while different species of *Cheiroseius* have been reported from both dry and humid environments, species of *Platyseius* Berlese have been found only in humid habitats. Species of both genera have large peritremes that function as plastron (Hinton 1971). These structures allow them to breathe underwater, with a respiratory efficiency reported to be comparable to that of many insects.

Cheiroseius species have been mentioned to feed on soil nematodes (Halliday et al. 1998). Their tweezer-like chelicerae, with a saw-shaped row of teeth on the distal section of the fixed digit, have been hypothesized to enable them to grasp nematodes in narrow spaces (Karg 1983).

Cheiroseius and *Platyseius* species have been mentioned as possible predators of mosquitoes. Smith (1983) reported *Platyseius subglaber* (Oudemans) on egg rafts of *Mansonia* (*Coquillettidia*) *perturbans* (Walker), attacking eggs and hatching larvae. No *Platyseius* but five *Cheiroseius* species have been described from insects (three species), rodent nests and bird nests (one species each) (Table 2.2). Species of both genera have been found on adult mosquitoes and crane flies (Reisen and Mullen 1978; Smith 1983; Halliday et al. 1998).

Platyseius species have not been evaluated experimentally and experimental information about *Cheiroseius* species refers only to reproduction. Mating has been mentioned as a pre-requisite for *Cheiroseius* nr. *jamaicensis* Evans and Hyatt, *Cheiroseius* nr. *mutilus* (Berlese), *Cheiroseius* nr. *neborealis* Evans and Hyatt and *Cheiroseius* nr. *parbatensis* Evans and Hyatt, while *Cheiroseius* nr. *curtipes* (Halbert) has been reported to reproduce by thelytoky (Walter and Lindquist 1995).

2.6.4 Lasioseius

This is one of the three most extensively studied genera of the ascid complex. A list of the studies about the life cycle and prey consumption of *Laseioseius* species is shown in Table 2.4. These mites show quite a wide variation in relation to items accepted as food, what has been related to the shape of the chelicerae (Walter and Lindquist 1989).

Karg (1983) reported some species of this genus to have teeth distributed along the whole edge of the fixed cheliceral digit, suggesting that this type of chelicera is most commonly found in nematophagous predators. Several species have been

Table 2.4 Studies on the life cycle and food consumption by *Lasioseius* species

Lasioseius species	Food source	Reference
L. aegypticus	Acarid mites, Collembola, *Penicillium viride*, *Fusarium oxysporum*	Afifi (1985)
L. africanus	*Rhizoglyphus robini*, *Tyrophagus putrescentiae*, *Musca domestica*, *Meloidogyne incognita*	Mowafi (2005)
L. athiasae	*R. robini*, *T. putrescentiae*, nematodes, *Aspergillus niger*, *F. oxysporum*	Nasr et al. (1990b)
L. athiasae	*Aceria dioscoridis*, *Tetranychus urticae*, *M. domestica*, *M. incognita*	Abou-Awad et al. (2001)
L. berlesei	*Acrobeloides* sp., *Cladosporium cladosporioides*	Walter and Lindquist (1989)
L. allii (= *L. bispinosus*)	*R. robini*, *T. putrescentiae*, *Drosophila melanogaster*, *M. domestica*	Nawar et al. (1990)
L. chaudhri	*Tarsonemus bilobatus*	Wu and Wang (1982), Zhang and Lin (1986, 1989)
L. scapulatus	*Aphelencus avenae* + *Rhizoctonia solani*	Mankau and Imbriani (1978), Imbriani and Mankau (1983)
L. scapulatus (= *dentatus*)	*Caloglyphus rodriguezi*, *T. putrescentiae*, *Penicillium digitatum*, *Meloidogyne javanica*	Fouly (1997)
L. floridensis (= *fimetorum*)	*T. putrescentiae*	Enkegaar and Broodsgaard (2000)
L. floridensis (= *fimetorum*)	*Heterodera trifolii*	van de Bund (1972)
L. floridensis	*Polyphagotarsonemus latus*, *T. putrescentiae*, *Rhabditella axei*, *Aspergilus flavus*, *Penicillium* sp., *Typha* sp. pollen	Britto et al. (2012)
L. kinikinik	*Acrobeloides* sp.	Walter and Lindquist (1989)
L. lindquisti	*A. dioscoridis*, *R. robini*, *T. urticae*, *F. oxysporum*, *Fusarium solani*, *Ricinus communis* pollen, *Phoenix dactylifera* pollen	Momen et al. (2011)
L. nr. *matthyssei*	*Raoiella indica*	Sheeja and Ramani (2009)
L. paraberlesei	*Steneotarsonemus spinki*	Chow and Liu (1984)
L. youcefi	*Acrobeloides* sp., *C. cladosporioides*	Walter and Lindquist (1989)
L. youcefi	*S. spinki, Tarsonemus* sp.	Lo and Ho (1984)
Lasioseius sp.	*T. putrescentiae*, *Bemisia tabaci*	Xu (2011)
Lasioseius sp.	*Carpoglyphus lactis*, *R. echinopus*, *Tetranychus kanzawai*, *T. putrescentiae*, *B. tabaci*, *M. domestica*	Wu et al. (2009)

shown able to develop when offered nematodes as prey, including *Lasioseius floridensis* Berlese (= *fimetorum* Karg) on *Heterodera trifolii* (Goffart) (van de Bund 1972) and *Lasioseius athiasae* Nawar and Nasr on *Meloidogyne incognita* (Kofoid and White) (Abou-Awad et al. 2001). In addition, experimental works have shown that several *Lasioseius* species are able to develop and reproduce when nematodes were offered as prey, namely *Lasioseius africanus* Nasr on *M. incognita* (Mowafi 2005), *L. athiasae* on unidentified free-living nematodes (Nasr et al. 1990a), *Lasioseius berlesei* (Oudemans) on *Acrobeloides* sp. (Walter and Lindquist 1989), *Lasioseius dentatus* (Fox) on *Meloidogyne javanica* (Treub) (Fouly 1997), *L. floridensis* on *Rhabditella axei* (Cobbold) (Britto et al. 2012), *Lasioseius kinikinik* Walter and Lindquist on *Acrobeloides* sp. (Walter and Lindquist 1989), *L. dentatus* (= *scapulatus* Kennett) on *Aphelencus avenae* (Imbriani and Mankau 1983) and *Lasioseius youcefi* Athias-Henriot on *Acrobeloides* sp. (Walter and Lindquist 1989).

Not much has been published about the relation between lepidopteran and *Lasioseius* species; Burkhardt (1920) reported development of *L. floridensis* (= *fimetorum*) on *A. kuehniella*. The same is true in relation to coleopteran species; *Lasioseius safroi* (Ewing) was mentioned as the most important egg predator of *Dendroctonus peseudotsugae* Hopkins (Bedard 1937). Apparently, no experimental work has been conducted on the effect of these mites on these insect groups. When offered as prey, several collembolan and insect species have been reported to allow development and reproduction of several *Lasioseius* species, including *Musca domestica* L. (housefly) for *L. africanus* (Mowafi 2005), *L. athiasae* (Abou-Awad et al. 2001) and *Lasioseius allii* Chant (= *bispinosus* Evans) (Nawar et al. 1990), as well as *D. melanogaster* for *L. allii* (= *bispinosus*) (Nawar et al. 1990) and *Bemisia tabaci* (Gennadius) for an unidentified *Lasioseius* species (Wu et al. 2009; Xu 2011).

Comparatively, more effort has been dedicated to the study of mite predation by *Lasioseius* species. Mites identified as *Lasioseius parberlesei* Bhattacharyya were reported to feed on *Steneotarsonemus spinki* Smiley (Chow and Liu 1984) and *Tetranychus urticae* Koch (De Leon-Facundo and Corpuz-Raros 2002); *Lasioseius* nr. *matthyssei* Chant was reported to feed on *Raoiella indica* Hirst (Sheeja and Ramani 2009), whereas *L. floridensis* (= *fimetorum*) was found to feed on eggs and larvae (but not on nymphs or adults) of *T. putrescentiae* (van de Bund 1972). Development and reproduction on *R. robini* were reported for *L. africanus* (Mowafi 2005), *L. athiasae* (Nasr et al. 1990b) and *L. allii* (= *bispinosus*) (Nawar et al. 1990; Lesna et al. 1995). Similarly, development and reproduction on *T. putrescentiae* were reported for *L. africanus* (Mowafi 2005), *L. athiasae* (Nasr et al. 1990a), *L. allii* (= *bispinosus*) (Nawar et al. 1990), *L. dentatus* (Fouly 1997), *L. floridensis* (= *fimetorum*) (Enkegaar and Broodsgaard 2000; Britto et al. 2012), *Lasioseius sewai* Nasr and Abou-Awad (Taha et al. 2007) and unidentified *Lasioseius* species (Wu et al. 2009; Xu 2011). Feeding and reproduction were reported for different *Lasioseius* species on mite groups from different habitats, namely *L. athiasae* on *Aceria dioscoridis* (Soliman and Abou-Awad) (Eriophyidae) and *T. urticae* (Abou-Awad et al. 2001), *Lasioseius chaudhrii* (Wu and Wang) on *Tarsonemus bilobatus* Suski (reported as *Tarsonemus fuzhouensis* Lin and Zhang) (Zhang and Lin 1989); *L. dentatus* on the astigmatid *Caloglyphus rodriguezi* Samsinak (Fouly 1997),

L. floridensis on *Polyphagotarsonemus latus* (Banks) (Tarsonemidae) (Britto et al. 2012), *Lasioseius lindquisti* Nasr and Abou-Awad on *A. dioscoridis* (Momen et al. 2011), an unidentified *Lasioseius* on *C. lactis*, *Rhizoglyphus echinopus* (Fumouze and Robin) and *Tetranychus kanzawai* Kishida (Wu et al. 2009) and *Lasioseius zaheri* Nasr on *T. putrescentiae* (Nawar and El-Sherif 1992). In a comparative study, Lo and Ho (1984) observed quicker development and higher oviposition and predation rates of *L. youcefi* on an unidentified *Tarsonemus* than on *S. spinki* Smiley, both tarsonemids from rice.

Fungus species have been shown to allow development and reproduction of several *Lasioseius* species, including *F. oxysporum* and *Hormodendrum viride* (Fresen.) Sacc. (mentioned as *Penicillium viride*) for *Lasioseius aegypticus* Afifi (Afifi 1985); *Aspergillus niger* Tiegh and *F. oxysporum* for *L. athiasae* (Nasr et al. 1990a); *Cladosporium cladosporioides* (Fresen.) G.A. de Vries for *L. berlesei* (Walter and Lindquist 1989); unidentified fungi for *L. chaudhrii*; *P. digitatum* for *L. dentatus* (Fouly 1997); *F. oxysporum* for *L. lindquisti* (Momen et al. 2011); *C. cladosporioides* for *L. youcefi* (Walter and Lindquist 1989) and *F. oxysporum* and *A. flavus* for *Lasioseius zaheri* Nasr (Nawar and El-Sherif 1992). Overall, oviposition was lower on fungi (usually fecundity between 10 and 20 eggs) than on other food sources (usually above 20 eggs); an outstanding exception was *L. lindquisti*, whose fecundity was slightly over 30 eggs per female on *F. oxysporum* [but not on *H. haematococca* (= *F. solani*)], according to Momen et al. (2011).

Three studies have evaluated the acceptance of pollen as food sources for these mites. *Lasioseius chaudhrii* was observed by Zhang and Lin (1989) on rice pollen and by Zhang and Lin (1991) on pollen of *Alopecurus aequalis* Sobol and of *Luffa cylindrica* Hoen. Pollen of *Phoenix dactylifera* L. was reported to allow development and reproduction of *L. lindquisti* (Momen et al. 2011). However, pollen of *Ricinus communis* L. did not promote the oviposition of *L. lindquisti* (Momen et al. 2011) or of *L. floridensis* (Britto et al. 2012), although it allowed immature development of the former species. Thus, acceptance of pollen may vary according to the *Lasioseius* species and, probably according to the type of pollen offered.

Available information has demonstrated the ability of many *Lasioseius* species to develop and reproduce on a range of different food items, suggesting at least some of them may be polyphagous. However, several species have been observed to refuse some potential food items. These include *L. aegypticus* on unidentified acarid mites and collembolans (Afifi 1985); *L. allii* (= *bispinosus*) on *T. putrescentiae* (Lesna et al. 1995); *L. floridensis* on *A. flavus*, *Penicillium* sp. and *Typha* sp. pollen (Britto et al. 2012); *L. lindquisti* on *R. robini*, *T. urticae*, *H. haematococca* (= *F. solani*) and *R. communis* pollen (Momen et al. 2011); and *Lasioseius* sp. on immature houseflies (Wu et al. 2009). In these cases, the species of *Lasioseius* either did not develop or developed but did not oviposit when offered those items as food.

Phoresis does not seem common for mites of this genus, although the type specimens of several *Lasioseius* species have been collected from rodents (15 species) or, more rarely, beetles (six species) and hummingbirds (five species) (Table 2.2). The relatively rare occurrence of phoresis suggests that these mites are well adapted to the usual changes in the niche they occupy, while matching the polyphagous habits of species of this group.

Mating has been shown necessary for oviposition of several *Lasioseius* species, namely *L. aegypticus* (Afifi 1985), *L. africanus* (Mowafi 2005), *Lasioseius boomsmai* Womersley (Walter and Lindquist 1995), *L. dentatus* (Fouly 1997), *L. floridensis* (Walter and Lindquist 1995; Britto et al. 2012), *L. kiniknik* Walter and Lindquist (1995), *L. lindquisti* (Momen et al. 2011), *Lasioseius queenslandicus* (Womersley) (Walter and Lindquist 1995), *Lasioseius subterraneus* (Chant) (Walter and Lindquist 1995) and *L. zaheri* (Nawar and El-Sherif 1992).

Thelytoky has been reported for other species or even other populations of species for which mating has been mentioned as required, namely *L. berlesei* (Walter and Lindquist 1995), *L. dentatus* (= *scapulatus*) (Mankau and Imbriani 1978; Imbriani and Mankau 1983; Walter and Lindquist 1995) and *L. youcefi* (Walter and Lindquist 1995). Variation in type of reproduction within a species has been reported for other mite groups (Norton et al. 1993). A possible cause for this variation is the presence of endosymbiont bacteria in some but not in other populations of the same species.

Larval feeding seems facultative for some *Lasioseius* species. Abou-Awad et al. (2001) observed larvae of *L. athiasae* to feed on some but not on other prey. Chow and Liu (1984) reported larvae of *L. parberlesei* to feed on *S. spinki*, whereas Zhang and Lin (1989, 1991) reported larvae of a species identified as *L. chaudhrii* to feed on *T. bilobatus* and on pollen. However, larval feeding was reported to be required by *L. dentatus* (= *scapulatus*) (Imbriani and Mankau 1983), *L. sewai* (Taha et al. 2007) and *L. zaheri* (Nawar and El-Sherif 1992), but not by *L. allii* (= *bispinosus*) on *R. robini*, *D. melanogaster*, *M. domestica* or *T. putrescentiae* (Nawar et al. 1990). Larval feeding was also observed for *L. youcefi* on *Tarsonemus* sp. and *S. spinki* (Lo and Ho 1984) and for *L. chaudhrii* on *T. bilobatus* (Zhang and Lin 1989).

Feeding habits of other blattisociid groups have been hardly studied. *Hoploseius* species have been reported from microdrosophilid flies (Lindquist 1970), whereas nematophagy was reported for *Arrhenoseius gloriosus* Walter and Lindquist (Blattisociidae) by Walter and Lindquist (2000), Walter (2003) and Beaulieu and Walter (2007).

2.7 Melicharidae

2.7.1 Melichares

A single reference was found regarding the life cycle of a species of this genus. Hughes (1976) reported the ability of *Melichares agilis* Hering to develop to adulthood when fed with *A. siro*.

2.7.2 Proctolaelaps

Proctolaelaps is one of the three most extensively studied genera of the ascid complex. A list of the studies about the life cycle and prey consumption of *Proctolaelaps* species is given in Table 2.5. Species of this broad genus also seem to be quite variable in relation to feeding habits.

Table 2.5 Studies on the life cycle and food consumption by *Proctolaelaps* species

Proctolaelaps species	Food source	Reference
P. bickleyi	Nematodes, *A. niger*, *Botrytis fabae*, *Rhizoctonia solani*	Nasr et al. (1990a)
P. bickleyi	*Aceria guerreronis*, *Steneotarsonemus furcatus*, *Tetranychus urticae*, *Rhizopus* aff. *stolonifer*, coconut pollen	Lawson-Balagbo et al. (2007)
P. bulbosus	*A. guerreronis*, *Tyrophagus putrescentiae*, *Rhizopus* aff. *Stolonifer*	Galvão et al. (2011)
P. deleoni	*Rhizoglyphus robini*, *T. putrescentiae*, *Rhabditis scanica*, *A. flavus*, *F. oxysporum*	Nawar (1992)
P. dendroctoni	*Rhabditis* sp. + *Cryptaphelenchus* sp.	Kinn (1983)
P. naggari	*Rhanditella muscicola*	Taha and Mohmoud (2007)
P. pygmaeus	*Alternaria tenuis*, *F. oxysporum*, *Penicillium* sp., *R. solani*, *Trichoderma viridis*	Ibrahim et al. (1989)
P. pygmaeus	*Meloydogine incognita*	Ahmed and Mowafi (1998)
P. pygmaeus	*R. robini*, *Panagrolaimus rigidus*	Metwalli et al. (1991)
P. striatus	*Aspergillus flavus*, *Aspergillus niger*, *Fusarium oxysporum*	Afifi et al. (1984)
P. subcorticalis	*Rhabditis* sp., *F. oxysporum*	Ibrahim et al. (1992)
P. yinchuanensis	*Dichromothrips corbetti*	Navasero et al. (2003)
P. yinchuanensis	*Suidasia pontifica*	Navasero et al. (2004)

Proctolaelaps lewisi (Garman and McGregor) [= *utahensis* (Chant)] was observed to feed on an unidentified nematode (Lindquist and Hunter 1965). Nematode predation was also reported by Hirschmann and Rühm (1953) for *Proctolaelaps fiseri* Samsinak (cited as *Lasioseius hystrix*) and by Krantz and Poinar (2004) for an undescribed *Proctolaelaps* species. The former was reported from galleries of an unidentified *Ips* sp., and the second, on the curculionid *Elaeidobius kamerunicus* (Faust), onto which it was reported to attack the bacteriophagous nematode *Cylindrocorpus* sp.

Several *Proctolaelaps* species have been shown to develop and reproduce on nematodes, namely *Proctolaelaps bickleyi* (Bram) on unidentified free living nematodes (Nasr et al. 1990b), *Proctolaelaps deleoni* Nawar, Childers and Abou-Awad on *Rhabditis scanica* (Nawar 1992), *Proctolaelaps dendroctoni* Lindquist and Hunter on an association of *Rhabditis* sp. and *Cryptaphelenchus* sp. (Kinn 1983), *Proctolaelaps naggari* (*nomen nudum*) on *Rhabditella muscicola* (Chitwood) (Taha and Mohmoud 2007), *Proctolaelaps pygmaeus* (Müller) on *M. incognita* (Ahmed and Mowafi 1998) and *Panagrolaimus rigidus* (Schneider) (Metwalli et al. 1991) and *Proctolaelaps subcorticalis* Lindquist on *Rhabditis* sp. (Ibrahim et al. 1992).

The association of *Proctolaelaps* and *Drosophila* Fallén species has been extensively reported in the literature, indicating the importance of these mites in laboratory rearings. *Proctolaelaps regalis* De Leon was reported to prey upon *Drosophila* species (Houck et al. 1991; Houck 1994). Given its behavior as a pseudo-parasite, often injuring eggs of different prey species found together in a particular habitat within a short time, without necessarily killing them, this mite has been mentioned as potentially implied in the horizontal transfer of *Drosophila* genes (Houck et al. 1991; Houck 1994). Increase in the population of *Proctolaelaps drosophilae* Karg, Baker and Jenkinson was reported (Karg et al. 1995) to be related to a crash in the population of *Drosophila* species, although actual feeding on the flies was not reported by the authors. *Proctolaelaps pygmaeus* [= *hypudaei* (Oudemans)] was suspected to cause depletion of a *Drosophila* colony by feeding on eggs of the insect (Cantelo and Boswell 1973; Ashburner 1989).

Association of *Proctolaelaps* species with bark beetles has also been mentioned. Predation on eggs and larvae were reported for *Proctolaelaps xyloteri* Samsinak and *P. dendroctoni* by Novak (1960) apud Lindquist (1969), and Lindquist and Hunter (1965), respectively. Predation on eggs and larvae of *Phloeotribus scarabaeoides* (Bernard), *Hylesinus fraxini* (Panzer) and *Hylesinus oleiperda* (Fabricius) by *Proctolaelaps eccoptogasteris* (Vitzthum) was reported by Russo (1938). The ability of *Proctolaelaps* to develop and reproduce when offered lepidopteran insects as prey has apparently not been reported.

A single publication refers to the ability of a *Proctolaelaps* species to prey on thrips. *Proctolaelaps yinchuanensis* Xue, Sui and Yi was reported by Navasero et al. (2003) to feed on immatures and adults of *Dichromothrips corbetti* (Priesner).

Insect parasitism has been reported for species of this genus. Egan and Moss (1969) and Egan and Hunter (1975) reported the parasitism of immatures of the cockroach *Nauphoeta cinerea* (Olivier) by *Proctolaelaps nauphoetae* (Womersley). Up to ten mites can be seen on a roach, but the activities of the host do not seem to be affected by the mites, as no significant damage was observed on the roaches. In addition to the blattisociid *B. patagiorum*, this is the only other member of the ascid complex known to truly parasitize arthropods. However, in *P. nauphoetae* the fixed digit is not attenuate, as usual in parasitic Mesostigmata (according to Halliday et al. 1998). The fixed digit of *P. regalis* is also not attenuate, but in the discussion in the legend of figure 1 of Houck et al. (1991), the authors state that obligatory parasites have reduced fixed digit, or edentate or reduced chelicerae, whereas omnivorous species retain the usual chelate-dentate type of chelicerae.

Even injury to humans has been attributed to *Proctolaelaps*. Andrews and Ramsay (1982) apud Halliday et al. (1998) reported *P. pygmaeus* to bite humans, causing dermatosis.

One of the first reports of mites of this genus on stored products was published by Bram (1956). The author described *P. bickleyi* in association with *S. cerealella* on corn ears, without specifying what it was feeding on. In two other occasions, this species was suspected to feed on fungi (Lindquist and Hunter 1965; Flechtmann 1976). The latter author reported to have been able to maintain several generations of this mite in a Petri dish containing an insect diet colonized by unidentified fungi.

Proctolaelaps pygmaeus [= *hypudaei*] has also been reported to feed on an unidentified fungus in the laboratory (Muma 1961), as well as on *A. flavus*, *F. oxysporum* and *H. viride* (= *P. viride*) (Shereef et al. 1980). Fungal and yeast-like inclusions were seen in the digestive system of this species by Halliday et al. (1998), indicating that unlike most members of the ascid complex this species ingests solid material. *Proctolaelaps euserratus* Karg was reported from animal and human decomposing tissues, probably preying upon necrophagous arthropods or feeding on fungi (Masan et al. 2013). The authors mentioned that the record of *P. epuraeae* on human remains in Spain (Saloña et al. 2010) refers to a misidentification of *P. euserratus*, and that records of a supposedly new species of *Proctolaelaps* on human remains in Belgium (Leclercq and Verstraeten 1988) and cat remains in Hawaii (Early and Goff 1986; Goff 1989) could also refer to *P. euserratus*. *Proctolaelaps regalis* can also feed on fungus and yeast (Houck et al. 1991; Houck 1994). Weisser and Zhang (1996) observed *Proctolaelaps* nr. *scolyti* Evans to be able to oviposit when fed with yeast or Kalmus medium with *Drosophila* larvae. They stated that although the mite did not feed on the eggs or larvae of that insect, the latter probably somehow promoted the development of fungi or yeast in the medium, upon which the mite fed.

Development and reproduction has been demonstrated under laboratory conditions for a few *Proctolaelaps* species on fungi, including *P. bickleyi* on *A. niger*, *Botrytis fabae* Sardiña and *R. solani* (Nasr et al. 1990b) and on *Rhizopus* aff. *stolonifer* (Ehrenb. ex Fr.) Vuill. (Lawson-Balagbo et al. 2007); *Proctolaelaps bulbosus* Moraes, Reis and Gondim Jr. on *R.* aff. *stolonifer* (Galvão et al. 2011) and *P. pygmaeus* as well as *Proctolaelaps striatus* Afifi, Hassan and El-Bishlawy on *F. oxysporum* (Ibrahim et al. 1989). However, fecundity was quite low (usually at most 10 eggs per female), except for *P. bickleyi* and *P. bulbosus* (each about 33 eggs per female), both feeding on *R.* aff. *stolonifer*. Although immature development could be completed, oviposition did not occur when some *Proctolaelaps* species were fed with some fungi, namely *P. deleoni* on *A. flavus* or *F. oxysporum* (Nawar 1992); *P. pygmaeus* on *Alternaria alternata* (Fr.) Keissl (mentioned as *A. tenuis*), *Penicillium* sp. and *Trichoderma viride* Pers. (Ibrahim et al. 1989) and *P. striatus* on *A. flavus* (Afifi et al. 1984). The latter species could not complete immature development on *A. niger* (Afifi et al. 1984).

A parasitic relationship of *Proctolaelaps kirmsei* Fain, Hyland and Aitken and its host plant *Hamelia patens* Jacq. in Costa Rica was mentioned by Paciorek et al. (1995), who calculated that it could consume about 16 % of the pollen available within the period corresponding to mite arrival and cessation of hummingbird visits to the flowers. Cruz et al. (2007) also reported *Proctolaelaps* species to significantly reduce the availability of nectar in flowers of *Heliconia* L. (Heliconiaceae), thus competing with hummingbirds for this source of food. However, in a laboratory study, *P. bickleyi* was not able to complete immature development on coconut pollen (Lawson-Balagbo et al. 2007).

Laboratory studies have shown the ability of a few *Proctolaelaps* species to develop and reproduce on mites. One of the oldest studies refers to predation of *T. urticae* by *P. pygmaeus* (= *hypudaei*) on fruit trees (Mathys and Tencalla 1959), although the authors did not consider the former to be an important predator of the latter. Other studies referred to the same predator on *R. robini* (Metwalli et al. 1991), to *P. bickleyi* and *P. bulbosus* on *Aceria guerreronis* Keifer (Lawson-Balagbo et al. 2007 and Galvão et al. 2011, respectively) and to *P. yinchuanensis* on *Suidasia pontifica* Oudemans (Navasero et al. 2004). However, some *Proctolaelaps* species were not able to complete immature development on certain mite species, namely *P. bickleyi* on *Steneotarsonemus furcatus* De Leon or *T. urticae* (Lawson-Balagbo et al. 2007), *P. bulbosus* on *T. putrescentiae* (Galvão et al. 2011) and *P. deleoni* on *R. robini* and *T. putrescenitae* (Nawar 1992).

Proctolaelaps regalis was reported to feed on free nutrients (Houck et al. 1991; Houck 1994). This is probably the only report of a member of the ascid complex feeding on inanimate substrate.

Karg (1988) considered species of *Proctolaelaps* to be characteristically phoretic on different insect groups. Twenty six species of this genus were described from specimens collected on insects, 13 on birds and 11 on mammals (Table 2.2). Treat (1969, 1975) reviewed data about phoresis of *Proctolaelaps* on moths, and on species that despite not been found phoreticly on moths, were found to feed on lepidopteran eggs and larvae. According to Halliday et al. (1998), *Proctolaelaps lobatus* De Leon has been reported in association with many lepidopteran species, while according to Krantz and Poinar (2004), an undescribed *Proctolaelaps* species was found to disperse phoretically on the curculionid *E. kamerunicus*. Karg (1988) referred to several previous records of association of *Proctolaelaps* species with different species of bark beetles.

The previous citations indicate that similar to what was reported for *Lasioseius* species, most *Proctolaelaps* species studied so far are able to develop and reproduce on several food items, indicating the polyphagous habit of members of this genus.

Mating has been shown as a pre-requisite for oviposition of *P. bickleyi* (Nasr et al. 1990b), *P. deleoni* (Nawar 1992), *Proctolaelaps aurora* (Vitzthum) [= *longipilis* (Chant)] (Walter and Lindquist 1995), *P. pygmaeus* (Metwalli et al. 1991; Walter and Lindquist 1995) and *P. striatus* (Afifi et al. 1984). However, the genetic system of *Proctolaelaps* has not been adequately studied. According to Norton et al. (1993), electrophoretic studies conducted by Kaliszewski (unpublished) suggested a haplodiploid genetic system of reproduction for *P. kirmsei* and *P. aurora* (= *longipilis*). Thelytoky has not been reported for species of this genus.

Egan and Moss (1969) and Egan and Hunter (1975) reported larviparity to occur in *P. nauphoetae*. Adaptations of females for this process include an enlarged idiosoma, which reaches an average of 670×314 µm when engorged, and reduced genital shield (Egan and Moss 1969). Larviparity is also suspected to occur in *Proctolaelaps vandenbergi* (Ryke) (Lindquist 1962).

Larval feeding was reported to be required by *P. bickleyi* (Lawson-Balagbo et al. 2007). However, larval feeding by *P. yinchuanensis* did not occur when it was fed with the thrips *D. corbetti* (Navasero et al. 2003). In a singular situation, Egan and Moss (1969) and Egan and Hunter (1975) showed experimentally that only adults of *P. nauphoetae* fed.

2.7.3 Rhinoseius *and* Tropicoseius

These mites are not predators, but feed on nectar (Colwell 1973; Colwell and Naeem 1994). It is not clear how they obtain protein, but it seems to be by uptaking nectar. Colwell (1973) summarized data of other authors indicating that some types of pollen may release free amino acids or enzymes when mixed with nectar and that some types of nectar contain amino acids. He also suggested that yeasts and bacteria could be other possible sources of nitrogen. Cruz et al. (2007) concluded that similar to species of *Proctolaelaps*, *Tropicoseius heliconiae* Baker and Yunker significantly reduced nectar availability in flowers of *Heliconia*, competing with hummingbirds for this food source.

Colwell and co-workers did an extensive investigation on mites of these genera, the interactions among themselves and their interaction with plants, hummingbirds and other organisms (Colwell 1979; Heyneman et al. 1991; Colwell and Naeem 1994). Colwell (1973) evaluated the competition between *Rhinoseius richardsoni* Hunter, *Tropicoseius colwelli* (Hunter) and other organisms with which they were related, discussing their co-existence. When males of some (but not all) species were found in the same flower, mortal combat occurred (Colwell 1986). In a different type of interaction involving a member of this group, Naeem et al. (1985) concluded that *Tropicoseius trinitatis* Fain, Hyland and Aitken could have been preyed upon by *Lasioseius elegans* Fain, Hyland and Aitken in heliconia flowers.

Given that transportation of *Tropicoseius* and *Rhinoseius* species between plants is done on the bills or the nasal cavities of hummingbirds (Colwell 1973; Colwell and Naeem 1994), the distribution of the mites is contained within the range of hummingbird occurrence (Colwell 1979). These mites can live on a variable number of plant substrates, but they do not seem to be specific in relation to the hummingbird species used as carriers (Colwell 1986). It is very rare to find more than one mite species in flowers of a given species, but very often more than one mite species are found on a single hummingbird (Colwell and Naeem 1979).

Colwell and Naeem (1979) mentioned *Tropicoseius chiriquensis* Baker and Yunker [= *epoecus* (Colwell and Naeem)] as possibly ovoviviparous, an egg-laying behavior compatible with the fact that the flowers in which this mite lives last less than a day (Colwell 1986).

2.8 Potential for Practical Use

Strategies for the biological control of pest organisms are usually classified as conservation, augmentation or classical biological control. The first two classes involve biological control agents already present in a given region, whereas the last involves their introduction in a new environment. In comparison with the phytoseiids, relatively little effort has been dedicated to evaluate the potential use of members of the ascid complex as biological control agents.

One of the first publications to critically evaluate the possible value of members of this complex for pest control under field conditions is that of Lindquist (1969). In that publication, the author reviewed the literature concerning the mites found in association with bark beetles, discussing the groups which seemed to influence the population levels of those organisms and proposing research themes to be emphasized so as to determine the levels of control they exerted, and ways to improve them. He highlighted the considerable importance of bark beetles, and the fact that controlling them by means other than biologically was usually difficult by different reasons, including their widespread occurrence on cultivated trees as well as on trees of the natural vegetation, often of discontinuous distribution or in inaccessible areas. As concluded by the author, improvement would require extensive investigation. Not much has changed after over 40 years since that paper was published. Unfortunately, research priorities listed then are still in need of being conducted. Thus, the determination of mites that are "primary associates" of bark beetles in different ecosystems still needs to be conducted before other actions can be proposed to improve the biological control of these organisms.

A discussion is subsequently presented about the potential of species of the ascid complex to serve as biological control agents for specific groups of harmful organisms, involving cases of the three biological control strategies.

2.8.1 Soil Nematodes

Potential in this regard was pointed out long ago. Recommendations by Karg (1967, 1983) correspond to the strategy known as conservation. They were based on observations of the author on the abundance and diversity of soil gamasid mites (including the ascid complex), emphasizing the importance of maintaining a protective litter layer on the soil surface, to reduce the effect of low temperature in the winter in temperate areas and of high temperature in the summer in tropical and subtropical areas, thus allowing the maintenance of nematophagous mites at adequate population levels. This is expected to result in better control of plant parasitic nematodes, as reported by van de Bund (1970, 1972) for mesostigmatid mites of the families Rhodacaridae and Laelapidae. Karg (1983) further suggested the use of stabilized manure, other types of organic matter, intercropping and crop rotation to promote the control of harmful nematodes by allowing better development of the predatory mites. The results of several of the previously reported studies are very encouraging, but more detailed field studies are needed to corroborate those findings.

Very promising results were reported by Cayrol (1970) in relation to the biological control of *Ditylenchus myceliophagus* J.B. Goodey, a pest of mushroom in Europe, using *Lasioseius thermophilus* Willmann. That nematode is a pest of the mushroom *Agaricus bisporus* (Lange) in France, where the predator was frequently found in mushroom cultivation. The author obtained experimental evidence of the efficiency of *L. thermophilus* in the control of *D. myceliophagus* and demonstrated that the predator did not affect the crop. Since Cayrol's publication, nothing has been reported regarding the use of that predator to control the pest. Further efforts seem warranted for use of *L. thermophilus* in the control of *D. myceliophagus*, or other predators of the same group for the control of similar nematodes. Given that most members of the ascid complex are found near the soil surface, it seems that better chances of success with those mites as control agents would be expected for plants with relatively shallow roots, such as horticultural or ornamental plants.

2.8.2 *Tenuipalpid and Tetranychid Mites*

In line with the scant information about the biology of *Asca* species, field studies on their possible role as control agents of damaging organisms have been few, despite the relatively common occurrence of these mites on plants. Unpromising results were obtained in a brief evaluation of an unidentified *Asca* species as a predator of *T. urticae* Koch and *Tetranychus evansi* Baker and Pritchard as prey (Rosa et al. 2005). However, Walter et al. (1993) reported *Asca* species to be involved in the natural control of eriophyoid and tenuipalpid mites in tropical areas of Australia. Without referring to any particular species, the authors also mentioned *Lasioseius* species to be of similar importance in the same region.

2.8.3 *Tarsonemid Mites on Rice*

Tarsonemid mites constitute important pests of rice in Asia and the Americas. Reviews of predatory mites, including members of the ascid complex, associated with *S. spinki*, one of the main tarsonemid species on rice, were provided by Hummel et al. (2009) and Zhang and Fan (2010). A species identified as *L. parberlesei* was considered by Tseng (1984) to be an important control agent of *S. spinki* on rice crops in Taiwan. According to the author, that predator overwinters mostly on weeds, moving to rice plants in the spring, reaching high levels in the summer (September – October).

However, Lo and Ho (1984) referred to better performance of a species identified as *L. youcefi* as predator of an unidentified *Tarsonemus* species than of *S. spinki* under laboratory conditions. That predator could actually be the same as the one reported by Tseng (1984) as *L. parberlesei*, given their great morphological similarity. They could also refer to the same species reported only as Ascidae by Lo and Ho (1979). Voucher specimens of the predator reported in those studies are not available for examination, but the illustration of the species reported by Tseng (1989) as commonly found in rice fields in Taiwan, mentioned in that paper as *L. parberlesei*, refers to a species different from *L. youcefi* and *L. parberlesei*, but belonging to the same *phytoseioides* species group (Lindquist 1964). In relation to the niche occupied on the host plant, there is a similarity between the behavior of *L. parberlesei* and *Typhlodromalus aripo* De Leon, a predaceous phytoseiid considered an effective control agent of the cassava green mite, *Mononychellus tanajoa* (Bondar) (Tetranychidae) in Africa (Yaninek and Hanna 2003). In both cases, the predator congregates at the meristematic region of the host plant, at the inner surface of the rice leaf sheath and the cassava tips, respectively.

Lasioseius chaudhrii was reported as an important biological control agent of another tarsonemid pest on rice in Asia. Wu and Wang (1982), Zhang and Lin (1986, 1989, 1991) mentioned it as the dominant natural enemy of *T. bilobatus* (reported as *T. fuzhouensis*) in the Fujian Province, southern China. Field studies reported by Zhang and Lin (1989, 1991) suggested the predator as able to reduce the population of the pest. Zhang and Lin (1991) stated that the predator provided obvious control of *T. bilobatus* on rice, and that the release of the predator in the early stage of rice development could provide effective control.

2.8.4 **Aceria guerreronis** *on Coconut*

Aceria guerreronis is one of the main pests of coconut in the Americas, Africa and parts of Asia (Navia et al. 2013). Major effort has been dedicated to the search for prospective predators in different parts of the world for its control. The predators most often found associated with it are phytoseiids. However, some species of the ascid complex have also been often found, especially *P. bickleyi* and *P. bulbosus*, but also some other species of this complex (Navia et al. 2013). Those two *Proctolaelaps* species were the most numerous predators associated with *A. guerreronis* in southeastern Brazil (Oliveira et al. 2012). It has been suggested that the exceedingly high feeding requirement of *P. bickleyi* may limit its performance as a control agent of *A. guerreronis* (Lima et al. 2012). Available information about the biology of *P. bulbosus* (Galvão et al. 2011) suggests the need for further studies to evaluate the possible efficiency of this species, determining whether its introduction to other sites would be warranted.

2.8.5 Mites on Stored Bulbs

The laelapid mite *Gaeolaelaps aculeifer* (Canestrini), often mentioned in the literature as *Hypoaspis aculeifer*, has been widely used for the biological control of *R. robini*, a serious mite pest of bulb producing plant species. Lesna et al. (1995) showed that under experimental conditions *L. allii* (= *bispinosus*) was also able to suppress the population of *R. robini*, but that the predator was not as effective as *H. aculeifer* in the propagation phase of lilies. However, they argued that by being smaller, *L. allii* could better search for bulb mites hidden inside lily bulbs. Thus, it is assumed that they could have a better performance on stored bulbs, reducing the chances of damage by *R. robini* when bulbs are planted. These findings suggest that the same or similar predators could be useful for the control of other pests of the same group in storage, as for example of *R. echinopus* on garlic.

2.8.6 Mite and Insect Pests in Storage of Other Items

Researchers have often considered the use of *Blattisocius* species promising for the control of pest mites and insects in storage, although Barker (1967) was inconclusive when considering *B. keegani* to control pest insects in storage. Emphasis has been placed on the study of *B. tarsalis*, as shown by the large number of papers dealing with different aspects of this mite. The main constraints in this case involve the fact that in an attempt to arrest seed germination process and disfavor development of associated undesirable organisms, stored products are usually kept at relatively low temperature and humidity levels, and that these conditions often hamper the performance of prospective predators.

Nielsen (1999a) studied the effect of different termperatures on *B. tarsalis*, finding this species to be active at 7 °C, to attack the prey even at 9 °C, and to attain high predation rates at 13 °C. This is of great importance, considering that the use of methyl bromide, a product that was commonly used in storage for pest control, has been prohibited in many countries (Nielsen 1999a).

In Greece, both *B. keegani* and *B. tarsalis* represented only a very small proportion of the predatory mites found in a commercial storage facility, in which another predator, *Cheyletus malaccensis* Oudemans was extremely more abundant (Athanassiou et al. 2005). The lower abundance of the former could at least in part be due to predation by the latter; Egan and Moss (1969) reported predation of all stages of *P. nauphoetae* by *Cheyletus eruditus* (Schrank). *Blattisocius tarsalis* was considered the major controlling agent of *E. cautella* in Kenya by Graham (1970), and sufficiently important to be considered as one of the parameters in the model developed by Throne et al. (1998) to simulate the population dynamics of that insect.

Hansen and Nielsen (2001) reported stimulating results of a study conducted in a commercial bran silo in Denmark. Releases of *B. tarsalis* from April to October 2000 allowed an apparent reduction in the population of *A. kuehniella* in comparison with average populations of the previous 4 years, when control was done with the application of a chemical pesticide. Nielsen (2003) was very optimistic about the possible use of *B. tarsalis* for the control of damaging organisms in storage, especially *A. kuehniella* and *S. cerealella*, considering the results he obtained in that and in his previous works. He referred to the need to evaluate the possible allergenic properties of these mites, both to humans and domesticated animals, before they could be recommended for practical use. These studies have apparently still not been conducted.

2.8.7 Other Insects

Dmoch (1995) was enthusiastic about the significant role of *A. semiscissus* in the control of sciarid larvae in mushroom houses in Poland. However, apparently nothing more has been published about the actual use of *A. semiscissus* in that country.

Bemisia tabaci is an insect pest of major importance worldwide. Phytoseiid species, especially *Amblyseius swirskii* Athias-Henriot, have been used to control it biologically. Xu (2011) referred to a new promising agent, of the ascid complex, for the control of this pest. He mentioned that effective control of *B. tabaci* was achieved by releases of an unidentified species of *Lasioseius* in screenhouses, with reductions of about 80 % of the pest population after 20 days of predator release. The use of mites of the ascid complex seems promising for the control of this pest, given that *B. tabaci* is favored by high level of relative humidity (Simmons and Mahroof 2011), which is also favorable to mites of the ascid complex.

Promising results were reported by Enkegaar and Broodsgaard (2000) in relation to predation of pupae of the thrips *Frankliniella occidentalis* (Pergande) by *L. floridensis* (= *L. fimetorum*). The authors estimated that the daily rate of consumption of the predator was similar to what had been determined for *Stratiolaelaps* (= *Hypoaspis*) *miles* (Berlese).

2.8.8 Fungi

Moustafa and El-Hady (2006) were optimistic about the biological control of the plant pathogenic fungi *Macrophomina phaseolina* (Tassi) Goid. (mentioned as *Sclerotium bataticola*) and *R. solani*. Based on laboratory studies, the authors reported that releases of the mites *P. pygmaeus* and *Ameroseius kosi* (El-Badry, Nasr and Hafez) and the fungus *Trichoderma harzianum* Rifai, together or in isolation, could be useful for the control of those fungi. Further efforts should be dedicated to confirm those results under different conditions.

2.9 Possible Limitations

The examples previously mentioned illustrate circumstances under which the use of members of the ascid complex for biological control seems more promising. However, based on biological studies of these mites, some possible limitation could be anticipated.

One of these refers to the seemingly low level of specificity of some of those predators. In perennial crops, the ability to consume other food sources in addition to the target prey favors the perpetuation of a predator in the environment, which in turn could favor biological control. However, specificity is desirable in annual crops in order to allow predators to control the pest more quickly, even if periodic predator releases are necessary. Thus, conducting studies to determine the range of acceptable prey by prospective predators seems of great importance.

Another possible limitation might refers to the use of some predators on perennial crops, because of their very high feeding requirements. That is the limitation referred by Lima et al. (2012) for *P. bickleyi* as a control agent of *A. guerreronis* on coconut palms, which may reach over 15 m in height, hampering eventual periodic releases of predators. On the other hand, the high feeding requirement could be a very positive aspect for use of this type of predators on crops on which periodic releases are feasible.

Humidity requirements for members of the ascid complex may also be a possible constraint. As reported by Rivard (1962a) for *B. dentriticus*, survival, speed of development and oviposition of this mite were greater at higher levels of humidity. Riudavets and Quero (2004) demonstrated experimentally the increased survival of *B. tarsalis* at higher humidity levels, being the effect more significant on the than on the egg stages. Population dynamic studies also support that hypothesis. de Sousa et al. (2005) found specimens of *B. tarsalis* throughout the year in food samples obtained from retailers. However, the levels of infestation were much higher in June, coinciding with the middle of the rainy season.

A determined limitation to the use of *B. tarsalis* for the control of *A. kuehniella* in storage may arise from the preferred microhabitat of the predator within the storage facility. Inefficient control was reported by Flanders and Badgley (1963) when the pest was more than 8 mm from the surface of the substrate (allotment of wheat flakes), and by Nielsen (1998) when the pest was 1 mm or 3 mm from the surface of the substrate (flour – most probably of wheat, although not stated by the author). This would suggest that the use of *B. tarsalis* would need to be associated with the use of complementary natural enemies that would move deeper into the substrate.

Although very little has been experimentally shown about the effect of humidity on other members of the ascid complex, their more common occurrence on plants in humid areas suggests that humidity may limit their use under open field conditions. Thus, the control of pests by these predators seems more promising on plants grown under high humidity conditions. This is the case, for example, of rice plants which are often grown under conditions of very high humidity, as well as of protected crops. We have also conducted observations on the effect of relative humidity on two species of the ascid complex. *Lasioseius floridensis* was shown to be more affected by low levels of relative humidity than values usually reported for phytoseiids. The effect of humidity was more pronounced on the adults, with about 50 % dying within 4 h of exposure to relative humidity of 67 % or lower (our unpublished data).

2.10 Conclusions

Differently from what has been observed for the phytoseiids, the potential for the use of predaceous species of the ascid complex still needs to be fully proven, although several promising species have been determined. However, it is certainly worthwhile not only to continue, but to enhance the studies of species of the ascid complex as biological control agents of harmful organisms.

These predators have been inadequately studied around the world. Most of the experimental studies conducted so far are laboratory evaluations of their life cycle and ability to feed on pre-determined food sources. Little has been done to attempt the practical use of members of the ascid complex. In order for these mites to be effectively used by growers, efforts should be concentrated on fieldwork to validate promising laboratory results.

At the same time, knowledge about the diversity of ascid species is still limited in many parts of the world, especially in the tropics. These studies should be supported, to increase the chances to find species potentially useful as biological control agents.

In parallel, specific aspects to be evaluated for each prospective species should include their response to different climatic conditions and the range of potential prey organisms in the habitat where the predator is proposed for release. The first aspect refers to one of potential limitations for mites of this group in practical use, whereas the second is ecologically relevant, and important for the future registration of the organism for commercialization in some countries.

Acknowledgements We are especially thankful to the following researchers who provided copies of the papers referred to in this publication: A.M. Afifi, C. G. Athanassiou, R.B. Halliday, F.M. Momen, P.S. Nielsen, J. Riudavets and R. Abo-Shnaf. We are also thankful to C.H.W. Flechtmann, for critically reviewing an earlier version of this document and providing us numerous suggestions.

References

Abou-Awad BA, Korayem AM, Hassan MF, Abou-Elela MA (2001) Life history of the predatory mite *Lasioseius athiasae* (Acari: Ascidae) on various kinds of food substances: a polypeptide analysis of prey consideration. J Appl Entomol 125:125–130

Afifi AM (1985) *Lasioseius aegypticus*, a new species from Egypt, with notes on its biology (Acari: Gamasida: Ascidae). Proceedings of Egypt's National Conference of Entomology 1:453–461, Cairo, 6–9 Dec 1982, pp 453–461

Afifi AM (1989) Biological studies and prey range of two predaceous soil mites (Acari: Gamasida). Bull Entomol Soc Egypt 68:175–180

Afifi AM, Hassan MF, El-Bilshlawy SO (1984) *Proctolaelaps striatus*, a new species from Egypt, with notes on its biology (Acari: Gamasida: Ascidae). Bull Fac Agr Univ Cairo 35:1215–1226

Afifi AM, Hassan MF, Nawar MS (1986) Notes on the biology feeding habits of *Protogamasellus minutus* Hafez, El-Badry & Nasr (Acari: Gamasida: Ascidae). Bull Entomol Soc Egypt 66:251–259

Ahmed MM, Mowafi MH (1998) Life tables of *Proctolaelaps pygmaeus* (Muller) (Gamasida: Ascidae) at different temperatures. Al-Azhar J Agr Res 27:338–348

Andrews JRH, Ramsay GW (1982) A case of papular dermatosis in man attributed to an ascid mite (Acari). J Med Entomol 19:111–112

Ashburner M (1989) *Drosophila*: a laboratory handbook. Cold Spring Harbor Laboratory Press, Cold Spring Harbor

Athanassiou CG, Palyvos NE (2006) Laboratory evaluation of two diatomaceous earth formulations against *Blattisocius keegani* Fox (Mesostigmata: Ascidae) and *Cheyletus malaccensis* Oudemans (Prostigmata: Cheyletidae). Biol Contr 38:350–355

Athanassiou CG, Kavallieratos NG, Palyvos NE, Sciarretta A, Trematerra P (2005) Spatiotemporal distribution of insects and mites in horizontally stored wheat. J Econ Entomol 98:1058–1069

Athias-Henriot C (1968) L'appareil d'insémination laelapoïde (Acariens anachtinotriches: Laelapoidea, ♀♀). Premières observations. Possibilité d'emploi à des fins taxonomiques. Bull Sic Borgogne 25:229–274

Baker EW, Wharton GW (1952) An introduction to acarology. The Macmillan Company, New York, p 465

Barker PS (1967) Bionomics of *Blattisocius keegani* (Fox) (Acarina: Ascidae), a predator on eggs of pests of stored grains. Can J Zool 45:1093–1099

Beaulieu F, Walter DE (2007) Predation in suspended and forest floor soils: observations on Australian mesostigmatic mites. Acarologia 47:34–54

Beaulieu F, Weeks AR (2007) Free-living mesostigmatic mites in Australia: their roles in biological control and bioindication. Australian J Entomol Agr 47:460–478

Beavers JB, Denmark HA, Sehime AG (1972) Predation by *Blattisocius keegani* on egg masses of *Diaprepes abbreviatus* in the laboratory. J Econ Entomol 65:1483–1484

Bedard WD (1937) Biology and control of the Douglas-fir beetle *Dendroctonus pseudotsugae* Hopkins (Coleoptera: Scolytidae) with notes on associated insects. Res Stud (State College of Washington) 5:103–105

Bhattacharyya AK (2001) Present state of knowledge of Indian ascid mites (Acarina: Mesostigmata) and their zoogeographical distribution. In: Second acarology symposium, abstracts, African Acarology Association, International Centre of Insect Physiology and Ecology, Nairobi. Available via DIALOG. http://www.africanacarologyassociation.com/symposiumII.htm. Accessed 31 Mar 2014

Binns ES (1972) *Arctoseius cetratus* (Sellnick) (Acarina: Ascidae) phoretic on mushroom sciarid flies. Acarologia 14:350–356

Binns ES (1974) Notes on the biology of *Arctoseius cetratus* (Sellnick) (Mesostigmata: Ascidae). Acarologia 16:577–582

Binns ES (1982) Phoresy as migration – some functional aspects of phoresy in mites. Biol Rev 57:571–620

Bram RA (1956) A new predatory mite from insect culture (Acarina: Phytoseiidae). Proc Entomoll Soc Washington 58:292–294

Britto EPJ, Gago E, Moraes GJ (2012) How promising is *Lasioseius floridensis* as a control agent of *Polyphagotarsonemus latus*? Exp Appl Acarol 56:221–231

Burkhardt F (1920) Zur Biologie der Mehlmotte (*Ephestia kuehniella* Zeller). Zeitschrift für Angewandte Entomologie 6:25–60

Cantelo WW, Boswell AL (1973) Mite control with chemicals: a *Drosophila* culture. Drosophila Information Service 50:163

Cayrol JC (1970) Action des autres composants de la biocénose du champignon de couche sur le nématode mycophage, *Ditylenchus myceliophagus* J. B. Goodey, 1958, et étude de son anabiose: Forme de survie en conditions défavorables. Revue d'écologie et de biologie du sol 7:409–440

Chow YS, Liu MA (1984) Preliminary studies on a parasite and a predatory mite of the tarsonemid mite *Steneotarsonemus spinki* Smiley. In: Griffiths DA, Bowman CE (eds) Acarology VI, vol 1. Ellis Horwood, Chichester, pp 323–327

Çobanoglu S, Bayram S, Denizhan E, Saglam HD (2007) The effect of different temperatures and food densities on development of *Blattisocius tarsalis* (Berlese) (Acari: Ascidae) reared on Mediterranean flour moth *Ephestia kuehniella* Zeller (Lepidoptera: Pyralidae). Turkiye Entomoloji Dergisi 31:7–20

Colwell RK (1973) Competition and coexistence in a simple tropical community. Am Nat 107:737–760

Colwell RK (1979) The geographical ecology of hummingbird flower mites in relation to their host plants and carriers. Recent Adv Acarol 2:461–468

Colwell RK (1986) Population structure and sexual selection for host fidelity in the speciation of hummingbird flower mites. In: Karlin S, Nevo E (eds) Evolutionary processes and theory. Academic, Orlando, pp 475–495

Colwell RK, Naeem S (1979) The first known species of hummingbird flower mite north of Mexico: *Rhinoseius epoecus* n. sp. (Mesostigmata: Ascidae). Ann Entomol Soc Am 72:485–491

Colwell RK, Naeem S (1994) Life-history patterns of hummingbird flower mites in relation to host phenology and morphology. In: Houck MA (ed) Mites – ecological and evolutionary analyses of life-history patterns. Chapman & Hall, New York, pp 23–44

Cruz DD, Abreu VHR, Sluys MV (2007) The effect of hummingbird flower mites on nectar availability of two sympatric *Heliconia* species in a Brazilian Atlantic forest. Ann Bot 100:581–588

Cruz WP, Sarmento RA, Pedro-Neto M, Teodoro AV, Rodrigues DM, de Moraes GJ (2014) Population dynamics of *Aceodromus convolvuli* (Acari: Mesostigmata: Blattisociidae) on spontaneous plants associated with *Jatropha curcas* in central Brazil. Exp Appl Acarol 64:309–319

Darst PH, King EW (1969) Biology of *Melichares tarsalis* in association with *Plodia interpunctella*. Ann Entomol Soc Am 62:747–749

Esteca F de CN, Pérez-Madruga Y, Britto EPJ, de Moraes GJ (2014) Does the ability of *Blattisocius* species to prey on mites and insects vary according to the relative length of the cheliceral digits? Acarologia 54(3):359–365

De Leon D (1967) Some mites of the Caribbean Area: Part II. The genus *Asca* in the Caribbean Area (Acarina: Ascidae). Allen Press, Inc. Lawrence, Kansas, pp 47–66

De Leon-Facundo JB, Corpuz-Raros LA (2002) Two new species and a new record of predatory mites of the genus *Lasioseius* Berlese (Acari: Ascidae) inhabiting ornamental crops in the Philippines. The Philippine Agr Sci 85:39–46

de Sousa JM, Gondim MGC Jr, Barros R, de Oliveira JV (2005) Ácaros em produtos armazenados comercializados em supermercados e feiras livres da cidade do Recife. Neotropical Entomol 34:303–309

Demite PR, Moraes GJ, McMurtry JA, Denmark HA, de Castilho RC (2012) Phytoseiidae Database. Available via DIALOG. www.lea.esalq.usp.br/phytoseiidae. Accessed 4 Jan 2014

Dmoch J (1995) *Arctoseius semiscissus* (Berlese 1892) (Acarina: Ascidae) foretic on mushroom sciarid flies in Poland – a possible agent for biological control of sciarids in mushroom houses. In: Elliot TJ (ed) Science and cultivation of edible fungi. Balkema, Rotterdam, pp 533–537

Early M, Goff ML (1986) Arthropod succession patterns in exposed carrion on the island of O'ahu, Hawaiian Islands, USA. J Med Entomol 23:520–531

Egan ME, Hunter PE (1975) Redescription of a cockroach mite, *Proctolaelaps nauphoetae*, with notes on its biology. Ann Entomol Soc Am 68:361–364

Egan ME, Moss WW (1969) The life cycle and behavior of a cockroach mite, *Proctolaelaps nauphoetae* (Acari: Mesostigmata: Ascidae). Notulae Naturae 420:1–9

Enkegaar A, Broodsgaard HF (2000) *Lasioseius fimetorum*: a soil-dwelling predator of glasshouse pests? Biol Contr 45:285–293

Flanders SE, Badgley ME (1963) Prey-predator interactions in self-balanced laboratory populations. Hilgardia 35(8):145–183

Flechtmann CHW (1976) Sobre dois ácaros (Arthropoda: Acari) frequentes em criações da broca da cana-de-açúcar (Insecta: Lepidoptera). Ecossistema 1(1):62–63

Fouly AH (1997) Influence of different nourishment on the biology of *Lasioseius dentatus* (Fox), a new record from Egypt (Acari: Gamasida: Ascidae). Egyptian J Biol Pest Contr 7:1–6

Galvão AS, Gondim MGC Jr, Moraes GJ (2011) Life history of *Proctolaelaps bulbosus* feeding on the coconut mite *Aceria guerreronis* and other possible food types occurring on coconut fruits. Exp Appl Acarol 53:245–252

Goff ML (1989) Gamasid mites as potential indicators of postmortem interval. In: Channabasavanna GP, Viraktamath CA (eds) Progress in acarology, vol 1. Oxford & IBH Publishing, New Delhi, pp 443–450

Graham WM (1970) Warehouse ecology studies of bagged maize in Kenya – II. Ecological observations on an infestation by *Ephestia* (*Cadra*) *cautella* (Walker) (Lepidoptera: Phycitidae). J Stored Product Res 6:157–167

Hafez SM, Mallawani MA, Taher SH (1988) Biological studies on *Blattisocius tarsalis* Keegan, a predatious mite inhabiting stored food in Egypt. Ann Agric Sci (Faculty of Agriculture, Ain Shams University of Cairo, Egypt) 33:1387–1393

Haines CP (1979) A revision of the genus *Blattisocius* Keegan (Mesostigmata: Ascidae) with special reference to *B. tarsallis* (Berlese) and the description of a new species. Acarologia 20:19–38

Haines CP (1981) Laboratory studies on the role of an egg predator, *Blattisocius tarsalis* (Berlese) (Acari: Ascidae), in relation to the natural control of *Ephestia cautella* (Walker) (Lepidoptera: Pyralidae) in warehouses. Bull Entomol Res 71:555–574

Halliday RB, Walter DE, Lindquist EE (1998) Revision of the Australian Ascidae (Acarina: Mesostigmata). Invertebr Taxon, East Melbourne 12:1–54

Hansen LS, Nielsen PS (2001) *Blattisocius tarsalis* as a control agent of *Ephestia kuehniella* in flour mills; initial results of a field trial. Proceedings of the first meeting of Working Group 4, Lisbon, 6–7 Sept 2001. Bio-control of arthropod pests in the stored products, pp 44–46

Hase A (1928) Insekten. In: Peterfi T (ed) Methodik der wissenschaftlichen Biologie, Allgemeine Physiologie, vol 2. Verlag von Julius Springer, Berlin, p 271

Hase A (1933) Über die Mehlmottenmilbe *Typhlodromus tineivorus* Oudems., 1929. Eine Berichtigung. Zoologischer Anzeiger, Leipzig 104:237–239

Heyneman AJ, Colwell RK, Naeem S, Dobkin DS, Hallet B (1991) Host plant discrimination: experiments with hummingbird flower mites. In: Price PW, Lewinsohn TM, Fernandes GW, Benson WW (eds) Plant-animal interactions: evolutionary ecology in tropical and temperate regions. John Wiley & Sons, New York, pp 45–485

Hinton HE (1971) Plastron respiration in the mite, *Platyseius italicus*. J Insect Physiol 17:1185–1199

Hirschmann W, Rühm W (1953) Milben und Fadenwürmer als Symphoristen und Parasiten des Buchdruckers. Mikrokosmos 43:7–10

Houck MA (1994) Mites as potential horizontal transfer vectors of eukaryotic mobile genes: *Proctolaelaps regalis* as a model. In: Houck MA (ed) Mites: ecological & evolutionary analyses of the life-history patterns. Chapman & Hall, London, pp 45–69

Houck MA, Clark JB, Peterson KR, Kidwell MG (1991) Possible horizontal transfer of *Drosophila* genes by the mite *Proctolaelaps regali*. Science 253:1125–1129

Hughes AM (1976) The mites of stored food and houses, 2nd edn, Technical Bulletin, 9. Ministry of Agriculture, Fisheries and Food, London, p 400

Hummel NA, Castro BA, McDonald EM, Pellerano MA, Ochoa R (2009) The panicle rice mite, *Steneotarsonemus spinki* Smiley, a re-discovered pest of rice in the United States. Crop Prot 28:547–560

Ibrahim GA, Afifi AM, Abdel-Halim SM (1989) Laboratory observations on the biology of *Proctolaelaps pygmaeus* (Muller) and *Proctogamasellus mica* (Athias-Henriot) (Acari: Gamasida: Asidae). Bull Entomol Soc Egypt 68:43–47

Ibrahim GA, Abdel-Samedi MA, El-Gazzar HF (1992) Biological aspects of predator mite *Proctolaelaps subcorticalis* Lindquist (Ascidae: Mesostigmata) with description to its immature stages. Minufiya J Agr Res 17:2015–2023

Imbriani JL, Mankau R (1983) Studies on *Lasioseius scapulatus*, a mesostigmatid mite predacious on nematodes. J Nematol 15:523–528

Karg W (1961) Oekologische Untersuchungen von edaphischen Gamasiden (Acarina: Parasitiformes) 2 Teil. Pedobiologia 1:77–98

Karg W (1967) Syno¨ kologische Untersuchungen von Bodenmilben aus forstwirtschaftlich und landwirtschaftlich genutzten Böden. Pedobiologia 7:198–214

Karg W (1983) Distribution and significance of predatory mites of the cohort Gamasina as antagonists of nematodes. Pedobiologia 25:219–232

Karg W (1988) To the knowledge of the genus *Proctolaelaps* Berlese, 1923 (Acarina: Parasitiformes). Zoologische Jahrbucher Systematik 115:441–454 [in German]

Karg W (1998) Neue Raubmilbenarten der Ascidae Oudemans, 1905 im tropischen Regenwald von Ecuador (Acarina: Parasitiformes). Zoologische Reihe 74:49–64

Karg W, Baker RA, Jenkinson L (1995) Mite infestation in *Drosophila* cultures and description of a new mite species of the genus *Proctolaelaps* Berlese with notes on its control in the laboratory (Acarina: Parasitiformes). Deutsche Entomologische Zeitschrift 42:187–190

Kinn DN (1983) The life cycle of *Proctolaelaps dendroctoni* Lindquist and Hunter (Acari: Ascidae): a mite associated with pine bark beetles. Int J Acarol 9:205–210

Krantz GW, Poinar GO Jr (2004) Mites, nematodes and the multimillion dollar weevil. J Nat Hist 38:135–141

Lawson-Balagbo LM, Gondim MGC Jr, de Moraes GJ, Hanna R, Schausberger P (2007) Life history of the predatory mite *Neoseiulus paspalivorus* and *Proctolaelaps bickleyi*, candidates for biological control of *Aceria guerreronis*. Exp Appl Acarol 43:49–61

Leclercq M, Verstraeten C (1988) Entomologie et médecine légale. Datation de la mort. Acariens trouvés sur des cadavres humains [Entomology and forensic medicine. Determination of the time of death. Acari found on human corpses]. Bull Annu Soc Royale Belge d'Entomol 124:195–200

Lesna I, Sabelis MW, Bolland HR, Conijn CGM (1995) Candidate natural enemies for control of *Rhizoglyphus robini* Claparède (Acari: Astigmata) in lily bulbs: exploration in the field and pre-selection in the laboratory. Exp Appl Acarol 19:655–669

Lima DB, Melo JWS, Gondim MGC Jr, Moraes GJ (2012) Limitations of *Proctolaelaps bickleyi* as control agents of *Aceria guerreronis*. Exp Appl Acarol 56:223–246

Lindquist EE (1961) Taxonomic and biological studies of mites of the genus *Arctoseius* Thor from Barrow, Alaska (Acarina: Aceosejidae). Hilgardia 30:301–350

Lindquist EE (1962) *Mucroseius monochami*, a new genus and species of mite (Acarina: Blattisociidae) symbiotic with sawyer beetles. Can Entomol 94:972–980

Lindquist EE (1963) A taxonomic review of the genus *Hoploseius* Berlese (Acarina: Blattisociidae). Can Entomol 95:1175–1185

Lindquist EE (1964) A revision of mites of subfamily Blattisociinae (Acarina: Blattisociidae) in American north of Mexico. Ph.D. dissertation, University of California, Berkeley. University Microfilms, Inc., Ann Arbor, p 402

Lindquist EE (1965) An unusual new species of *Hoploseius* Berlese (Acarina: Blattisociidae) from Mexico. Can Entomol 97(11):1121–1131

Lindquist EE (1969) Mites and the regulation of bark beetle populations. In: Evans GO (ed) Proceedings of the 2nd international congress of acarology, Sutton Bonington, 19–25 July 1967. Akadémiai Kiadó, Budapest, pp 389–399

Lindquist EE (1970) Relationships between mites and insects in forest habitats. Can Entomol 102:978–983

Lindquist EE (1971) New species of Ascidae (Acrina: Mesostigmata) associated with forest insect pests. Can Entomol 103:919–942

Lindquist EE (1995) Remarkable convergence between two taxa of ascid mites (Acari: Mesostigmata) adapted to living in pore tubes of bracket fungi in North America, with description of *Mycolaelaps* new genus. Can J Zool 73:104–128

Lindquist EE, Evans GO (1965) Taxonomic concepts in the Ascidae, with a modified setal nomenclature for the idiosoma of the Gamasina (Acarina: Mesostigmata). Mem Entomol Soc Canada 47:1–64

Lindquist EE, Hunter PE (1965) Some mites of the genus *Proctolaelaps* (Berlese) (Acarina: Blattisociidae) associated with forest insect pests. Can Entomol 97:15–32

Lindquist EE, Walter DE (1989) *Antennoseius* (*Vitzthumia*) *janus* n.sp. (Acari: Ascidae), a mesostigmatic mite exhibiting adult female dimorphism. Can J Zool, Ottawa 67:1291–1310

Lindquist EE, Krantz GW, Walter DE (2009) Order Mesostigmata. In: Krantz GW, Walter DE (eds) A manual of acarology, 3rd edn. Texas Tech University Press, Lubbock, pp 124–232

Lo KC, Ho CC (1979) Ecological observations on rice tarsonemid mite, *Steneotarsonemus spinki* (Acarina: Tarsonemidae). J Agr Res China 28:181–192

Lo KC, Ho CC (1984) Studies on the life history and predation of *Lasioseius* and its tolerance to some pesticides. Special Issue Taiwan Agr Res Inst 16:235–240 [in Chinese, with English abstract]

Makarova OL (1995) On the association of gamasid mites of the genus *Arctoseius* (Mesostigmata, Ascidae) with winter crane flies (Diptera, Trichoceridae) in the northern Taimyr Peninsula. Arctic Insect News 6:2–5

Mankau R, Imbriani JL (1978) Studies on the mite *Lasioseius scapulatus* a predator on soil nematodes. Nematropica 8:15

Masan P, Perotti MA, Saloñas-Bordas MI, Braig HR (2013) *Proctolaelaps euserratus*, an ecologically unusual melicharid mite (Acari: Mesostigmata) associated with animal and human decomposition. Exp Appl Acarol 61:415–429

Mashaya N (2002) Predation on the booklouse *Liposcelis entomophila* (Enderlein) by *Blattisocius dentriticus* (Berlese) and their susceptibility to deltamethrin and fenitrothion. Inscet Sci Appl 22:75–79

Mathys G, Tencalla Y (1959) Note préliminaire sur la biologie et la valeur prédatice de *Proctolaelaps hypudaei* Oudms (Acarien: Mesostigmata: Aceosejidae). Stations Fédérales D'Essais Agricoles, Lausanne 600:645–654

McMurtry JA, Moraes GJ, Sourassou NF (2013) Revision of the lifestyles of phytoseiid mites (Acari: Phytoseiidae) and implications for biological control strategies. Syst Appl Acarol 18:297–320

Metwalli AM, Abbassy MRA, Montaser SA, Mowafy MH (1991) Life history of the ascid mite *Proctolaelaps pygmaeus* Muller when fed on two different preys. Al-Azhar J Agr Res 13:237–246

Momen FM, Abou-Elela AM, Metwally AM, Naser AK, Saleh KHM (2011) Biology and feeding habits of the predacious mite, *Lasioseius lindquisti* (Acari: Ascidae) from Egypt. Acta Phytopathol Entomol Hungarica 46:151–163

Moore JC, Walter DE (1988) Arthropod regulation of micro- and mesobiota in below-ground detrital food webs. Annu Rev Entomol 33:419–439

Moraza ML, Lindquist EE (2011) A new genus of fungus-inhabitting blattisociid mites (Acari: Mesostigmata: Phytoseioidea) from Middle America, with a key to genera of the subfamily Blattisociinae. Zootaxa 2758:1–25

Moustafa MA, El-Hady MM (2006) Biocontrol of some sclerotia forming fungi attacking green beans plant using two fungivorus mites and *Trichoderma harzianum*. J Agr Sci Mansoura Univ 31:3239–3245

Mowafi MH (2005) Effect of food type on development and reproduction of mite species *Lasioseius africanus* Nasr (Acari: Ascidae). Egyptian J Biol Pest Contr 15:93–95

Muma MH (1961) Mites associated with citrus in Florida. Bull Univ Fla Agr Exp Stn 640:39

Naeem S, Dobkin DS, O'Connor BM (1985) *Lasioseius* mites (Acari: Gamasida: Ascidae) associated with hummingbird-pollinated flowers in Trinidad, West Indies. Int J Entomol 27:338–353

Narita JPZ, Pédelabat M, Moraes GJ (2013) A new species of *Neocypholaelaps* Vitzthum (Acari: Ameroseiidae), with notes on the cheliceral lobes and ventral pore-like structures of mites of this family. Zootaxa 3666:1–15

Nasr AK, Nawar MS, Mowafi MH (1990a) Biological studies and feeding habits of *Lastoseius* [sic] *athiasae* Nawar & Nasr (Acari: Mesostigmata: Ascidae) in Egypt. Bull Zool Soc Egypt 39:75–88

Nasr AK, Nawar MS, Mowafi MH (1990b) Biological studies on *Proctolaelaps bickleyi* Bram (Acari: Gamasida: Ascidae). Bull Zool Soc Egypt 39:89–100

Navasero MM, Hirao GA, Navasero MV, Santiago DR (2003) Predatory behavior of *Proctolaelaps yinchuanensis* Xue, Sui & Yi (Ascidae: Acarina). Philippine Entomol 17(2):185

Navasero MM, Hirao GA, Santiago DR, Navasero MV, Raros LC (2004) Laboratory rearing technique for the predatory *Proctolaelaps yinchuanensis* Xue, Sui & Yi (Ascidae: Gamasida: Acari) using *Suidasia pontifica* Oudemans (Suidasiidae: Acaridae: Acari). Philippine Entomol 18(2):180

Navia D, Gondim MGC Jr, Aratchige NS, Moraes GJ (2013) A review of the status of the coconut mite, *Aceria guerreronis* (Acari: Eriophyidae), a major tropical mite pest. Exp Appl Acarol 59:67–94

Nawar MS (1992) Life tables of *Proctolaelaps deloni* Nawar, Childers and Abou-Setta (Gamasida: Ascidae) at different temperatures. Exp Appl Acarol 13:281–285

Nawar MS, El-Sherif AA (1992) Biological studies and description of developmental stages of *Lasioseius zaheri* Nasr (Acari: Ascidae). Ann Agr Sci Moshtohor 30:581–589

Nawar MS, Nasr AK (1989) Biology of the ascid mite *Protogamasellus primitivus similis* Genis, Loots and Ryke with description of immature stages (Acari: Mesostigmata: Ascidae). Bull Fac Agr Cairo Univ 68:85–94

Nawar MS, Rakha MA, Ali FS (1990) Laboratory studies on the predaceous mite, *Lasioseius bispinosus* Evans (Acari: Mesostigmata: Ascidae) on various kinds of food substances. Bull Entomol Soc Egypt 69:247–255

Nesbitt HHJ (1951) A taxonomic study of the Phytoseiinae (family Laelaptidae) predaceous upon Tetranychidae of economic importance. Zoologische Verhandelingen, p 64 + 32 plates

Nielsen PS (1998) *Blattisocius tarsalis* (Berlese), would this predatory mite be effective against moth eggs in Scandinavian flour mills? Integ Prot Stored Products IOBC Bull 21:83–87

Nielsen PS (1999a) The impact of temperature on activity and consumption rate of moth eggs by *Blattisocius tarsalis* (Acari: Ascidae). Exp Appl Acarol 23:149–157

Nielsen PS (1999b) The use of *Blattisocius tarsalis* (Acari: Ascidae) for biological control in flour mills. Proceedings of the 7th international working conference on stored-product protection, Beijing, 14–19 Oct 1998, vol 2, pp 1265–1268

Nielsen PS (2001) Development of *Blattisocius tarsalis* (Acari: Ascidae) at different temperatures. Exp Appl Acarol 25:605–608

Nielsen PS (2003) Predation by *Blattisocius tarsalis* (Berlese) (Acari: Ascidae) on eggs of *Ephestia kuehniella* Zeller (Lepidoptera: Pyralidae). J Stored Products Res 39:395–400

Norton RA, Kethley JB, Johnston DE, O'Connor BM (1993) Phylogenetic perspectives on genetic systems and reproductive modes of mites. In: Wrensch DL, Ebbert MA (eds) Evolution and diversity of sex ratio of insects and mites. Chapman & Hall Publications, New York, pp 8–99

Novak V (1960) Die natürlichen Feinde und Krankheiten des gemeinen Nutzholzborkenkäfers *Trypodendron lineatum* Oliv. Zoologicke Listy Folia Zool 9:309–322

O'Farrell AF, Butler PM (1948) Insects and mites associated with the storage and manufacture of foodstuffs in Northern Ireland. Econ Proc Royal Dublin Soc 3:343–407

Oliveira DC, Moraes GJ, Dias CTS (2012) Status of *Aceria guerreronis* Keifer (Acari: Eriophyidae) as a pest of coconut in the state of Sao Paulo, southeastern Brazil. Neotropical Entomol 41:315–323

Oudemans AC (1929) Acarologische Aanteekeningen C. Entomologischen Berichten 8:28–36
Paciorek CJ, Moyer BR, Levin RA, Halpern SL (1995) Pollen consumption by the hummingbird flower mite *Proctolaelaps kirmsei* and possible fitness effects on *Hamelia patens*. Biotropica 27:258–262
Pan JW (1985) Discovery of *Blattisocius dentriticus* and *Blattisocius keegani* in salted meet in stored houses. Entomol Knowledge 1985:263–264
Reisen WK, Mullen GR (1978) Ecological observations on acarine associates (Acari) of Pakistan mosquitoes (Diptera: Culicidae). Entomol Exp Appl 7:769–776
Rezk HA (2000) Mites associated with stored dried-dates in Egypt and the role of *Blattisocius keegani* Fox as biological control agent. Alexandria J Agr Res 45:179–191
Riudavets J, Quero R (2003) Prey preference of the predatory mite *Blattisocius tarsalis* (Acari: Ascidae). Proceedings of the 8th international working conference on stored product protection, York, 22–29 July 2002, pp 297–299
Riudavets J, Quero R (2004) Effect of relative humidity on the preimaginal development of *Blattisocius tarsalis* (Acari: Ascidae). Integ Prot Stored Products, IOBC Bull/wprs 27(9):177–180
Riudavets J, Maya M, Monserrat M (2002a) Predation by *Blattosocius tarsalis* (Acari: Ascidae) on stored product pests. Bull OILB/SROP 25(3):121–126
Riudavets J, Pons MJ, Lucas E (2002b) Insects and mites of stored products in the northeast of Spain. Integ Prot Stored Products, IOBC Bull 25:41–44
Rivard I (1960) A technique for individual rearing of the predacious mite *Melichares dentriticus* (Berlese) (Acarina: Aceosejidae) with notes on its life history and behaviour. Can Entomol 92:834–839
Rivard I (1962a) Influence of humidity on the predaceous mite *Melichares dentriticus* (Berlese) (Acarina: Aceosejidae). Can J Zool 40:761–766
Rivard I (1962b) Some effects of prey density on survival, speed of development, and fecundity of the predaceous mite *Melichares dentriticus* (Berl.) (Acarina: Aceosejidae). Can J Zool 40:1233–1236
Rosa AA, Gondim MGC Jr, Fiaboe KKM, Moraes GJ, Knapp M (2005) Predatory mites associated with *Tetranychus evansi* Baker & Pritchard (Acari: Tetranychidae) on native solanaceous plants of coastal Pernambuco State, Brazil. Neotropical Entomol 3494:689–692
Rudzinska M (1998) Life history of the phoretic predatory mite *Arctoseius semiscissus* (Acari: Ascidae) on a diet of sciarid fly eggs. Exp Appl Acarol 22:643–648
Ruf A (1996) Life-history patterns in soil-inhabiting mesostigmatid mites (Dermanyssina: Parasitina). In: Mitchell R, Horn DJ, Needham GR, Welbourn WC (eds) Acarology IX. Proceedings, Columbus, 17–22 July 1994, vol 1, pp 621–628
Russo G (1938) VI. Contributo alla conoscenza dei coleotteri scolitidi. Fleotribo: *Phloeotribus scarabaeoides* (Bern.) Fauv. Parte Seconda. Biografia, simbionti, danni e lotta. Bollettino del Laboratorio di Entomologia Agraria di Portici 2:1–420
Saber SA, El-Laithy AYM, Korayem AM (2007) Biology of the predaceous mite, *Protogamasellus denticus* Nasr (Mesostigmata: Ascidae) feeding on different nematode diets. Int J Nematol 17:17–20
Saloña MI, Moraza L, Carles-Tolrá M, Iraola V, Bahillo P, Yélamos T, Outerelo R, Alcaraz R (2010) Searching the soil: forensic importance of edaphic fauna after the removal of a corpse. J Forensic Sci 55:1652–1655
Sardar MA, Murphy PH (1987) Feeding tests of grassland soil-inhabiting gamasine predators. Acarologia 28:117–121
Sheeja UM, Ramani N (2009) Feeding potential of *Lasioseius* sp. (Acari: Mesostigmata), a promising predator of the red palm mite, *Raoiella indica* Hirst (Acari: prostigmata) ecosystem. Karnataka J Agr Sci 22:698–700
Shereef GM, Zaher MA, Afifi AM (1980) Biological studies and feeding habits of *Proctolaelaps pygmaeus* (Muller) (Mesostigmata: Ascidae) in Egypt. First Conference of the Plant Protection Research Institute, Cairo 3:169–181

Simmons AM, Mahroof RM (2011) Response of *Bemisia tabaci* (Hemiptera: Aleyrodidae) to vapor pressure deficit: oviposition, immature survival, and body size. Ann Entomol Soc Am 104:928–934

Smith BP (1983) The potential of mites as biological control agents of mosquitoes. In: Hoy MA, Cunningham GL, Knutson L (eds) Proceedings of biological control of pests by mites. University of California, Berkeley, 1982, pp 79–85

Stein W (1960) Untersuchungen über die Möglichkeit einer Bekämpfung von Raubmilben (*Blattisocius tenivorus*) in Zuchten der Getreidemotte (*Sitotroga cerealella* (Oliv.)) durch Anwendung von Akariziden. Zeitschrift fur Pflanzenkrankheiten, Pflanzenpathologie und Pflanzenchutz 67:77–87

Sznajder B, Sabelis MW, Egas M (2010) Response of predatory mites to a herbivore-induced plant volatile: genetic variation for context-dependent behaviour. J Chem Ecol 36:680–688

Taha HA, Mohmoud MA (2007) Description and biological studies on the predatory mite *Proctolaelaps naggari* n.sp (Acari: Ascidae: Gamasida). In: Abstracts, Fourth African Acarology Symposium, Hammamet, Tunisia, 22–26 Oct, African Acarology Association. http://www.africanacarologyassociation.com/symposiumIII.pdf. Accessed 18 Apr 2014

Taha HA, Metwally AM, Atwa W, El-Sanady M (2007) Biological and prediction studies on the two acarine predators *Lasioseius sewai* Nasr & Abou-Awad and *Blattisocius keegani* Fox fed on the grain mite *Tyrophagus putrescentiae* (Acari: Acaridae). Egyptian J Agr Res 85:1659–1668

Thind BB, Ford HL (2006) Laboratory studies on the use of two new arenas to evaluate the impact of the predatory mites *Blattisocius tarsalis* and *Cheyletus eruditus* on residual populations of the stored product mite *Acarus siro*. Exp Appl Acarol 38:167–180

Thomas HQ, Zalom FG, Nicola NL (2011) Laboratory studies of *Blattisocius keegani* (Fox) (Acari: Ascidae) reared on eggs of navel orangeworm: potential for biological control. Bull Entomol Res 101:499–504

Throne JE, Hagstrum DW, Nawrot J (1998) Computer model for simulating almond moth (Lepidoptera: Pyralidae) population dynamics. Environ Entomol 27:344–354

Treat A (1966) A new *Blattisocius* from noctuid moths. J NY Entomoll Soc 74:143–159

Treat A (1969) Behavorial aspects of the association of the mites with noctuid moths. In: Proceedings of the 2nd international congress of acarology, Sutton Bonington, Akadémiai Kiadó, Budapest, 19–25 July 1967, pp 275–286

Treat A (1975) Mites of moths and butterflies. Comstock Publishing Associates, Ithaca, p 362

Trivedi TP, Rajagopal D, Tandon PL (1994) Life table for establishment of potato tuber moth, *Phthorimaea operculella*. J Indian Potato Assoc 21:97–105

Tseng YH (1984) Mites associated with weeds, paddy rice, and upland rice fields in Taiwan. In: Griffiths DA, Bowman CE (eds) Acarology VI, vol 2. Ellis Horwood Publisers, Chichester, pp 770–780

Tseng YH (1989) Taxonomical studies on mites associated with weeds, paddy rice and upland rice fields in Taiwan (I). Chinese J Entomol 3:1–35

van de Bund CF (1970) Gamasides as predators of phytophagous nematodes. In: Bozek J, Suski ZW, Jakubowska J (eds) Problems of acarology, symposium. Polska Akademia Nauk, Warszawa, p 280

van de Bund CF (1972) Enkele waarnemingen aan *Lasioseius fimetorum* Karg, 1971 in een gezelschap van mijten, springstaarten en nematoden tussen de wortels van witte klaver. Entomologische Berichten 32:6–12

Voigts H, Oudemans AC (1905) Zur Kenntnis der Milben-Fauna von Bremen. Abhandlungen. Naturwissenschaftlichen Verein zu Bremen, Bremen 18:199–253

Walter DE (1987) Life history, trophic behavior, and description *of Gamasellodes vermivorax* n. sp. (Mesostigmata: Ascidae), a predator of nematodes and arthropods in semiarid grassland soils. Can J Zool Ottawa 65:1689–1695

Walter DE (2003) The genus *Gamasellodes* (Acari: Mesostigmata: Ascidae): New Australian and North American species. Syst Appl Acarol (Special Publications, London) 15:1–10

Walter DE, Ikonen EK (1989) Species, guilds, and functional groups: taxonomy and behavior in nematophagous arthropods. J Nematol 21:315–327
Walter DE, Kaplan DT (1990) A guild of thelytokous mites (Acari: Mesostigmata) associated with citrus roots in Florida. Environ Entomol 19:1338–1343
Walter DE, Lindquist EE (1989) Life history and behavior of mites in the genus *Lasioseius* (Acari: Mesostigmata: Ascidae) from grassland soils in Colorado, with taxonomic notes and description of a new species. Can J Zool 67:2797–2813
Walter DE, Lindquist EE (1995) The distributions of parthenogenetic ascid mites (Acari: Parasitiformes) do not support the biotic uncertainty hypothesis. Exp Appl Acarol 19:423–442
Walter DE, Lindquist EE (2000) *Arrhenoseius gloriosus* n.g., n.sp. (Acari: Mesostigmata: Ascidae), an arrhenotokous mite from rainforests in Queensland, Australia. Acarologia 41:53–68
Walter DE, Halliday RB, Lindquist EE (1993) A review of the genus *Asca* (Acarina: Ascidae) in Australia, with descriptions of three new leaf-inhabiting species. Invertebr Taxon 7:1327–1347
Weis-Fogh T (1948) Ecological investigations on mites and collemboles in the soil. Natura Jutlandic 1:135–270
Weisser WW, Zhang Z-Q (1996) Infestation of *Drosophila* cultures (Diptera: Drosophilidae) by mites of *Proctolaelaps* (Acari: Ascidae) and experiments on mite feeding habits. Syst Appl Acarol 1:41–44
Wu W-N, Wang C-L (1982) A new species of *Gnorimus* from Fujian (Acarina: Phythoseiidae). Wuyi Sci J 2:134–136
Wu Z-Y, Zhang X-H, Luo J, Fang K-H (2009) Effects of five preys on growth and reproduction of *Lasioseius* sp. J Fujian Agr For Univ (Nat Sci Ed) 38:581–584
Xu X (2011) Study on the biology and application of *Lasioseius* sp.. MS thesis presented to Fujian Agriculture and Forestry University. Available via DIALOG. www.dissertationtopic.net/doc/324688. Accessed 14 Apr 2014
Yaninek JS, Hanna R (2003) Cassava green mite in Africa – a unique example of successful biological control of a mite pest on a continental scale. In: Borgemeister C, Langewald J, Neuenschwander P (eds) Biological control in IPM systems in Africa. CABI Publishing, Wallingford, pp 61–75
Zhang Z-Q, Fan Q-H (2010) Blattisociidae of China. Zoosymposia 4:280–287
Zhang YX, Lin J-Z (1986) Identification of phytoseiids—the four species of natural enmies [sic] for controling [sic] tarsonemids on rice. Entomol Knowledge 23:83–85
Zhang Y-X, Lin J-Z (1989) Role of *Gnorimus chaudhrii* (Acari: Phytoseiidae) in controlling tarsonemid mite in paddy fields in China. Indian J Acarol 10:83–86
Zhang Y, Lin J (1991) Studies of *Gnorimus chaudhrii* (Acari: Phytoseiidae), an important predator of tarsonemid mites in rice fields. Chinese J Biol Contr 7:163–165

Chapter 3
The Potential of Free-Living Laelapid Mites (Mesostigmata: Laelapidae) as Biological Control Agents

Grazielle Furtado Moreira and Gilberto José de Moraes

3.1 Introduction

Methods for pest control without the use of pesticides have been extensively studied since the 1950s, after the first registered cases of pest resistance to pesticides. One of these alternative methods, biological control, has been implemented extensively in the last decades, with the establishment of private companies dedicated to the mass production of biological control agents.

Predatory mites are among these agents, especially those of the family Phytoseiidae, order Mesostigmata, widely used for the control of phytophagous mites and small insects inhabiting the aerial plant parts. For the control of edaphic pests or pests that spend part of their life in the soil, four predatory mites of the family Laelapidae have also been used, namely *Androlaelaps casalis* (Berlese), *Gaeolaelaps aculeifer* (Canestrini), *Stratiolaelaps miles* (Berlese) and *Stratiolaelaps scimitus* (Womersley).

Laelapidae is one of the 15–16 families of Dermanyssoidea, a superfamily that includes free-living and parasitic forms (Lindquist et al. 2009). Described slightly over 120 years ago (Berlese 1892), this family includes about 1,300 species classified in slightly over 90 genera (Beaulieu et al. 2011). Recent evidences indicate Laelapidae to be a paraphyletic group (Dowling and O'Connor 2010). Thus, important taxonomic changes in this group are expected in the near future.

G.F. Moreira (✉)
Departamento de Fitossanidade, Faculdade de Ciências Agrárias e Veterinárias, Universidade Estadual Paulista "Júlio de Mesquita Filho", 14884-900 Jaboticabal, SP, Brazil
e-mail: grabiologia@yahoo.com.br

G.J. de Moraes
Departamento de Entomologia e Acarologia, Escola Superior de Agricultura "Luiz de Queiroz", Universidade de São Paulo, 13418-900 Piracicaba, SP, Brazil
e-mail: moraesg@usp.br

D. Carrillo et al. (eds.), *Prospects for Biological Control of Plant Feeding Mites and Other Harmful Organisms*, Progress in Biological Control 19,
DOI 10.1007/978-3-319-15042-0_3

Laelapidae has been divided in 11 subfamilies, including free-living and arthropod, mammal and bird associates. Free-living and arthropod associated species have been considered (e.g. Radovsky 1969; Casanueva 1993) to be part of Hypoaspidinae and Melittiphinae. The first of these has been ranked by some authors (e.g. Karg 2003) at the family level (Hypoaspididae). Hypoaspidinae is the group of interest in this chapter, given the role of its members as natural enemies of pest species; Melittiphinae are usually associated with non-pestiferous Hymenoptera.

Many genera of Hypoaspidinae have been treated in the literature (e.g. Karg 1993, 1994, 1997, 2003, 2006) as subgenera of *Hypoaspis* Canestrini and *Pseudoparasitus* Oudemans, but in this publication they are considered at the generic level. Consideration of those groups as genera is consistent with Casanueva (1993) and recent revisions published by different authors (e.g. Walter and Campbell 2003; Beaulieu 2009; Joharchi and Halliday 2011; Joharchi et al. 2011; Moreira et al. 2014).

The hypoaspidin *Euandrolaelaps* Bregetova is considered a junior synonym of *Alloparasitus* Berlese (Karg 1993). *Androlaelaps* Berlese, considered by Casanueva as hypoaspidin, is currently placed in Laelapinae (Lindquist et al. 2009), but it is considered in this chapter because of the potential of some species to control some undesirable organisms.

3.2 Brief Morphological and Biological Characterization

Adult females of hypoaspidins range in color from light to dark brown, and in length from about 0.3 to about 0.8 mm. Mounted females of most hypoaspidins are easily distinguished from other mite groups for their flask-shaped genital shield and a well separated subtriangular anal shield; genital shield occasionally broad, abutting the subtriangular anal shield (e.g. *Pseudoparasitus*), although a ventrianal shield is present in *Ololaelaps* Berlese and *Oloopticus* Karg. The dorsal shield is sclerotized, undivided, with more than 20 pairs of setae, usually 37–39 pairs, often including extra unpaired or extra paired setae in relation to the pattern shown by Lindquist and Evans (1965) for the Blattisociidae (then considered as Ascidae), and extensively used for other mesostigmatid groups. Peritrematic shield not fused with exopodal shield. Tibia I usually with two ventral setae, including *av2*; genu IV usually with two posterolateral setae, including *pl2*; subcapitulum with internal malae usually well developed, with fimbriate lateral margins and as long as or longer than corniculi (Lindquist et al. 2009).

As in other Mesostigmata, hypoaspidins have five developmental stages: egg, larva, protonymph, deutonymph and adults (females and males). The reproduction in Laelapidae can be by haplodiploidy, arrhenotoky or thelytoky (Norton et al. 1993).

At 25 °C, development from egg to adult takes about 7 days. The adults feed on small insects, other mites, nematodes, annelids and springtails. Some of them have been demonstrated to survive 20–150 days without food (Ignatowicz 1974; Wright and Chambers 1994; our personal observation), provided environmental humidity is high (Ignatowicz 1974; our personal observation). Each female may lay 1–7 eggs a day at around 25 °C.

3.3 Trends in the Efforts to Understand the Hypoaspidins

Most of the taxonomic studies of the Laelapidae refer to mammal associated subfamilies, given their medical and veterinarian importance and the relatively large number of researchers interested in these groups. A catalog of the free-living and arthropod associated laelapids, under preparation by G.F. Moreira, R.A.P. Freire, H. Klompen and G.J. de Moraes, lists nearly 750 papers about the taxonomy of the laelapids of the subfamilies Hypoaspidinae, Mellithiphinae and Laelapinae; about 200 of these papers refer to the hypoaspidins. The pattern of variation in the number of papers published in the last ca. 215 years has been quite erratic, with few papers published until 1950, followed by increasing numbers until 2000 (with an outlying high number in the 1960s) and an apparent decline since then (Fig. 3.1). This pattern can be directly related to the publications of the following authors, who greatly contributed to the understanding of the taxonomy of these mites: A. Berlese from Italy (1890s–1920s), S.K. Bhattacharyya from India (1960s–1970s), M. Costa from Israel (1960s–1980s), W. Karg from Germany (last five decades) and L.M. Ma from China (1990s and 2000s).

About 40 % of the hypoaspidins were described from the Paleartic region, where most of the authors cited in the previous paragraph conducted their studies, followed by the Neotropical (29 %), Australasian (9 %), Ethiopian (9 %), Oriental (8 %) and Neartic (5 %) regions (Fig. 3.2).

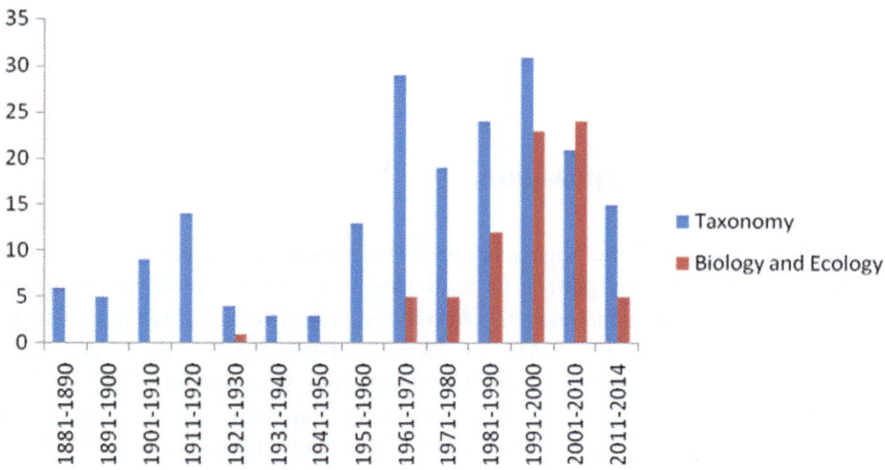

Fig. 3.1 Number of publications on biology/ecology and taxonomy of species of Hypoaspidinae since 1881

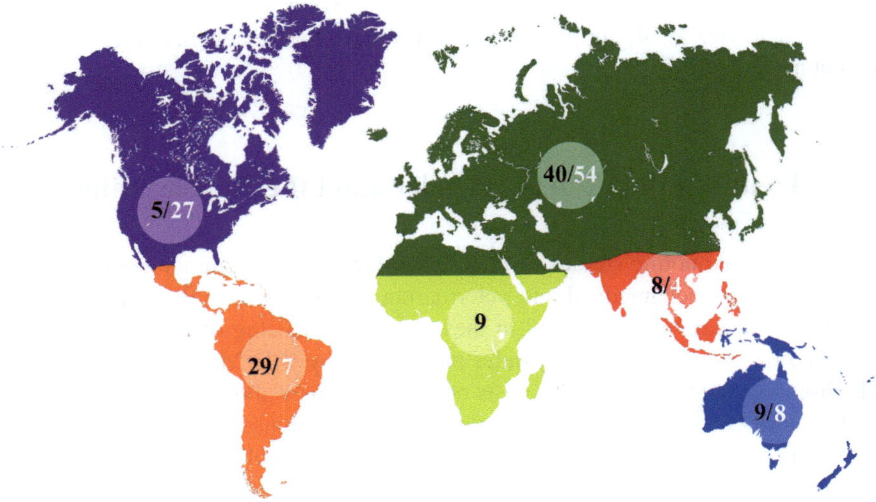

Fig. 3.2 Percentage of publications on taxonomy (*black numbers*) and biology/ecology (*white numbers*) of species of Laelapidae on the different regions. Nearctic (*purple*), Neotropical (*orange*), Palearctic (*green*), Ethiopian (*yellow*), Oriental (*red*) and Australasian (*blue*) (Color figure online)

Papers concerning the biology, ecology and practical use of hypoaspidins are less numerous than those about taxonomy, despite the recognized potential of those mites for the control of soil pests. About 75 papers refer to the biology, ecology, or practical use of mites of this group, most of which published in the last two decades (Fig. 3.2), with major contributions by D. Walter (Australia/North America). As reported for the taxonomic studies, papers about the biology and ecology of this group refer mostly to the Paleartic region (54 %). Papers about other regions refer to: 27 % for the Neartic, 8 % for the Australasian, 7 % for the Neotropical and 4 % for the Oriental regions.

3.4 Diversity of Hypoaspidins

In the last revision of the supraspecific classification of this group, Casanueva (1993) considered Hypoaspidinae to be divided in the tribes Pseudoparasitini and Hypoaspidini. The first included the free-living, soil-dwelling, predatory genera *Alloparasitus* Berlese, *Cosmolaelaps* Berlese, *Gaeolaelaps* Evans and Till, *Ololaelaps* Berlese, *Pseudoparasitus* Oudemans and *Stratiolaelaps* Berlese. The second included paraphage genera associated with Blattaria (*Blaberolaelaps* Costa and *Gromphadorholaelaps* Till) and Coleoptera (*Coleolaelaps* Berlese, *Dyscinetonyssus* Moss and Funk and *Hypoaspis* Canestrini). Since them, *Chapalania* Hoffman and López-Campos and *Mumulaelaps* Clark, both Coleoptera associates, have been included by their authors in Hypoaspidinae, without any attempt to place them in tribes.

Raymentia Womersley was considered by Walter et al. (2002) to be related to Hypoaspidinae and Melittiphinae. In this chapter, this genus is considered to belong to Hypoaspidinae, as are also *Angosomaspis* Costa, *Dicrocheles* Krantz and Khot, *Dynastaspis* Costa, *Lucanaspis* Costa, *Promacrolaelaps* Costa and *Oloopticus* Karg (the latter considered as Hypoaspididae by Karg 2003).

With the inclusion of *Raymentia*, in total 20 genera containing about 422 species are considered to constitute the Hypoaspidinae (Table 3.1). Nine genera are mainly associated with Coleoptera and two are associated with Blattodea. *Dicrocheles* species are parasitic on Noctuidae moths (Lepidoptera), while *Raymentia* are associated with bees. The largest hypoaspidin genera *Cosmolaelaps*, with 114 described species, and *Gaeolaelaps*, with 103 species, are usually found in soil and litter, although frequently reported in association with arthropods (Blattodea, Formicidae, Scarabaeidae and Termitididae). Six genera are monotypic and five have less than ten described species. *Alloparasitus*, *Ololaelaps*, *Oloopticus* and *Pseudoparasitus* are usually found in soil and litter, while *Stratiolaelaps* is frequently found on mammals and their nests and in litter.

3.5 Biological Information

As summarized by Beaulieu and Walter (2007), the habitat occupied by edaphic Mesostigmata is strikingly different from that occupied by plant inhabiting mites of the same order in at least the following aspects: scarcity of alternative pollen and honeydew (commonly used as food for the plant inhabiting Mesostigmata); high availability of a diversity of potential invertebrate prey (including other Mesostigmata), which are partitioned in two functional groups, namely those dependent on water-film for movement and feeding, as the nematodes, and those found in air-space, most arthropods. Edaphic Mesostigmata are mostly generalists, specialization usually consisting of preference for the consumption of members of a single general prey type, as nematodes or arthropods.

Despite the free-living behavior of most hypoaspidins, a large number of species has been described dispersing phoretically on arthropods, mainly Coleoptera and Blattodea. The phoretic species usually have morphological characteristics that apparently make them adapted to particular microhabitats on their hosts. Walter and Moser (2010) mentioned loss of setae (especially those usually added later in ontogeny) in arthropod-associated *Gaeolaelaps* (*J1* usually added in the protonymphal stage; *z3*, *R2-x* and *UR*-series usually added in the deutonymphal stage) as a supposed result of accelerated development or reduced interactions with other organisms in nests or burrows occupied by their transporters. The lack of setae in arthropod associated species was also mentioned by Faraji and Halliday (2009) and Beaulieu (2009). Costa (1971) mentioned the presence of macrosetae in arthropod associated mesostigmatids. He stated that macrosetae can be thigmotactic and concerned with finding the right location on the host, or else can serve to avoid overcrowding on the host, assuring each mite its own space. However, leg macrosetae are also found in most phytoseiid mites, a group usually not carried by arthropods.

Table 3.1 Diversity of genera and habitats from which species of each genus of Hypoaspidinae were described (numbers in parentheses indicate number of species described from the respective habitat)

Genus	Approx. nr. species	Habitats (number of species)
Alloparasitus	10	Soil (6), rocky meadow (1), mammal nest (1), mammal (1), unspecified (1)
Angosomaspis	1	Dynastidae (1)
Blaberolaleaps	2	Blaberidae (2)
Chapalania	1	Melolonthidae (1)
Coleolaelaps	29	Scarabaeidae (16), Dynastidae (2), Cerambycidae (1), Passalidae (1), centipede (1), soil (1), rodent nest (1), others (1), unspecified (4)
Cosmolaelaps	114	Soil (67), Formicidae nest (13), mammals (5), Formicidae (4), Termitidae nest (4), Passalidae (2), beetle (1), Blaberidae (1), Paradoxosomatidae (1), rotten fruit (1), unspecified (14)
Dicrocheles	2	Noctuidae (2)
Dynastaspis	2	Dynastidae (2)
Dyscinetonyssus	1	Scarabaeidae (1)
Gaeolaelaps	103	Soil (55), Blaberidae (7), Passalidae (6), mammals (5), Formicidae (2), mammal nest (2), Poaceae (2), Asparagacea (1), beetle (1), bird (1), Bird nest (1), Blattidae (1), Carabidae (1), Cerambycidae (1), Formicidae nest (1), Gryllidae (1), millipede (1), spider (1), Termitidae nest (1), others (2), unspecified (5)
Gromphadorholaelaps	1	Blaberidae (1)
Hypoaspis s. strictu	43	Scarabaeidae (28), Dynastidae (8), soil (2), Formicidae nest (1), Geotrupidae (1), insect larva (1), mangrove (1)
Lucanaspis	1	Scarabaeidae
Mumulaelaps	1	Scarabaeidae
Ololaelaps	27	Soil (16), unspecified plant species (2), algae (1), mammal (1), Termitidae nest (1), unspecified (6)
Oloopticus	12	Soil (12)
Promacrolaelaps	2	Scarabaeidae (2)
Pseudoparasitus	39	Soil (18), unspecified plant species (3), mammal (3), ant (1), boring dust (1), flowers (1), leaf mould (1), mangrove (1), under bark (1), under rocks (1), unspecified (4)
Raymentia	3	Bee (3)
Stratiolaelaps	28	Soil (9), mammal (6), algae (1), ant nest (1), bird nest (1), Orthoptera (1), poultry house (1), stable (1), tree hole (1), unspecified (6)

Soil: including grasses, humus, litter, manure, mosses, rotting trunks on the soil surface, sphagnum and turf

Representative biological and ecological studies concerning each genus are subsequently presented.

3.5.1 Androlaelaps

Predation of insects and mites by species of this genus has been reported. Studies conducted by Mihm and Chiang (1976) showed predation of an unidentified species of *Androlaelaps* on eggs and first instar larvae of the beetles *Diabrotica longicornis* (Say) and *Diabrotica virgifera* Le Conte (Chrysomelidae), important corn pests in the USA. *Androlaelaps casalis* (Berlese) was observed in laboratory experiments to feed and oviposit on the poultry red mite, *Dermanyssus gallinae* (De Geer) (Dermanyssidade) (Lesna et al. 2009). Some *Androlaelaps* have been reported as ectoparasites of mammals and birds, but no evidence of haematophagy of *A. casalis* was found by Lesna et al. (2009) when hatchlings of starlings (*Sturnus vulgaris* L.) were exposed to *A. casalis*.

Androlaelaps casalis was reported by Chen and Meng (1987) to reproduce by haplodiploidy.

3.5.2 Chapalania

The single species of this genus, *Chapalania cifuentesi* Hoffman and López-Campos, was described by Hoffmann and López-Campos (1995) from Mexico, in association with *Strategus aloeus* (L.) (Coleoptera: Melolonthidae). The authors suggested a case of multiple phoresy, with the presence of *Histiostoma* sp. (Astigmatina) on the legs of female *C. cifuentesi* that was phoretically associated with *S. aloeus*. Nothing has been reported about the biology of this laelapid mite.

3.5.3 Coleolaelaps

This genus consists of species commonly associated with scarab beetles (Scarabaeidae). Grandi (1925) suggested *Coleolaelaps inopinatus* Grandi to feed on *Tyrophagus mycophagus* Mégnin (Astigmatina), both associated with the scarab *Anoxia matutinalis* var. *suturalis* Reitter. Hinckley (1967) mentioned an unidentified *Coleolaelaps* in association with all stages of another scarab beetle, *Oryctes rhinoceros* (L.), an important palm pest on some Pacific Islands. The author reported this mite to be harmless to the adult beetle, but to be a possible egg predator. In laboratory studies, the mite could not attack eggs with hardened chorion, suggesting that the laelapid left the carrier to feed on recently laid eggs.

3.5.4 Cosmolaelaps

Two *Cosmolaelaps* species were reported in association with passalid (Passalidae) beetles, *Cosmolaelaps trifidus* (Pearse, Patterson, Rankin and Wharton) and *Cosmolaelaps euventricosus* (Karg). Hunter and Mollin (1964) observed adults of *C. trifidus* (mentioned as *Cosmolaelaps passali*, its junior synonym) to use the chelicerae to attach to the passalid *Odontotaenius disjunctus* (Illiger). Non-feeding larvae and protonymphs were found deep in frass and fecal material, while deutonymphs were found on the surface of the frass and fecal material of the same beetle.

Feeding capacity and fecundity were evaluated for *Cosmolaelaps claviger* (Berlese) on house fly [*Musca domestica* (L.) (Muscidae)] eggs and larvae (Afifi and van der Geest 1984), and for *Cosmolaelaps barbatus* Moreira, Klompen and Moraes, and *Cosmolaelaps jaboticabalensis* Moreira, Klompen and Moraes on pupae of the thrips *Frankliniella occidentalis* (Pergande) (Thripidae) (our unpublished observation). *Cosmolaelaps* sp. nr. *multisetosus* was reported to feed on larvae of *Helicoverpa armigera* (Hübner) (Noctuidae) (Beaulieu and Walter 2007). Complete development of *Cosmolaelaps vacuus* (Michael) was observed when feeding on eggs of house fly (Abou-Awad et al. 1989). Feeding and oviposition on nymphs of Collembola was reported by Afifi and van der Geest (1984) for *C. claviger*, by Walter (1987, 1988) for *C. vacuus* and by Beaulieu and Walter (2007) for *C.* sp. nr. *multisetosus.*

Mites of several groups have also been reported to be attacked by *Cosmolaelaps* species. *Cosmolaelaps barbatus* and *C. jaboticabalensis* were observed to feed and oviposit on *Rhizoglyphus echinopus* (Fumouze and Robin) (Astigmatina) (our unpublished observation). Complete development on the same prey was reported for *C. claviger* (Afifi and van der Geest 1984) and *C. vacuus* (Abou-Awad et al. 1989), and on *Tyrophagus putrescentiae* Schrank (Astigmatina) for *Cosmolaelaps paulista* Freire and Moraes (Freire and Moraes 2007a), *C. jaboticabalensis* (Moreira et al. 2015) and *Cosmolaelaps qassimensis* Fouly and Al-Rehyiani (Fouly and Abdel-Baky 2015).

Afifi and van der Geest (1984) reported *C. claviger* to oviposit on immature stages of *Tetranychus urticae* Koch (Prostigmata). Walter (1987, 1988) observed complete development and oviposition of *C. vacuus* on *Oppiella nova* (Oudemans) (Oribatida) and *T. putrescentiae. Cosmolaelaps simplex* Berlese was able to complete development on *Caloglyphus rodriguezi* Samsinak (Astigmatina), while its larval stage was observed to be completed without feeding (Al Rehiayani and Fouly 2005) and *C. qassimensis* on *Caloglyphus rodriguez* Samsinak (Fouly and Abdel-Baky 2015).

Predation of *Cosmolaelaps* on nematodes has been widely studied. *Cosmolaelaps claviger* and *Cosmolaelaps* sp. nr. *multisetosus* were observed to feed on free-living nematodes under laboratory condition (Afifi and van der Geest 1984; Beaulieu and Walter 2007, respectively). All post-embryonic stages of *Cosmolaelaps indicus* Bhattacharyya (mentioned as *Hypoaspis calcuttaensis*, its synonym) were observed to feed on vermiform stages and free eggs (but not on egg masses) of *Meloidogyne javanica* (Treub) (Meloidogynidae) by Walia and Mathur (1995). Bilgrami (1997) reported the same predator (also mentioned as *H. calcuttaensis*) to be able to prey on 31 species of free-living Rhabditida nematodes, Enoplida [*Acrobeles* sp., *Acrobeloides* sp., *Cephalobus* sp., *Chiloplacus symmetricus* (Thorne), *Mesorhabditis* sp., *Rhabditis* sp. and *Tobrilus* sp.], plant parasitic Tylenchida, Aphelenchida, and Dorylaimida [*Anguina tritici* (Stainbuch), *Aphelenchoides* sp., *Basiria* sp., *Helicotylenchus indicus* Siddiqi, *Hemicriconemoides mangiferae* Siddiqi, *Hemicycliophora dhirendri* Husain and Khan, *Heterodera mothi* Khan, *Hirchmanniella oryzae* (Haan), *Hoplolaimus indicus* Sher, *Longidorus* sp., *Meloidogyne incognita* (Kofoid and White), *Paratrichodorus* sp., *Paralongidorus citri* (Siddiqi), *Scutellonema* sp., *Tylenchorhynchus mashhoodi* Siddiqi and Basir and *Xiphinema basiri* Siddiqi] or predaceous Dorylaimida, and Diplogasterida [*Dorylaimus stagnalis* Dujardin, *Aporcelaimellus nivalis* (Altherr), *Aquatides thornei* (Schneider), *Discolaimus silvicolus* Sauer and Annells, *Mononchoides fortidens* (Schuurmans Stekhoven), *Mononchoides longicaudatus* Khera, *Mononchus aquaticus* Coetzee and *Mylonchulus dentatus* Jairajpuri]. The same author reported that in the absence of prey *C. indicus* was observed to feed on its own fecal matter. Cannibalism was not observed and increased predation with increasing prey numbers suggested a density dependent behavior.

Al Rehiayani and Fouly (2005) reported *C. simplex* to develop and oviposit on egg masses of *M. javanica* and *Tylenchulus semipenetrans* Cobb (Tylenchulidae) under laboratory conditions. *Cosmolaelaps barbatus* and *C. jaboticabalensis* were observed to oviposit on the entomopathogenic nematodes, *Heterorhabditis amazonensis* GL Andaló, Nguyen and Moino Jr. (Heterorhabditidae), *Steinernema brazilense* CB06 Nguyen, Ginarte, Leite, Santos and Harakava (Steinernematidae) and *Protorhabditis* sp. (Rhabditidae) (our personal observation). *Cosmolaelaps vacuus* completed development and oviposited on the free-living nematodes *Panagrolaeimus* sp. (Panagrolaimidae), *Aphelencus* sp. (Aphelenchoididae) and *Acrobeloides* sp. (Cephalobidae), but not on the entomopathogenic *Steinernema feltiae* Filipjev (Steinernematidae) or *Heterorhabditis heliothidis* (Khan, Brooks and Hirschmann) (Heterorhabditidae) (Walter 1987, 1988).

Populations of *C. vacuus* from Colorado, USA, and *C. jaboticabalensis* from São Paulo, Brazil, were observed to be thelytokous (Walter and Oliver 1989; Moreira et al. 2015, respectively), although males have been reported in Russian and Polish populations of *C. vacuus* (Bregetova 1977; Gwiazdowicz 2004). The matting process of *C. claviger* was reported to last 15 min, while unmated females were observed to reproduce by arrhenotoky (Afifi and van der Geest 1984). *Cosmolaelaps* sp. nr. *weeversi* was also mentioned to reproduce by arrhenotoky (Walter and Kaplan 1990).

3.5.5 Dyscinetonyssus

Based on morphological characteristics, Moss and Funk (1965) suggested *Dyscinetonyssus hystricosus* Moss and Funk to be a beetle parasite. According to these authors, strongly thickened setae *av1* and *av3* on tarsus II, short peritreme, serrated labrum, movable digit edentate or with a single small tooth and with a ventral basal notch, and larviparous habit are common characteristics of parasitic dermanyssoids (mentioned as laelaptoids). Nothing else was published about the biology or ecology of mites of this group.

3.5.6 Gaeolaelaps

As summarized by Walter and Oliver (1989), *Gaeolaelaps* are aggressive predators of a large variety of arthropod and nematode prey. Most biological studies of this genus refer to *G. aculeifer*, species seemingly well distributed in Europe and North America.

This species was observed to feed on injured or dead *Acheta domesticus* L. (Gryllidae), *M. domestica*, *Drosophila melanogaster* Morgan (Drosophilidae) and *Culex* sp. (Ignatowicz 1974); a mixture of *D. melanogaster* and the mite *T. urticae* (Lobbes and Schotten 1980); and dipteran larvae (Sardar and Murphy 1987). It has also been reported to feed (Sardar and Murphy 1987) and to develop (Murph and Sardar 1991) on eggs of the beetle *Tribolium* sp. (Tenebrionidae), as well as to oviposit on second instar larvae, prepupae and pupae of *F. occidentalis* (Berndt et al. 2004b).

Springtails can also be preyed by *G. aculeifer*. This predator was observed to feed on undisclosed springtails (Ignatowicz 1974; Sardar and Murphy 1987) and on *Folsomia candida* Willem and *Folsomia fimetaria* L. (Heckmann et al. 2007). It was also reported to reproduce on *Sinella coeca* (Schott) and *Hypogastrura denticulata* (Bagnell) (Usher and Davis 1983), and to develop and reproduce on an unidentified springtail (Ragusa et al. 1986), on *Onychiurus* sp. (Murph and Sardar 1991) and on *Isotoma notabilis* (Schaefer) (Kevan and Sharma 1964).

Mites have also been observed to serve as prey for this predator, especially the Astigmatina, as *R. echinopus* (Ignatowicz 1974); *Acarus siro* L., *Schwiebea talpa* Oudemans and *T. putrescentiae* (Sardar and Murphy 1987); and *Caloglyphus* cf. *michaeli* Oudemans (Heckmann et al. 2007). In addition, it was observed to develop and reproduce on *T. putrescentiae* (Kevan and Sharma 1964; Barker 1968; Lobbes and Schotten 1980; Murph and Sardar 1991), *Glycyphagus domesticus* (de Geer) (Barker 1968), *Tyrophagus* sp. (Ragusa et al. 1986), an unidentified Histiostomatidae (Ragusa et al. 1986), *R. echinopus* (Ragusa and Zedan 1988; Amin et al. 2014), *Caloglyphus* sp. (Ruf 1991) and *Tyrophagus similis* Volgin (Kasuga et al. 2006). There is also a report about its ability to feed on the Oribatida *Liebstadia similis* Michael and *Ceratozetes gracilis* (Michael) (Sardar and Murphy 1987).

Gaeolaelaps aculeifer was reported to feed on the nematode *Turbatrix silusiae* de Man (Heckmann et al. 2007) and to develop on *Rhabditis* sp. (Murph and Sardar 1991). Inserra and Davis (1983) mentioned a species identified as *G.* nr. *aculeifer* to feed on the plant parasitic nematodes *Heterodera schachtii* Schmidt, *Meloidogyne chitwoodi* Golden, O'Bannon, Santo and Finley and *Meloidogyne hapla* Chitwood.

Small oligochaets of the family Enchytraeidae have also been reported to be attacked by *G. aculeifer*. Predation has been reported on unidentified species of this family (Ignatowicz 1974; Sardar and Murphy 1987) as well as on *Enchytraeus crypticus* Westheide and Graefe (Heckmann et al. 2007). This is not a desirable attribute from the standpoint of the use of this predator for biological control purposes, given the known beneficial action of enchytraeids in the soil. The extent of the predation is unknown and should be evaluated.

In addition to its predatory habit, it has also been reported to feed on pollen of *Acanthus mollis* L., *Carbobrotus edulis* (L.) and *Ipomoea purpurea* (L.), despite the lower oviposition and immature survivorship on *C. edulis* than on arthropod prey (Ragusa et al. 1986).

Copulation of adult males with deutonymphs of *G. aculeifer* was reported by Ignatowicz (1974). The effect of predator density on oviposition was described by Ruf (1991). The author reported daily production of two eggs per isolated female and of one egg per female when groups of five females were kept together. He also observed that during population growth the number of males remains stable or reduced with increasing numbers of female. Cannibalism was reported by Ignatowicz (1974), Ruf (1991) and Berndt et al. (2003).

Larvae of *G. aculeifer* were observed not to feed, whereas protonymphs were observed to survive for about 20 days without feeding at 23–25 °C (Ignatowicz 1974). Eggs did not developed and females did not oviposit at 0 °C, 4 °C or 8 °C (Kevan and Sharma 1964). Usher and Davis (1983) suggested a collaborative predation behavior for *G. aculeifer* when associated with the springtail *S. coeca*, which is larger than the predator. This behavior was not expressed when the predator was associated with the springtail *H. denticulada*, smaller than the predator.

Walter and Oliver (1989) studied biological aspects of *Gaeolaelaps oreithyiae* Walter and Oliver. They observed this species to feed on larvae of the moth *Plodia interpunctella* (Hubner) (Pyralidae) and of the beetle *Oryzaephilus surinamensis* (L.) (Silvanidae), as well as on aphids, scale crawlers and thrips, as well as on mites and millipedes. The predator was observed to oviposit when fed with the collembolans *Hypogastura scotti* Yosi and *Tullbergia granulata* (Mills), and to oviposit on the nematode *Acrobeloides* sp.. *Gaeolaelaps oreithyiae* was observed to reproduce by thelitoky.

Gaeolaelaps invictianus Walter and Moser is known to be phoretic on female and male alate forms of the ant *Solenospsis invicta* (Formicidae), acting probably as a generalist predator that inhabits ant colonies (Walter and Moser 2010). The phoretic behavior on both ant sexes has been interpreted as an indication of a recent evolution of the process between these organisms, given that only alate females are able to found new colonies.

3.5.7 Gromphadorholaelaps

Studies conducted by Yoder and Barcelona (1995) and Yoder (1996, 1997) suggested that the type species of this monotypic genus, *Gromphadorholaelaps schaeferi* (Till), found in association with the hissing-cockroach *Gromphadorhina portentosa* (Schaefer and Peckham), is not parasitic, but should have a mutualistic association with the cockroach. The mite feeds on cockroach saliva (ptyalophagy, unique among Acari) and consumes the cockroach parasitic mites *Proctolaelaps nauphaetae* (Womersley) (Melicharidae) and *Locustacarus* sp. (Podapolipidae).

Cannibalism has been reported in *G. schaeferi*, larvae being consumed by adult females. As a possible morphological adaptation, the larva has unusually long opisthonotal setae that could play a role in avoiding cannibalism (Gerdeman and Klompen 1998).

In a recent study, Yoder et al. (2009) also showed regulation of the external mycoflora of *G. portentosa* by *G. schaeferi*. The mite was observed to feed on the fungi *Alternaria* sp., *Aspergillus* sp., *Cladosporium* sp., *Geotrichum* sp., *Mucor* sp., *Penicillum* sp., *Rhizopus* sp. and *Trichoderma* sp. The authors suggested this behavior to represent a commensalistic process, with no clear benefits for the cockroach, as indicated by laboratory studies.

3.5.8 Hypoaspis

In the strict taxonomic sense, *Hypoaspis* species are beetle associates. Costa (1971) reported *Hypoaspis rhinocerotis* Oudemans to feed on eggs of the scarab *O. rhinoceros*, while Al-Deeb et al. (2012) reported the presence of the same mite on the related scarab beetle *Oryctes agamemnon arabicus* (Burmeister). The latter authors could not observe any effect of the mite on the beetle, suggesting the former to feed on deutonymphs of the astigmatin *Sancassania* sp., a species commonly found phoretically in association with that beetle. Predation by *Hypoaspis* species on small invertebrates underneath the elitra of scarab beetle has also been suggested by Joharchi and Halliday (2011).

Çobanoğlu et al. (2003) mentioned *Hypoaspis krameri* (Canestrini and Canestrini) as an ectoparasite of larvae of the scarab beetle *Anoxia orientalis* Kryn., an important pest of different crops, mainly in nurseries. According to these authors, the activity of the beetle larvae is reduced as a consequence of microorganisms that penetrate the larvae through the wounds produced by the mite. *Hypoaspis polyphyllae* Khanjani and Ueckermann was mentioned by Khanjani and Ueckermann (2005) to be parasitic on the scarab *Polyphylla olivieri* Laporte, a pest of fruit trees in Iran.

A large number of species described under *Hypoaspis* might belong to different genera, as reported in a recent revision of the genus by Joharchi and Halliday (2011). These constitute a group containing an array of morphologically poorly known species, described in *Hypoaspis* but whose conclusive generic placement remains uncertain. One of those species, *Hypoaspis solimani* Shereef, Mawar and Ahmed (not a true *Hypoaspis*) was observed to feed and oviposit on larvae of the house fly and of the vinegar fly (*D. melanogaster*), as well as on the edaphic fungi *Macrophomina phaseoli* (Maubl.) and *Fusarium oxysporum* Schlecht (Shereef et al. 1992). Oviposition was shown to be higher on the first two food sources.

Hypoaspis solimani was mentioned by Shereef et al. (1992) to be arhenotokous, although the authors mentioned mating to be essential for egg laying. Larvae were not observed to feed by those authors.

3.5.9 Mumulaelaps

The single species of this genus, *Mumulaelaps ammochostos* Clark, was found on larvae of the scarab *Pericoptus truncatus* Fabricius. Signs of piercing by the mite were not seen on the cuticle of the larvae, which is consistent with the fact that hypoaspidin chelicerae do not seem adapted for hemolymph uptake. Analyses involving the use of stable isotope indicated no sign of ingestion of hemolymph or cuticular lipids, leading Clark and Hawke (2012) to suspect that *M. ammochostos* either fed on a species of *Sancassania* often found on the same phoront beetle, or on entomopathogenic or free-living nematodes.

3.5.10 Ololaelaps

Ololaelaps bregetovae Shereef and Soliman was reported by Shereef and Soliman (1980) to develop when offered a combination of the fungus *Aspergillus flavus* Link and the mite *R. echinopus*, but not on any of these offered in isolation. *Ololaelaps bregetovae* could not survive for more than 8 days under dry conditions (the exact humidity level was not specified by the authors) but was able to survive for 13 days without food.

Epski et al. (1988) reported the phoretic behavior of the entomopathogenic nematode *S. feltiae* on *Ololaelaps placidus* (Banks) (mentioned as *Ololaelaps veneta* Berlese, its junior synonym), under laboratory condition. They also observed that the predator could not complete development when offered the nematodes *S. feltiae* or *H. heliothidis* as prey. Unmated females were reported to be unable to oviposit.

3.5.11 Stratiolaelaps

Walter and Campbell (2003) evaluated different Australian populations of *Stratiolaelaps* initially identified as *S. miles*, concluding that the apparently cosmopolitan *S. miles* could actually be composed of a complex of cryptic species, each with a potentially different use in applied biological control. This finding is of particular importance because of the relatively large number of studies about this putative species, and the fact that populations identified as *S. miles* have been widely used commercially for pest control. According to Cabrera et al. (2005), the species reported as *S. miles* by Chambers et al. (1993), Wright and Chambers (1994) and

Ydegaard et al. (1997) was actually *S. scimitus*. Regardless of that conclusion, in this chapter we refer to the species names as originally mentioned in the respective papers.

Quite a few studies have been conducted to evaluate the ability of mites of this genus to prey on insects. Mihm and Chiang (1976) showed the predation of an unidentified *Stratiolaelaps* species on eggs and larvae of the beetles *D. longicornis* and *D. virgifera. Stratiolaelaps miles* was observed to feed on all larval instars of *Bradysia paupera* (Tuomikoski) (Wright and Chambers 1994). It was able to complete development and to oviposit on larvae of the fly *Heteropeza pygmaea* (Winnertz) (Cecidomyiidae) (Ali and Brennan 1997), and to oviposit when fed with second instar larvae, prepupae and pupae of *F. occidentalis* (Berndt et al. 2004b). A study about the effect of temperature on the development of *S. miles* fed on a combination of larvae of the fungus gnats *B. paupera* and *Bradysia tritici* (Coquillet) was conducted by Ydegaard et al. (1997). Although the predator was observed to develop and reproduce at temperatures between 15 and 30 °C, a high mortality rate (52.3 %) was observed at 15 °C; mortality was lower at intermediate temperatures (11.6 and 3.2 % at 20 and 25 °C) and again tended to increase, reaching 23.9 %, at 30 °C.

Stratiolaelaps scimitus was observed by Brødsgaard et al. (1996) to be able to complete development and to reproduce on pupae of the leafminer *Liriomyza bryoniae* Kalt (Agromyzidae), on larvae of the fungus gnat *Lycoriella solani* Winnertz (Sciaridae), as well as on pupae of the gall-midge *Aphidoletes aphidimyza* Rondani (Cecidomyiidae). This species was shown to prefer *L. solani* over *F. occidentalis*, which in turn was preferred over the springtail *Isotomurus* sp. as prey (Brødsgaard et al. 1996). *Stratiolaelaps scimitus* was also observed to complete development on *L. solani* (Ali and Brennan 1997; Enkegaard et al. 1997) and on *Bradysia* sp. (Cabrera et al. 2005).

Stratiolaelaps species were also reported to feed on different astigmatin mites. Hoda et al. (1986) also reported the development of *Stratiolaelaps miles* on *R. echinopus*. Development and oviposition of *S. miles* were reported by Shereef et al. (1980) on *R. echinopus* or *T. putrescentiae*, by Wright and Chambers (1994) and Ali and Brennan (1997) on *A. siro* and by Brødsgaard et al. (1996) and Enkegaard et al. (1997) on *T. putrescentiae*. Laboratory colonies of *S. miles* were maintained by Steiner et al. (1999) using *T. putrescentiae* as prey, while colonies of *S. scimitus* were maintained by Freire and Moraes (2007b) with the use of the same prey. However, *S. scimitus* was not able to complete development on *Sancassania* aff. *sphaerogaster* (Cabrera et al. 2005).

Stratiolaelaps miles was also observed to feed and oviposit on the mesostigmatid *D. gallinae* (Tuovinen 2008; Ali et al. 2012). Castilho et al. (2009) mentioned the apparent consumption of *Pygmephorus* sp. (Pygmephoridae) by *S. scimitus* in mushroom houses.

Messelink and van Holstein-Saj (2006) demonstrated occasional predation of *S. scimitus* (mentioned as *Hypoaspis miles*) on *Steneotarsonemus latipes* (Halbert). However, a greenhouse experiment later conducted by the same authors (Messelink and van Holstein-Saj 2011) showed that *S. scimitus* affected the establishment of

Neoseiulus barkeri (Phytoseiidae), a potential predator of *S. latipes*, on *Hippeastrum* sp. (Amaryllidaceae) crop.

Stratiolaelaps lamington Walter and Campbell was mentioned to be able to complete development on the springtail *F. candida* (Beaulieu and Walter 2007) and *S. miles* to develop and reproduce on the springtail *I. notabilis* (Shereef et al. 1980).

Few studies have shown the ability of *Stratiolaelaps* species to feed on nematodes. Brødsgaard et al. (1996) observed *S. miles* to be able to complete development on *S. feltiae*, while Berndt et al. (2003) observed the same species to oviposit when feeding on *T. silusiae* (Panagrolaimidae).

As a consequence of its generalist feeding habit, *S. scimitus* was observed to feed on potworms (Enchytraeidae) (Cabrera et al. 2005), but the extent of the predation is unknown.

The effect of temperatures ranging between 15 and 28 °C on the biology of *S. miles* was studied by Wright and Chambers (1994). They observed that at 10 °C eggs did not hatch, adults were unable to feed and to lay eggs.

Protonymphs, deutonymphs and adults of *S. miles* were observed to survive for respectively about 12, 22 and 24 days at 20 °C without feeding (Wright and Chambers 1994). Enkegaard et al. (1997) observed that the oviposition period of unmated females was significantly longer than of mated females. Despite the longer oviposition period, virgin females had lower oviposition rates.

Except for the non-feeding larvae, cannibaslitic behaviour was determined by Berndt et al. (2003) to be exerted by all post-embryonic stages of *S. miles* after 14 days of starvation. Nymphs were able to reach adulthood and to oviposit when feeding on con-specific mites, although the number of eggs produced increased tenfold when the nematode *T. silusiae* was offered as prey.

Larvae of *S. miles* were observed not to feed when exposed to the mites *A. siro*, *R. echinopus* or *T. putrescentiae* or to larvae of the fly *H. pygmaea* (Ali and Brennan 1997; Shereef et al. 1980).

DeJong et al. (1981) reported *S. miles* to be haplodiploid. Concurrently, arrhenotoky was reported for this species by Shereef et al. (1980), Walter and Kaplan (1990) and Enkegaard et al. (1997).

3.6 Practical Use of Laelapids for Biological Control

Use of predatory mites for applied biological control is dependent on a chain of scientific and technological research. First, prospective species are determined by initial exploratory surveys; usually, these are later screened under controlled laboratory conditions by taking into account a set of biological data, often including life table parameters. Selected species then usually evaluated under semi-field or field condition before a final decision about their possible practical use can be reached. Few studies have been conducted to validate the potential of laelapid species in field experiments. Most of these studies refer to the use of *G. aculeifer*.

Practical use of predatory mites for augmentative biological control, involving periodic releases, is highly dependent upon the availability of efficient mass production systems for the selected species. That means that regardless of the efficiency of a given candidate, its practical use may not be possible unless it can be produced in large numbers at competitive cost.

Laelapid mites can be mass produced with the adoption of methods that have also been widely used the production of phytoseiid mites. Steiner et al. (1999) presented a simple method for the mass rearing of *S. miles* using *T. putrescentiae* as prey. The method was similar to what had been proposed by Ramakers and van Leiburg (1982) for the production of phytoseiids, allowing a 35 fold increase of the predator population in 21 days. Using essentially the same procedure, Freire and Moraes (2007b) obtained a 42 fold increment of the population of *S. scimitus* within a period of 30 days.

Four laelapid species are currently produced commercialized as biological control agents of pest and parasitic arthropods (Table 3.2). Most of the producers are from Europe where greenhouse crops occupy a large area and where the target pests of these predatory mites are commonly found. Selected examples of specific groups of harmful organisms targeted for control with the use of predatory laelapids are subsequently presented.

3.6.1 Thrips

Some thrips species are major pests of several crops. *Frankliniella occidentalis* is one of these. The life cycle of this and other thrips comprises an edaphic phase corresponding to about a third of its total duration. In this phase, laelapid mites, which are essentially edaphic, can prey on the pest. Satisfactory control of *F. occidentalis* has been reported with the release of the commercially available *G. aculeifer* and *S. miles*. An inundative introduction of 1,600 specimens per plant of a species identified as *Gaeolaelaps* sp. nr. *aculeifer* reduced in about 30 % the emergence of *F. occidentalis* adults in a cucumber crop (Gillespie and Quiring 1990). Experimental releases of *G. aculeifer* and *S. miles* have shown reduction of respectively 80.5 and 61 % of the *F. occidentalis* populations in bean crops (Berndt et al. 2004a).

When in combination with entomopathogenic nematodes, *G. aculeifer* reduced thrips emergence in up to 82 % (Premachandra et al. 2003) on bean crops. The same predador was reported to provide better thrips control when in combination with *Neoseiulus cucumeris* (Oudemans) than when released in isolation (Wiethoff et al. 2004). However, *G. aculeifer* was reported not be sufficiently efficient to control *F. occidentalis* in production of the ornamentals *Saintpaulia ionantha* Wendl (Glockemann 1992). This negative result could be due to the insufficient number of predators released. Private companies producing those predators commercially recommend the release of 100–500 predators per m^2, depending on the level of the pest population.

Table 3.2 Laelapidae species currently commercialized for pest control

Species	Producer	Country/region	Target pest
Androlaelaps casalis	Koppert	Netherlands	*Dermanyssus gallinae*
Gaeolaelaps aculeifer	Koppert	Europe	*Bradysia* spp.
	Biological Services	Australia	*Rhizoglyphus* spp.
			Steneotarsonemus laticeps
			Frankliniella occidentalis
			Thrips spp.
Stratiolaelaps miles	Defenders	England	*Bradysia* spp.
	Interior Landscaping Products	England	*Bradysia* spp.
	Ladybird Plantcare	Great Britain	*Bradysia* spp.
	Syngenta	England	*Bradysia* spp.
			Thrips
	Buglogical Control Systems	USA	*Bradysia* spp.
			Frankliniella occidentalis
	Biobest	Belgium	*Bradysia* spp.
			Thrips
			Shore fly
	BioProduction	Denmark	Thrips
			Bradysia spp.
Stratiolaelaps scimitus	PROMIP	Brazil	*Bradysia* spp.
	Evergreen Growers Supply	USA	*Bradysia* spp.
			Frankliniella occidentalis
	Koopert	Europe	*Bradysia* spp.
			Frankliniella occidentalis
			Thrips tabaci
			Thrips fuscipennis
			Echinothrips americanus

3.6.2 Sciarid Flies

Sciarid flies, commonly referred to as fungus gnats, are pests of mushrooms and several cultivated plants of different groups, especially in nurseries. Larvae of these species attack the underground plant structures, but in mushrooms the larvae can cause direct damage the epigeous part of the fungus, opening galleries and turning them unsuitable for commercialization. The main fungus gnats mentioned as pests belong to the genera *Bradysia* and *Lycoriella. Gaeolaelaps aculeifer*, *S. miles* and *S. scimitus* are currently commercialized to control these pests.

Gillespie and Quiring (1990) reported satisfactory reduction of the population of *Bradysia* spp. with the inoculative release of 125 specimens of a species identified as *Gaeolaelaps* sp. nr. *aculeifer* per cucumber plant. Chambers et al. (1993) released *S. miles* on potted cyclamen and poinsettia in glasshouses to control *Bradysia* spp., at a rate of 55 mites per pot, obtaining satisfactory control. The predator was observed on the compost for up to 7 weeks after the release.

The introduction of *S. miles* in association with the use of the insecticide methoprene, an insect growth regulator, increased the yield of *Agaricus bisporus* (Lange) when compared with the use of the insecticide alone (Ali et al. 1999). The insecticide did not affect the predatory mite, whose population increased during the experimental period. The introduction of *S. miles* to the compost during its colonization by the mycelium of *A. bisporus* reduced the emergence of adult sciarids by 87 % (Jess and Kilpatrick 2000).

Cabrera et al. (2004) reported the compatibility of some pesticides with *S. scimitus* under laboratory conditions. The acaricide dicofol, the insecticide pyriproxyfen and the fungicides fosetyl-Al and mefenoxam did not affect development and reproduction of that predator. In laboratory trials, releases of that predator reduced infestation of *Bradysia matogrossensis* (Lane) in mushroom houses by about 80 % within 30 days (Freire et al. 2007). In a subsequent evaluation in mushroom houses, Castilho et al. (2009) compared the effectiveness of two and a single release of *S. scimitus* to control the pest, observing that a single release of the predator, immediately after casing, was enough to reduce the number of flies, to improve quality and increase yield.

3.6.3 Shore Fly

The shore fly *Scatella tenuicosta* Collin (Ephydridae) is an important pest of several cultivated plants, mainly in nurseries. Releases of *G. aculeifer* at a rate of 5,000 mites per m^2 of a mint (*Mentha piperita* L.) crop reduced the population of *S. tenuicosta* in over 83 %. Curative releases of *S. miles* at the same rate provided up to 16–65 % control of the pest, whereas preventive releases was more efficient, reducing pest population by over 68 % (Vänninen and Koskula 2004).

3.6.4 Nematodes

Plant parasitic nematodes cause damage to many crops. Although a considerable number of studies to show the potential of predatory soil mites, mainly *Cosmolaelaps* species, to control nematodes have been conducted, until now no laelapid species has been commercialized for that propose.

The association between the plant-parasitic nematode *T. semipenetrans* and the laelapids *Cosmolaelaps keni* Hafez, El-Badry and Nasr and *O. bregetovae* was reported by El-Banhawy et al. (2006) in citrus orchards in Egypt. Promising results were reported by Al Rehiayani and Fouly (2005) with the release of *C. simplex* for the control of that nematode, an important parasite of different crops. The authors reported a reduction of 75 % in the population of *T. semipenetrans* in citrus roots with the release of the predator in comparison with a control treatment, leading the authors to conclude that *C. simplex* has potential as a biological control agent of *T. semipenetrans*. Further efforts to confirm that conclusion as well as to evaluate the possible use of laelapids against other nematode species are warranted, especially in crops with shallow root systems.

3.6.5 Bulb Mite

The bulb mite *Rhizoglyphus robini* Chaparède is an important pest of cultivated lilies, mainly in propagation phase (Lesna et al. 1996). In a preliminary study to find potential predators of *R. robini*, Lesna et al. (1995) demonstrated that differently from *S. miles*, *G. aculeifer* was able to feed and reproduce on *R. robini*. Subsequently, Lesna et al. (1996) showed the suppression of *R. robini* population on different spatial scales, including in plastic bags with lily bulbs and vermiculite, used in the propagation of lily bulbs. In a greenhouse experiment, Lesna et al. (2000) reported the elimination of bulb mites when the predator:prey ratio was 3:1 using intact bulbs in peat soil.

3.6.6 Poultry Red Mite

The poultry red mite is a worldwide parasite of chicken and many other bird species. It is controlled basically with the use of chemical pesticides, but there is a major interest today in the use of natural enemies for that purpose. Since 2010, *A. casalis* and *S. scimitus* have been commercialized by private Dutch companies for the control of that parasite. Lesna et al. (2012) referred to the possible efficiency of those species as control agents of the poultry red mite. Both predators were able to reduce the prey population level in a small scale experiment, with a better performance of *A. casalis*.

Ali et al. (2012) observed *S. miles* to feed on adults of the poultry red mite under laboratory conditions. Predation rates were higher at 22 or 28 than at 15 °C. The authors reported that the presence of dust and refuges in the experimental units did not affect predation, but the presence of another prey (*T. putrescentiae*) reduced predation rate.

Given the positive results of the studies conducted so far, further efforts evaluating other laelapid mite species seems worthwhile, especially for use in countries where *A. casalis* and *S. scimitus* are not available.

3.7 Conclusions

The success obtained so far with the use of laelapid mites for the control of undesirable organisms should stimulate further efforts to make their use even more widespread and intense. One of the most positive characteristics of some of the laelapid species refer to their high tolerance to starvation. Even when the pest is reduced to very low levels, these predators are able to remain alive for relatively long periods (up to nearly a month or more), ready to affect a new pest outbreak. This characteristic certainly leads to higher survivorship of the released predators, reducing control cost and consequently making the technique more readily adopted.

As indicated in the previous sections of this chapter, knowledge about the potential of laelapids as biological control agents is still relatively small. In extensive regions of the globe, especially in the tropics, little or nothing is known about the prevailing laelapid species. A large number of species, quite different from the ones known to date, is expected to be found, and some of these perhaps could be used for periodic releases in their region of natural occurrence as well as in other regions. Thus, emphasis should be given to basic inventory efforts in areas where faunistic surveys have not been adequately conducted.

Selected species found in newly explored areas should be scrutinized in relation to their potential as biological control agents of edaphic pests, first in laboratory investigations and later in complementary field evaluations. Possible reward from these types of studies is well exemplified by the current use of predatory mites for the biological control of the poultry red mite. Studies conducted in The Netherlands to turn the use possible initiated with the evaluation of the mite fauna associated with the European starling (*S. vulgaris*) (Lesna et al. 2009), which was followed by an investigation to determine the potential of two of the predator species collected (Lesna et al. 2012). Another example refers to the use of *S. scimitus* in Brazil. Interest in the use of this predator arouse from its incidental collection under natural conditions leading to the studies conducted by Freire et al. (2007) and Castilho et al. (2009). These studies confirmed the results of previous evaluations conducted in other countries in relation to the ability of *S. scimitus* to control species of fungus gnats on different crops.

The reduced number of experiments conducted under field conditions, despite the increase in the number of papers about the biology and ecology of hypoaspidins, leads to a still limited use of these mites for the control of undesirable organisms. Species of *Cosmolaelaps*, for example, have shown good potential to control plant nematodes but conclusive research must be conducted.

Also of interest would be the more conclusive evaluation of laelapid species as biocontrol agents of *Diabrotica* species. Differently from the positive results obtained by Mihm and Chiang (1976), in the study conducted by Prischmann-Voldseth and Dashiell (2013), releases of *G. aculeifer* under field condition failed to provide effective control of *Diabrotica virgifera virgifera* LeConte. According to the authors, the generalist behavior of that predator probably reduced the natural soil invertebrate fauna, decreasing the naturally occurring biocontrol.

Additional effort seems also warranted in relation the mass production of laelapid predators. The improvement of mass rearing techniques, the development of artificial diets, as implemented with some phytoseiid mites (Nguyen et al. 2013, 2014), and the search for alternative prey, may optimize their mass production, reducing production costs and making biological control competitive with other forms of pest control.

References

Abou-Awad BA, Nasr AK, Gomaa EA, Abou-Elela M (1989) Feeding, development and reproduction of the predatory mite, *Hypoaspis vacua* on various kinds of food substances (Acari: Laelapidae). Int J Trop Insect Sci 10:503–506

Afifi A, van der Geest L (1984) Notes on the development and biology of the predaceus soil mite *Cosmolaelaps claviger* (Berlese 1883) (Gamasida: Laelapidae). In: Griffiths D, Bowman C (eds) Acarology IV, vol I. Ellis Horwood Limited Publishers, Chichester, pp 585–590

Al Rehiayani SM, Fouly AH (2005) *Cosmolaelaps simplex* (Berlese), a polyphagous predatory mite feeding on root-knot nematode *Meloidogyne javanica* and citrus nematode *Tylenchulus semipenetrans*. Pakistan J Biol Scis 8:168–174

Al-Deeb MA, Muzaffar SB, Sharif EM (2012) Interactions between phoretic mites and the Arabian rhinoceros beetle, *Oryctes agamemnon arabicus*. J Insect Sci 12:1–14

Ali O, Brennan P (1997) Development, feeding and reproduction of the predatory mite, *Hypoaspis miles* (Acari: Mesostigmata: Laelapidae) on different prey. Syst Appl Acarol 2:81–88

Ali O, Dunne R, Brennan P (1999) Effectiveness of the predatory mite *Hypoaspis miles* (Acari: Mesostigmata: Hypoaspidae) in conjunction with pesticides for control of the mushroom fly *Lycoriella solani* (Diptera: Sciaridae). Exp App Acarol 23:65–77

Ali W, George DR, Shiel RS, Sparagano OAE, Guy JH (2012) Laboratory screening of potential predators of the poultry red mite (*Dermanyssus gallinae*) and assessment of *Hypoaspis miles* performance under varying biotic and abiotic conditions. Veterinary Parasitol 187:341–344

Amin MR, Khanjani M, Zahiri B (2014) Estimating development of preimaginal and fecundity of *Gaeolaelaps aculeifer* (Acari: Laelapidae) feeding on *Rhizoglyphus echinopus* (Acari: Acaridae) at constant temperatures. J Crop Prot 3(Suppl):581–587

Barker PS (1968) The response of a predator, *Hypoaspis aculeifer* (Canestrini) (Acarina: Laelapidae), to two species of prey. Can J Zool 47:343–345

Beaulieu F (2009) Review of the mite genus *Gaoelaelaps* Evans & Till (Acari: Laelapidae), and description of a new species from North America, *G. gillespiei* n. sp. Zootaxa 2158:33–49

Beaulieu F, Walter DE (2007) Predation in suspended and forest floor soils: observations on Australian mesostigmatic mites. Acarologia 47:34–54

Beaulieu F, Dowling APG, Klompen H, Moraes GJ, Walter D (2011) Superorder Parasitiformes Reuter, 1909. In: Zhang, Z-Q (ed) Animal biodiversity: an outline of higher-level classification and survey of taxonomic richness (Addenda 2013). Zootaxa 3703:123–128

Berlese A (1892) Acari, Miriapoda et Scorpiones Hucusque in Italia Reperta. Ordo Mesostigmata (Gamasidae) 11:1–144

Berndt O, Meyhöfer R, Poehling H-M (2003) Propensity towards cannibalism among *Hypoaspis aculeifer* and *Hypoaspis miles*, two soil-dwelling predatory mite species. Exp Appl Acarol 31:1–14

Berndt O, Meyhöfer R, Poehling H-M (2004a) The edaphic phase in the ontogenesis of *Frankliniella occidentalis* and comparison of *Hypoaspis miles* and *Hypoaspis aculeifer* as predators of soil-dwelling thrips stages. Biol Contr 30:17–24

Berndt O, Poehling HM, Meyhöfer R (2004b) Predation capacity of two predatory laelapid mites on soil-dwelling thrips stages. Entomol Exp Appl 112:107–115

Bilgrami AL (1997) Evaluation of the predation abilities of the mite *Hypoaspis calcuttaensis*, predaceous on plant and soil nematodes. Fundam Appl Nematol 20:96–98

Bregetova NG (1977) Laelaptidae. In: Ghilyarov MS, Bregetova NG (eds) Key to the soil-inhabiting mites. Mesostigmata. Nauka, Leningrad, pp 169–226 [in Russian]

Brødsgaard HF, Sardar MA, Enkegaard A (1996) Prey preference of *Hypoaspis miles* (Berlese) (Acarina: Hypoaspididae): non-interference with other beneficials in glasshouse crops. IOBC/WPRS Bull 19:23–26

Cabrera AR, Cloyd RA, Zaborski ER (2004) Effects of greenhouses pesticides on the soil-dwelling predatory mite *Stratiolaelaps scimitus* (Acari: Mesostigmata: Laelapidae) under laboratory conditions. Biol Microbial Contr 97:793–799

Cabrera A, Cloyd R, Zaborski E (2005) Development and reproduction of *Stratiolaelaps scimitus* (Acari: Laelapidae) with fungus gnat larvae (Diptera: Sciaridae), potworms (Oligochaeta: Enchytraeidae) or *Sancassania* aff. *sphaerogaster* (Acari: Acaridae) as the sole food source. Exp Appl Acarol 36:71–81

Casanueva ME (1993) Phylogenetic studies of the free-living and arthropod associated Laelapidae (Acari: Mesostigmata). Gayana Zool 57:21–46

Castilho RC, Moraes GJ, Silva ES, Freire RAP, Da Eira FC (2009) The predatory mite *Stratiolaelaps scimitus* as control agent of the fungus gnat *Bradysia matogrossensis* in commercial production of the mushroom *Agaricus bisporus*. Intl J Pest Mgt 55:181–185

Chambers RJ, Wright EM, Lind RJ (1993) Biological control of glasshouse sciarid flies (*Bradysia* spp.) with the predatory mite, *Hypoaspis miles*, on cyclamen and poinsettia. Biocontrol Sci Tech 3:285–293

Chen C, Meng Y (1987) Studies on karyotype and C-banded, G-banded chromosomes of *Haemolaelaps casalis*. Zool Res 8:143–148

Clark J, Hawke D (2012) A new epizoic laelapid mite from the New Zealand sand scarab *Pericoptus truncatus* larvae and its isotopic ecology. New Zealand J Zool 39:187–199

Çobanoğlu S, Çakmak I, Başpinar H (2003) *Hypoaspis krameri* (Canestrini, 1881) (Mesostigmata: Laelapidae) an ectoparasitic mite associated with *Anoxia orientalis* Kryn. (Col., Scarabaeidae) from Turkey. Entomol Monthly Mag 139:97–101

Costa M (1971) Mites of the genus *Hypoaspis* Canestrini, 1884 s. str. and related forms (Acari: Mesostigmata) associated with beetles. Bull British Mus (Nat Hist) Zool 21:69–98

DeJong JH, Lobbes PV, Bolland HR (1981) Kariotypes and sex determination in two species of laelapid mites (Acari: Gamasida). Genetica 55:187–190

Dowling APG, O'Connor BM (2010) Phylogenetic relationships within the suborder Dermanyssina (Acari: Parasitiformes) and a test of dermanyssoid monophyly. Intl J Acarol 36:299–312

El-Banhawy E, Nasr A, Afia S (2006) Survey of predacious soil mites (Acari: Mesostigmata) in citrus orchards of the Nile Delta and Middle Egypt with notes on the abundance of the citrus parasitic nematode *Tylenchulus semipenetrans* (Tylenchida: Tylenchulidae). Int J Trop Insect Sci 26:64–69

Enkegaard A, Sardar M, Brødsgaard H (1997) The predatory mite *Hypoaspis miles*: biological and demographic characteristics on two prey species, the mushroom sciarid fly, *Lycoriella solani*, and the mould mite, *Tyrophagus putrescentiae*. Entomol Exp Appl 82:135–146

Epski ND, Walter DE, Capinera JL (1988) Potential role of nematophagous microarthropods as biotic mortality factors of entomogenous nematodes (Rhabditida: Steinernematidae: Heterorhabditidae). J Econ Entomol 81:821–825

Faraji F, Halliday B (2009) Five new species of mites (Acari: Laelapidae) associated with large Australian cockroaches (Blattodea: Blaberidae). Intl J Acarol 35:245–264

Fouly AH, Abdel-Baky NF (2015) Influence of prey types on the biological characteristics of *Cosmolaelaps qassimensis* (Acari: Laelapidae). J Entomol 12:21–29

Freire R, Moraes G (2007a) Description of a new species of *Cosmolaelaps* Berlese (Acari: Laelapidae: Hypoaspidinae) from Brazil and its biological cycle. Intl J Acarol 33:353–358

Freire RAP, Moraes G (2007b) Mass production of the predatory mite *Stratiolaelaps scimitus* (Womersley) (Acari: Laelapidae). Syst Appl Acarol 12:117–119

Freire R, Moraes G, Silva E, Vaz A, Castilho R (2007) Biological control of *Bradysia matogrossensis* (Diptera: Sciaridae) in mushroom cultivations with predatory mites. Exp Appl Acarol 42:87–93

Gerdeman B, Klompen JY (1998) The larva of *Gromphadorholaelaps schaeferi* Till (Acari: Laelapidae), an associate of the Madagascar hissing-cockroach, *Gromphadorhina portentosa* (Schaum). Intl J Acarol 24:301–305

Gillespie DR, Quiring DMJ (1990) Biological control of fungus gnats, *Bradysia* spp. (Diptera: Sciaridae), and western flower thrips, *Frankliniella occidentalis* (Pergande) (Thysanoptera: Thripidae), in greenhouses using a soil-dwelling predatory mite, *Geolaelaps* sp. nr. *aculeifer* (Canestrini) (Acari: Laelapidae). Can Entomol 122:975–983

Glockemann B (1992) Biological control of *Frankliniella occidentalis* on ornamental plants using predatory mites. Bull OEPP/EPPO 22:397–404

Grandi G (1925) Contributo alla conoscenza biologica e morphologica di alcuni Lamellicorni fillofagi (*Amphiphallus assimilisobscurus* Brenske; *Haplidia etrusca* Kraatz.; *Anoxia matutinalis suturalis* Rttr.), e descrizione di una nuova species di acaro (*Coleolaelaps inopinatus* Grnd.). Bollettino Del Laboratorio di Zoologia Generale e Agraria del Reale Istituto Superiore Agrario in Portici 18:159–224

Gwiazdowicz DJ (2004) Records of heteromorphic males of *Hypoaspis* (*Cosmolaelaps*) *vacua* (Michael, 1891) (Acari: Mesostigmata: Laelapidae) from Poland. J Acarol Soc Japan 13:181–184

Heckmann L-H, Ruf A, Nienstedt KM, Krogh PH (2007) Reproductive performance of the generalist predator *Hypoaspis aculeifer* (Acari: Gamasida) when foraging on different invertebrate prey. Appl Soil Ecol 36:130–135

Hinckley AD (1967) Associates of the coconut rhinoceros beetle in Western Samoa. Pacific Insects 9:505–511

Hoda FM, Taha HA, Ibrahim GA, El-Beheri MM (1986) Biological observations on the predator mite, *Hypoaspis miles* Berlese (Acarina: Laelapidae). Bull Soc Entomol d'Egypte 66:103–106

Hoffmann A, López-Campos G (1995) Género y especie nuevos de Hypoaspidinae (Acarida: Laelapidae) en un caso de foresia multiple. Ann Inst Biol Univ Nac Autónoma Mexico 66:33–46

Hunter PE, Mollin K (1964) Mites associated with the passalus beetle. I. Life stages and seasonal abundance of *Cosmolaelaps passali* n. sp. (Acarina: Laelaptidae). Acarologia 6:247–256

Ignatowicz S (1974) Observations on the biology and development of *Hypoaspis aculeifer* Canestrini, 1885 (Acarina: Gamsides). Zool Poloniae 24:41–59

Inserra RN, Davis DW (1983) *Hypoaspis* nr. *aculeifer*: a mite predacious on root-knot and cyst nematodes. J Nematol 15:324–325

Jess S, Kilpatrick M (2000) An integrated approach to the control of *Lycoriella solani* (Diptera: Sciaridae) during production of the cultivated mushroom (*Agaricis bisporus*). Pest Mgt Sci 56:477–485

Joharchi O, Halliday B (2011) New species and new records of mites of the family Laelapidae (Acari: Mesostigmata) associated with Coleoptera in Iran. Zootaxa 2883:23–38

Joharchi O, Halliday B, Saboori A, Kamali K (2011) New species and new records of mites of the family Laelapidae (Acari: Mesostigmata) associated with ants in Iran. Zootaxa 2972:22–36

Karg W (1993) Acari (Acarina), Milben Parasitiformes (Anactinochaeta) Cohors Gamasina Leach, Raubmilben. Die Tierwelt Deutschlands 59:1–523

Karg W (1994) Raubmilben der cohors Gamasina Leach (Acarina: Parasitiformes) vom Galapagos–Archipel. Mitteilungen aus dem Zoologischen Museum in Berlin 70:179–216

Karg W (1997) Zur Kenntnis der Raubmilbencohors Gamasina Leach (Acarina: Parasitiformes) tropischer Gebiete. Mitteilungen aus dem Zoologischen Museum in Berlin 73:63–88

Karg W (2003) Die raubmilbengattungen *Afrogamasellus* Loots et Ryke und *Oloopticus* Karg mitzwei neuen Arten – Ein Beitrag zur Evolution der Bodenmilben (Acarina: Gamasina). Abhandlungen und Berichte des Naturkundemuseums Görlitz 75:23–33

Karg W (2006) The systematics of Parasitiformes, especially of Gamasina Leach (Acarina), with new species from Ecuador. Mitteilungen aus dem Museum für Naturkunde in Berlin, Zoologische Reihe 82:140–169

Kasuga S, Kanno H, Amano H (2006) Development, oviposition, and predation of *Hypoaspis aculeifer* (Acari: Laelapidae) feeding on *Tyrophagus similis* (Acari: Acaridae). J Acarol Soc Japan 15:139

Kevan DKM, Sharma GD (1964) Observations on the biology of *Hypoaspis aculeifer* (Canestrini, 1884), apparently new to North America (Acarina: Mesostigmata: Laelaptidae). Acarologia 6:647–658

Khanjani M, Ueckermann EA (2005) *Hypoaspis* (*Hypoaspis*) *polyphyllae* n. sp. (Mesostigmata: Laelapidae) parasitic on larvae of *Polyphylla olivieri* Castelnau (Coleoptera: Scarabaeidae) in Iran. Intl J Acarol 31:119–122

Lesna I, Sabelis MW, Bolland HR, Conijn CGM (1995) Candidate natural enemies for control of *Rhizoglyphus robini* Claparède (Acari: Astigmata) in lily bulbs: exploration in the field and pre-selection in the laboratory. Exp Appl Acarol 19:655–669

Lesna I, Sabelis M, Conijn C (1996) Biological control of the bulb mite, *Rhizoglyphus robini*, by the predatory mite, *Hypoaspis aculeifer*, on lilies: predator- prey interactions at various spatial scales. J Appl Ecol 33:369–376

Lesna I, Conijn CGM, Sabelis MW, van Straalen NM (2000) Biological control of the bulb mite, *Rhizoglyphus robini*, by the predatory mite, *Hypoaspis aculeifer*, on lilies: predator-prey dynamics in the soil, under greenhouse and field conditions. Biocontrol Sci Tech 10:179–193

Lesna I, Wolfs P, Faraji F, Roy L, Komdeur J, Sabelis MW (2009) Candidate predators for biological control of the poultry red mite *Dermanyssus gallinae*. Exp Appl Acarol 48:63–80

Lesna I, Wolfs P, Faraji F, Roy L, Komdeur J, Sabelis MW (2012) Laboratory tests for controlling poultry red mites (*Dermanyssus gallinae*) with predatory mites in small 'laying hen' cages. Exp Appl Acarol 58:371–383

Lindquist EE, Evans GO (1965) Taxonomic concepts in the Ascidae, with a modified setal nomenclature for the idiosoma of the Gamasina (Acarina: Mesostigmata). Mem Entomol Soc Canada 47:1–64

Lindquist EE, Krantz GW, Walter DE (2009) Order Mesostigmata. In: Krantz GW, Walter DE (eds) A manual of acarology, 3rd edn. Texas Tech University Press, Texas, pp 124–232

Lobbes P, Schotten C (1980) Capacities of increase of the soil mite *Hypoaspis aculeifer* Canestrini (Mesostigmata: Laelapidae). J Appl Entomol 90:9–22

Messelink GJ, van Holstein-Saj R (2006) Potential for biological control of the bulb scale mite (Acari: Tarsonemidae) by predatory mites in amaryllis. Proc Netherlands Entomol Soc Mtg 17:113–118

Messelink GJ, van Holstein-Saj R (2011) Generalist predator *Stratiolaelaps scimitus* hampers establishment of the bulb scale mite predator *Neoseiulus barkeri* in Hippeastrum. Proc Netherlands Entomol Soc Mtg 22:67–73

Mihm J, Chiang H (1976) Laboratory studies of the life cycle and reproduction of some soil- and manure-inhabiting mites (Acarina: Laelapidae). Pedobiologia 16:353–363

Moreira GF, Klompen H, Moraes GJ (2014) Redefinition of *Cosmolaelaps* Berlese (Acari: Laelapidae) and description of five new species from Brazil. Zootaxa 3764:317–346

Moreira GF, Morais MR, Busoli AC, Moraes GJ (2015) Life cycle of *Cosmolaelaps jaboticabalensis* (Acari: Mesostigmata: Laelapidae) on *Frankliniella occidentalis* (Thysanoptera: Thripidae) and two factitious food sources. Exp Appl Acarol 65:219–226

Moss WW, Funk RC (1965) Studies on the developmental chaetotaxy of *Dyscinetonyssus hystricosus* n. g., n. sp. (Acari: Mesostigmata: Laelaptoidea). Acarologia 7:235–267

Murph PW, Sardar MA (1991) Resource allocation and utilization contrasts in *Hypoaspis aculeifer* (Can.) and *Alliphis halleri* (G. & R. Can.) (Mesostigmata) with emphasis on food source. In: Schuster R, Murphy PW (eds) The Acari. Reproduction, development and life-history strategies. Chapman & Hall Publications, New York, pp 301–311

Nguyen DC, Vangansbeke D, Lü X, Clercq P (2013) Development and reproduction of the predatory mite *Amblyseius swirskii* on artificial diets. BioControl 58:369–377

Nguyen DC, Vangansbeke D, Clercq PD (2014) Artificial and factitious foods support the development and reproduction of the predatory mite *Amblyseius swirskii*. Exp Appl Acarol 62:181–194

Norton RA, Kethley JB, Johnston DE, O'Connor BM (1993) Phylogenetic perspectives on genetic systems and reproductive modes of mites. In: Wrensch DL, Ebbert MA (eds) Evolution and diversity of sex ratio of insects and mites. Chapman & Hall Publications, New York, pp 8–99

Premachandra WTSD, Borgemeister C, Berndt O, Ehlers R-U, Poehling H-M (2003) Combined releases of entomopathogenic nematodes and the predatory mite *Hypoaspis aculeifer* to control soil-dwelling stages of western flower thrips *Frankliniella occidentalis*. BioControl 48:529–541

Prischmann-Voldseth DA, Dashiell KE (2013) Effects of releasing a generalist predator (Acari: *Gaeolaelaps aculeifer*) on a subterranean insect herbivore (Coleoptera: *Diabrotica virgifera virgifera*). Biol Contr 65:190–199

Radovsky RJ (1969) Adaptive radiation in the parasitic Mesostigmata. Acarologia 11:450–483

Ragusa S, Zedan MA (1988) Biology and predation of *Hypoaspis aculeifer* (Canestrini) (Parasitiformes: Dermanyssidae) on *Rhyzoglyphus echinopus* (Fum. & Rob.) (Acariformes: Acaridae). Redia 71:213–225

Ragusa S, Zedan MA, Sciacchitano MA (1986) The effects of food from plant and animal sources on the development and egg production of the predacious mite *Hypoaspis aculeifer* (Canestrini) (Parasitiformes: Dermanyssidae). Redia 69:481–488

Ramakers PMJ, van Leiburg MJ (1982) Start of commercial reproduction and introduction of *Amblyseius mckenzie* Sch. & Pr. (Acarina: Phytoseiidae) for the control of *Thrips tabaci* Lind. (Thysanoptera: Thripidae) in glasshouses. Mededelingen van de Faculteit der Landbouwwetenschappen. Rijksuniversiteit Gent 47:541–545

Ruf A (1991) Do females eat males? Laboratory studies on the population development of *Hypoaspis aculeifer* (Acari: Parasitiformes). In: Dusbábek F, Bukva V (eds) Modern acarology, vol 2. Academia, Prague and SPB Academic Publishing bv, The Hague, pp 487–492

Sardar MA, Murphy PW (1987) Feeding tests of grassland soil-inhabiting gamasine predators. Acarologia 28:117–121

Shereef G, Soliman Z (1980) Biological studies on *Ololaelaps bregetovae* sp. n. and *Kleemania plumosus* Oud. in Egypt. Bull Zool Soc Egypt 30:81–86

Shereef GM, Soliman ZR, Afifi AM (1980) Economic importance of the mite *Hypoaspis miles* (Berlese) (Mesostigmata: Laelapidae) and its life history. Bull Zool Soc Egypt 30:103–108

Shereef G, Nawar M, Ahmed M (1992) Effect of food on development and reproduction of *Hypoaspis solimani* n.sp. (Acari: Laelapidae). Egypt J Agr Res 70:1123–1136

Steiner M, Goodwin S, Wellham T (1999) A simplified rearing method for *Stratiolaelaps (Hypoaspis) miles* (Acari: Laelapidae). IOBC Bull 22:241–242

Tuovinen T (2008) Predatory arthropods as biocontrol agents of *Dermanyssus gallinae*? In: BSP Spring, trypanosomiasis/leishmaniasis & malaria meetings, 30 Mar–2 Apr 2008, Newcastle. Abstract Booklet: 170

Usher MB, Davis PR (1983) The biology of *Hypoaspis aculeifer* (Canestrini) (Mesostigmata): is there a tendency towards social behavior? Acarologia 24:243–250

Vänninen I, Koskula H (2004) Biocontrol of the shore fly *Scatella tenuicosta* with *Hypoaspis miles* and *H. aculeifer* in peat pots. BioControl 49(2):137–152

Walia KK, Mathur S (1995) Predatory behavior of two nematophagous mites, *Tyrophagus putrescentiae* and *Hypoaspis calcuttaensis*, on root-knot nematode, *Meloidogyne javanica*. Nematol Mediterranea 23:255–261

Walter D (1987) The influence of prey type on the development and reproduction of some predatory soil mites. Pedobiologia 30:419–424

Walter D (1988) Nematophagy by soil arthropods from the Shortgrass Steppe, Chihuahuan Desert and Rocky Mountains of the Central United States. Agr Ecosyst Environ 24:307–316

Walter DE, Campbell N (2003) Exotic vs endemic biocontrol agents: would the real *Stratiolaelaps miles* (Berlese) (Acari: Mesostigmata: Laelapidae), please stand up? Biol Contr 26:253–269

Walter DE, Kaplan DT (1990) A guild of thelytokous mites (Acari: Mesostigmata) associated with citrus roots in Florida. Environ Entomol 39:1338–1343

Walter DE, Moser JC (2010) *Gaeolaelaps invictianus*, a new and unusual species of hypoaspidin mite (Acari: Mesostigmata: Laelapidae) phoretic on the red imported fire ant *Solenopsis invicta* Buren (Hymenoptera: Formicidae) in Louisiana, USA. Intl J Acarol 36(5):399–407

Walter D, Oliver J Jr (1989) *Geolaelaps oreithyiae*, n. sp. (Acari: Laelapidae), a thelytokous predator of arthropods and nematodes, and a discussion of clonal reproduction in the Mesostigmata. Acarologia 4:293–303

Walter DE, Beard JJ, Walker KL, Sparks K (2002) Of mites and bees: A review of mite-bee associations in Australia and a revision of *Raymentia* Womersley (Acari: Mesostigmata: Laelapidae), with the description of two new species of mites from *Lasioglossum* (*Parasphecodes*) spp. (Hymenoptera: Halictidae). Australian J Entomol 41:128–148

Wiethoff W, Poehling H, Meyhöfer R (2004) Combining plant- and soil-dwelling predatory mites to optimise biological control of thrips. Exp Appl Acarol 34:239–261

Wright EM, Chambers RJ (1994) The biology of the predatory mite *Hypoaspis miles* (Acari: Laelapidae), a potential biological control agent of *Bradysia pauper* (Diptera: Sciaridae). Entomophaga 29(2):225–235

Ydegaard S, Enkegaard A, Brødsgaard H (1997) The predatory mite *Hypoaspis miles*: temperature dependent life table characteristics on a diet of sciarid larvae, *Bradysia paupera* and *B. tritici*. Entomol Exp Appl 85:177–187

Yoder J (1996) The Madagascar hissing-cockroach mite (*Gromphadorholaelaps schaeferi*): first observation of its larva and ptyalophagy in Acari. Intl J Acarol 22(2):141–148

Yoder J (1997) Exterminator-mites (Acari: Dermanyssidae) on the giant Madagascar hissing-cockroach. Intl J Acarol 23(4):233–236

Yoder J, Barcelona J (1995) Food and water resources used by the Madagascar hissing-cockroach mite, *Gromphadorholaelaps schaeferi*. Exp Appl Acarol 19:259–273

Yoder J, Chambers M, Condon MB, Zettler L (2009) Regulation of the external mycoflora of the giant Madagascar hissing-cockroach, *Gromphadorhina portentosa*, by its mite associate, *Gromphadorholaelaps schaeferi*, and its implication on human health. Symbiosis 47:93–98

Chapter 4
Macrochelid Mites (Mesostigmata: Macrochelidae) as Biological Control Agents

Letícia H. de Azevedo, Rowan M. Emberson, Fernanda de C.N. Esteca, and Gilberto José de Moraes

4.1 Introduction

Biological control of plant pests using predatory mites has been extensively practiced in many countries, especially in protected crops. The predators most often used for that purpose belong to the Phytoseiidae (Gerson et al. 2003; Beaulieu and Weeks 2007), especially because these are usually by far the most common plant inhabiting group of predatory mites.

The family Macrochelidae Vitzthum (Fig. 4.1), of the order Mesostigmata, is one of the most abundant and diverse groups of predatory mites living in excrements or decomposing animals (Vitzthum 1930; Koehler 1999; Behan-Pelletier 2003; Gerson et al. 2003; Braig and Perotti 2009; Lindquist et al. 2009; Perotti and Braig 2009). Filipponi (1964) divided the macrochelids into three main groups, fimicolous, humicolous and insecticolous, referring to their association with excrements, humus or insects (mainly as carriers of these mites). They have also been found in nests or galleries of mammals, birds and social insects, as summarized by Lindquist et al. (2009), where they find an ample assortment of food sources, represented by different organisms found in those microhabitats (Krantz and Whitaker 1988).

Macrochelids have a potential as biological control agents of organisms with which they have been found in association, including different species of flies, and other groups of organisms.

L.H. de Azevedo (✉) • F.C.N. Esteca • G.J. de Moraes
Departamento de Entomologia e Acarologia, Escola Superior de Agricultura "Luiz de Queiroz", Universidade de São Paulo, 13418-900 Piracicaba, SP, Brazil
e-mail: leticiahenrique2806_br@yahoo.com.br; fernanda.esteca@usp.br; moraesg@usp.br

R.M. Emberson
Department of Ecology, Faculty of Agriculture and Life Sciences, Lincoln University, 84, 7647 Canterbury, New Zealand
e-mail: rockview@actrix.co.nz

D. Carrillo et al. (eds.), *Prospects for Biological Control of Plant Feeding Mites and Other Harmful Organisms*, Progress in Biological Control 19,
DOI 10.1007/978-3-319-15042-0_4

Fig. 4.1 Adult female of *Glyptholaspis* sp. (Macrochelidae) preying larvae of *Musca domestica*

4.2 Morphological Characterization

In addition to the original descriptions, morphological characterizations of Macrochelidae were provided by several authors, including Evans (1956), Evans and Browning (1956), Krantz (1962), Krauss (1970), Bregetova (1977), Hyatt and Emberson (1988), Mašán (2003) and Lindquist et al. (2009).

Within the Mesostigmata, diagnostic characteristics of females of this family are as follows: most frequently well sclerotized, with undivided dorsal shield bearing at least 28 pairs of setae (sometimes with additional unpaired, or asymmetrically inserted setae); sternal shield with three pairs of setae; genital shield with one pair of setae and with a pair of conspicuous accessory sclerites beneath lateral margins; ventrianal shield with 1–5 pairs of setae in addition to circumanal setae (except in *Neopodocinum* Oudemans and *Reductholaspis* Emberson, with a small anal shield only); peritreme generally looped around the stigma (except in some species of *Neopodocinum*). Arthrodial membrane at base of movable cheliceral digit with an elongate fimbriate or brush-like process; palp apotele 3-tined; tarsus I usually without pretarsus (except in some species of *Neopodocinum*); paired paradactyli of pretarsi II-IV usually broad, extending to apex of claws and distally divided or serrated (not always distinct because of position in mounted specimens). Males with a holoventral shield or separate sternigenital and ventrianal shields.

4.3 Trends in Understanding the Morphology, Taxonomy, and Biology of Macrochelids

Knowledge about the morphology and taxonomy of macrochelids has increased greatly in the last 60 years. Most important contributions include the detailed study of *Glyptholaspis confusa* (Foà) by Van der Hammen (1964), body chaetotaxy

(Lindquist and Evans 1965; Halliday 1986, 1987), leg chaetotaxy (Evans 1963; Evans and Hyatt 1963), cribrum (Krantz and Redmond 1989), dorsal porotaxy (Van der Hammen 1964; Krantz and Redmond 1987; Moraza 2004), comparative studies of the female reproductive tract (Petrova 1960; Costa 1966, 1967; Athias-Henriot 1968), etc.

Filipponi (1959) was the first author to study the effect of environmental conditions on the morphology of macrochelids. A few years later, he (Filipponi 1964) proposed the use of experimental studies to separate closely related macrochelid species, including biological information and phenotypic and genotypic variation resulting from the variation of environmental factors. This initiative, and the detailed conduct of these studies, represent an important contribution to the taxonomy of mites in general, not only of macrochelids.

A comprehensive work to pull together information and references on the taxonomy of the macrochelids, in the form of a catalog, was initiated by R. M. Emberson in the early 2000s. Other collaborators have joined with him since then, but the work is still ongoing. Until now, almost 250 papers related to the taxonomy of this group have been detected. This work already allows some interpretation of the trends in the taxonomic knowledge of the family.

It seems that the first paper dealing with the taxonomy of this group was published almost 250 years ago (Fig. 4.2), by Scopoli (1772) in describing *Acarus muscaedomesticae*. For about 100 years after that description, the number of publications on macrochelid taxonomy remained very low. Between 1880 and 2000, the number of new publications increased, with outlying higher numbers between 1901 and 1920, and between 1961 and 1970, probably reflecting contributions of A. Berlese in Italy and G.W. Krantz in the USA, respectively. However, a reduction in the number of published papers has been noticed since 2003.

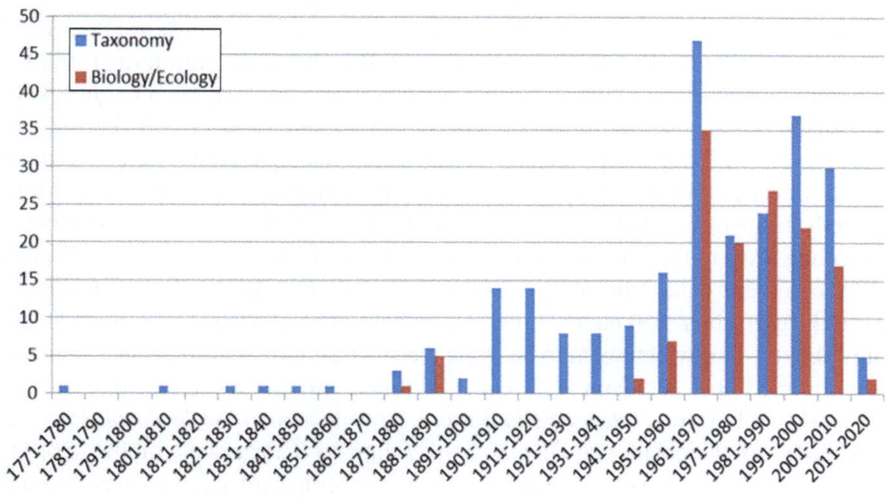

Fig. 4.2 Number of publications on taxonomy and biology/ecology of macrochelids per decade

Quite a different progression was observed in relation to knowledge about the biology and ecology of the macrochelids (Fig. 4.2). The first study was published about 135 years ago, and apparently very little was published in the subsequent 70 years (until the early 1950s). The first publication concerning the predatory behavior of these mites was published only about 70 years ago (Pereira and Castro 1945). Soon after that, the number of publications increased considerably up to the mid 1970s, mainly due to the efforts by A. Filipponi and R.C. Axtell. The number of publications then reduced considerably. We have been unable to determine the cause of this lack of interest in the study of the macrochelids in the last decades, but it could be related to the fact that not enough effort has been dedicated to the promotion of the practical use of these mites and to the general lack of funding sources for basic science.

4.4 Diversity of the Macrochelidae

The current concept of this family is largely based on the work of Evans (1956), who divided Vitzthum's (1930) concept of the family into the subfamilies Macrochelinae and Parholaspinae, and Krantz (1960), who raised those subfamilies to the family level. The Macrochelidae is in need of a thorough revision because of insufficient detail provided in many of the original species descriptions. This situation has led to the creation of many synonyms, a number of homonyms and problems in ascertaining the correct name and authority for many species.

The first revision of the family at the generic level was conducted by Krantz (1962), who considered it to include ten genera. He mentioned that *Macrocheles* Latreille was by far the largest genus, with more than 100 species. He also provided keys to the species of *Geholaspis* Berlese, *Grafia* Krantz and *Lordocheles* Krantz, and listed relevant publications which included taxonomic keys: Valle (1953) for *Geholaspis* species; Balogh (1958) for European species of *Longicheles*; Evans and Browning (1956) for British macrochelids; and Bregetova and Koroleva (1960) for Russian macrochelids. Other important regional taxonomic papers providing keys to genera and species are those of Krauss (1970) and Karg (1993) for Europe; Bregetova (1977) for the former Soviet Union; Krantz and Ainscough (1990) for North America; Halliday (2000) for Australia; and Mašán (2003) for Slovakia. More recently, Emberson (2010) provided a reappraisal of some basal macrochelid groups, providing a key to the genera and subgenera of Geholaspini, a newly proposed tribe.

According to data gathered by one of the authors of this chapter (R. M. Emberson, unpublished), about 550 named mite species have been assigned to Macrochelidae, only about 470 of which seem to be valid; these are currently arranged in 23 genera. He observed that the largest diversity is found in *Macrocheles*, followed by *Holocelaeno* Berlese, *Holostaspella* Berlese, *Neopodocinum*, *Nothrolaspis* Berlese, *Macrholaspis* Oudemans and *Glyptholaspis* Filipponi & Pegazzano (about 320, 41, 37, 33, 20, 18 and 15 species, respectively). Ten other genera are monospecific, and the remaining genera contain between 2 and 11 species each.

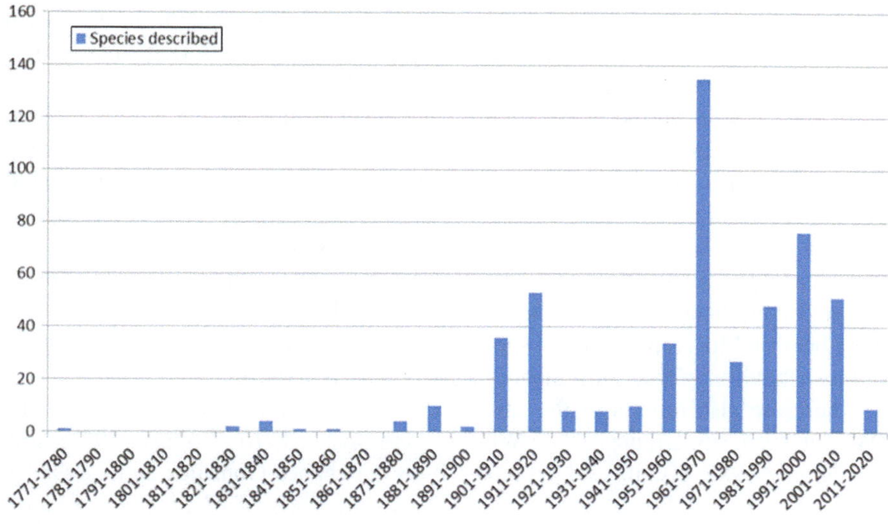

Fig. 4.3 Number of macrochelid species described per decade

The first macrochelid species to be described (Scopoli 1772) was named *Acarus muscaedomesticae* Scopoli. The second species was described only almost 50 years later. The number of species described along the years (Fig. 4.3) has the same pattern of variation reported for the number of taxonomic papers published.

4.5 Associations with Other Organisms

The association of mesostigmatid mites with other arthropods seems to have evolved several times, with the development of a wide range of attachment mechanisms that facilitate mite phoretic dispersal, as summarized by Hunter and Rosario (1988). According to those authors, the complexity of these associations, particularly with insects, indicates a long history of evolution. Phoresy is usually necessary for mite species specialized in living in patchy habitats, which become periodically inadequate for their survival, as in the case of animal excreta or decomposing animal cadavers.

Much has been published about the phoretic behavior of macrochelids because of their common occurrence in patchy habitats. A comprehensive discussion of phoresy in macrochelids was published by Krantz (1998). This process may be facultative or obligatory, but in some cases phoretic specificity in macrochelids has evolved to a degree as intimate as in true parasitism (Krantz 1967). A close association between three macrochelid species, *Aethosoma burchellestes* Krantz, *Macrocheles rettenmeyeri* Krantz and *Macrocheles dibamos* Krantz, and species of army ants (*Eciton* spp.) was reported by Rettenmeyer (1962). He presumed all three species to be ectoparasites, as opposed to other macrochelid species also associated with ants acting as coprophages or scavengers in ant colonies, or by using ants to disperse (phoresy).

Mated macrochelid females are the usual dispersants, attaching to insects when the habitat becomes unfavorable and detaching when the insects visit suitable new mite habitat, to feed or oviposit (Farish and Axtell 1971). Based on the literature taken into account in the preparation of this chapter, most of the reports refer to *Macrocheles* species, but all species of *Holocelaeno* and *Neopodocinum*, and many species of *Holostaspella*, together with species of some of the smaller macrochelid genera, are also known to be phoretic. As carriers, most reports refer to Coleoptera species, mainly those whose larvae feed on dung (the dung beetles, or Scarabaeinae, as well as Aphodiinae, the small dung beetles), but also to Diptera. Krantz (1983a) reported that about 280 macrochelid species have been found in association with dung beetles. Most reports do not specify the family of the dipteran carrier, but most of those that do, refer to Muscidae, especially *Musca domestica* L.

There are also reports of the association of these mites with small mammals (Womersley 1955; Krantz 1983b; Whitaker and Spicka 1976; Whitaker 1977; Adis and Krantz 1985; Krantz and Whitaker 1988; Estébanes-González and Cervantes 2009). Krantz and Whitaker (1988) summarized reports of *Macrocheles* species associated with small mammals in North America and added new records of this type of association. Altogether, 32 *Macrocheles* species were reported in association with 53 mammal species. It is thought that the mites attach to the host mammals when the latter visit the substrate where the mites live, in search of food. A curious example of this type of association is of *Macrocheles lukoschusi* Krantz, known to live in the rectum of the sloth *Bradypus tridactylus* L., apparently feeding on nematodes (Krantz 1983b; Adis and Krantz 1985).

Facultative parasitism has been reported by *Macrocheles muscaedomesticae* on adult houseflies (*Musca domestica* L.) and by *Macrocheles subbadius* (Berlese) on adult *Drosophila nigrospiracula* Patterson and Wheeler, by Jalil and Rodriguez (1970) and Polak (1996), respectively.

Phoresy in *Glyptholaspis* Filipponi and Pegazzano is less common, usually involving a beetle as carrier (Krantz 1983a). It has been reported for *G. americana* and *G. confusa* on stable fly (*Stomoxys calcitrans* L.) (Axtell 1961), some species of Scarabaeidae (Halliday 1986; Glida et al. 2003) and the housefly (Axtell 1964; Niogret et al. 2006).

According to Krantz (1998), a phylogenetic analysis of selected phoretic and non-phoretic macrochelid taxa has shown a strong correlation between phoresy and evolutionary position, indicating that an increasing commitment to phoresy is correlated with an advance from early derivative to more evolved taxa.

Some studies support the hypothesis that the phoretic behavior in *Macrocheles* is closely related with the stage of decomposition, temperature and relative humidity (Singh et al. 1967; Jalil and Rodriguez 1970; Farish and Axtell 1971; Stafford and Bay 1987; Ho 1990; Rodrigueiro and Prado 2004). According to Farish and Axtell (1971), the decomposition stage of the substrate interferes with moisture content. With humidity loss at increasing temperatures, the substrate becomes progressively less suitable for mite development, leading to dispersal by phoresy. Mite tolerance to temperature and relative humidity may have an indirect effect on the phoretic rate. Singh et al. (1967) reported *M. muscaedomesticae* to be more sensitive to fluctuation of temperature than of relative humidity. As ambient temperature rises above their preferred range, mites leave for more favourable niches (Ho 1990).

As discussed by Hunter and Rosario (1988) for mites in general, Krantz (1998) considered that phoresy, as a survival mechanism, offered many macrochelids ample reason for development of special structures to assure their successful transfer to new habitats. According to that author, most phoretic *Macrocheles,* and species of other phoretic macrochelid genera (for example *Holocelaeno*, *Holostaspella* and *Neopodocinum*), have a special mechanism to grasp an individual seta of their beetle or fly carrier. This mechanism involves the use of a set of two closely associated teeth [referred to by Krantz (1998) as the subterminal bidentate tooth] located near the middle of the inner edge of the movable cheliceral digit and an opposing ribbed section of the inner edge of the fixed digit. A different specialized structure is found in *M. rettenmeyeri*, which has a close association with the army ant *Eciton dulcis crassinode* Borgmeier (Rettenmeyer 1962). Legs IV of this mite are greatly enlarged, serving as functional claws for the ant worker, by the attachment of the mite to the pulvillus of the ant hind leg.

4.6 Predation Potential and Life Cycle

Experimental data on predation potential and life cycle have only been published for species of two macrochelid genera. By far, most detailed information is available for *Macrocheles* species, with a few studies conducted on *Glyptholaspis* species.

Several macrochelid species have been reported as predators of eggs and first instar larvae of the housefly, as first noted by Pereira and Castro (1945). However, they are also known to feed on other fly species, nematodes, collembola and other small arthropods, as subsequently described.

An important aspect interfering with the predation ability of macrochelids refers to the location of their prey in the substrate. Perotti (2001) evaluated the effect of prey (housefly eggs) location on the predatory ability of *M. muscaedomesticae* and *G. confusa*. This author showed that *M. muscaedomesticae* was an effective predator and located both hidden and exposed eggs of *M. domestica*, while *G. confusa* only showed a significant predation on hidden eggs.

4.6.1 Macrocheles

Most studies on the life cycle and prey consumption of species of this genus used fly and nematode species as prey (Table 4.1). Many studies refer to predation of housefly without providing details about the process (Pereira and Castro 1945, 1947; Filipponi 1959; Filipponi and Petrelli 1967; Ito 1973; Filipponi et al. 1971; Manning and Halliday 1994).

Table 4.1 Studies on the life cycle and prey consumption by *Macrocheles* species and respective food source

Macrocheles species	Food source	Reference
M. boudreauxi	Nematodes: Rhabditidae	Kinn and Witcoskjy (1977)
M. glaber	Flies: *Lucilia* sp., *Musca domestica*, *Musca vetustissima*	Filipponi (1959), Manning and Halliday (1994), Al-Dulaimi (2002)
	Nematodes: *Panagrellus* sp., Rhabditidae	Filipponi and Delupis (1963), Manning and Halliday (1994)
	Other soil invertebrates: *Enchytraeus albidus* (Enchytraeidae) and Poduromorpha Collembola	Filipponi and Delupis (1963)
M. insignitus	Unidentified flies	Filipponi and Ilardi (1959)
M. matrius	Fly: *M. domestica*	Filipponi and Cicolani (1974), Cicolani (1978), Soliman et al. (1978)
	Nematodes: Rhabditidae	Filipponi and Cicolani (1974), Cicolani (1978)
M. merdarius	Flies: *Lucilia* sp., *M. domestica*, *M. vetustissima*	Filipponi and Francaviglia (1964), Manning and Halliday (1994)
	Nematodes: *Panagrellus* sp., Rhabditidae	Filipponi and Francaviglia (1964), Manning and Halliday (1994)
M. muscaedomesticae	*M. domestica*	Pereira and Castro (1945, 1947), Filipponi (1959), Wade and Rodriguez (1961), Rodriguez et al. (1962), Filipponi and Delupis (1963), Wallwork and Rodriguez (1963), Filipponi and Francaviglia (1964), O'Donnell and Axtell (1964), Kinn (1966), Singh et al. (1966), Filipponi and Petrelli (1967), Willis and Axtell (1968), Filipponi (1971), Filipponi et al. (1971), Ito (1971, 1973, 1977), Al-Dulaimi (1986), Ho (1989), Geden and Axtell (1988), Geden et al. (1988, 1990), Almeida (1994), Perotti (2001), Guanilo and Cordero (2003)
	Other flies: *Fannia canicularis*, *Haematobia irritans*, *Musca autumnalis*, *Phormia regina*, *Stomoxys calcitrans*	Kinn (1966), Singh et al. (1966), O'Donnell and Nelson (1967), Perotti (2001)
	Nematodes: *Aphelenchus avenae*, *Aphelenchoides composticora*, *Diplogasteroides* sp., *Diplogaster* sp., *Panagrellus redivivus*, *Panagrolaimus* sp., *Rhabditella* sp., Rhabditidae, *Rhabditis* (*Rhabditella*) *axei*, *Rhabditis elongata*, *R. teres*, *Rhitis inermiformis*	Rodriguez et al. (1962), Filipponi and Delupis (1963), Filipponi and Francaviglia (1964), Filipponi and Petrelli (1967), Filipponi (1971), Filipponi et al. (1971), Ito (1971, 1973, 1977), Geden and Axtell (1988), Royce and Krantz (1991)

	Other invertebrates: *Enchytraeus albidus* (Enchytraeidae), *Phthorimaea operculella*, Poduromorpha Collembola	Filipponi and Delupis (1963), Filipponi and Petrelli (1967), Hassan et al. (2002)
M. mycotrupetes	Flies: *M. domestica*	Krantz and Royce (1994)
	Nematode: *Rhitis inermiformis*	Royce and Krantz (1991), Krantz and Royce (1994)
M. parapisentii	Nematodes: Rhabditidae	Costa (1967)
M. penicilliger	Flies: *Lucilia* sp., *M. domestica*, *M. vetustissima*	Filipponi (1971), Manning and Halliday (1994)
	Nematodes: *Panagrellus* sp., Rhabditidae	Filipponi (1971), Manning and Halliday (1994)
M. peniculatus	Flies: *Lucilia* sp., *M. domestica*, *Musca vetustissima*	Filipponi and Delupis (1963), Filipponi and Francaviglia (1964), Filipponi (1971), Manning and Halliday (1994)
	Nematodes: *Panagrellus* sp., Rhabditidae	Filipponi and Delupis (1963), Filipponi (1971), Manning and Halliday (1994)
	Other invertebrates: *Enchytraeus albidus*, Poduromorpha Collembola	Filipponi and Delupis (1963), Filipponi and Francaviglia (1964)
M. peregrinus	Flies: *Haematobia irritans exigua*, *Haematobia thirouxi potans*	Doube et al. (1986), Roth et al. (1988)
M. perglaber	Flies: *M. domestica*	Filipponi and Delupis (1963), Filipponi and Francaviglia (1964), Filipponi and Petrelli (1969)
	Nematodes: Rhabditidae, *Rhitis inermiformis*	Filipponi and Delupis (1963), Filipponi and Francaviglia (1964), Filipponi and Petrelli (1969), Royce and Krantz (1991)
	Other invertebrates: *Enchytraeus albidus* (Enchytraeidae) and Poduromorpha Collembola	Filipponi and Delupis (1963), Filipponi and Petrelli (1969)
M. pisentii	Nematode: *Rhitis inermiformis*	Royce and Krantz (1991)
M. robustulus	Flies: *Bradysia* spp., *Lucilia* sp., *M. domestica*, *Musca vetustissima*	Filipponi and Delupis (1963), Filipponi and Mosna (1968), Manning and Halliday (1994), Grosman et al. (2011)
	Nematodes: *Panagrellus* sp., Rhabditidae.	Filipponi and Delupis (1963), Costa (1966), Manning and Halliday (1994)
	Other invertebrates: *Enchytraeus albidus* (Enchytraeidae), *Frankliniella occidentalis*, Poduromorpha Collembola	Filipponi and Delupis (1963), Messelink and Holstein-Saj (2008)

(continued)

Table 4.1 (continued)

Macrocheles species	Food source	Reference
M. rodriguezi	Acarid mites (Acaridae)	Oliver and Krantz (1963)
M. saceri	Nematode: *Rhitis inermiformes*	Royce and Krantz (1991)
M. scutatus	Fly: *M. domestica*	Filipponi and Delupis (1963)
	Nematodes: Rhabditidae	Filipponi and Delupis (1963)
	Other invertebrates: *Enchytraeus albidus* (Enchytraeidae) and Poduromorpha Collembola	Filipponi and Delupis (1963)
M. similis	*Tyrophagus similis*	Saito and Takaku (2013)
M. subbadius	Flies: *Lucilia* sp., *M. domestica*, *M. vetustissima*	Filipponi and Francaviglia (1964), Filipponi (1971), Cicolani et al. (1977), Manning and Halliday (1994)
	Nematodes: *Panagrellus* sp., Rhabditidae, *Rhitis inermiformis*	Filipponi and Francaviglia (1964), Filipponi (1971), Royce and Krantz (1991), Manning and Halliday (1994)
M. superbus	Flies: Diptera	Blaszak et al. (1990)
	Other invertebrates: Enchytraeidae, Poduromorpha Collembola	Blaszak et al. (1990)
M. vernalis	Nematode: *Rhitis inermiformis*	Royce and Krantz (1991)
M. virgo	Flies: *Lucilia* sp., *Musca vetustissima*	Manning and Halliday (1994)
	Nematodes: *Panagrellus* sp.	Manning and Halliday (1994)

Predation rates were reported for *M. muscaedomesticae* on housefly (Filipponi and Delupis 1963; Wallwork and Rodriguez 1963; O'Donnell and Axtell 1964; Kinn 1966; Singh et al. 1966; Willis and Axtell 1968; Ito 1977; Soliman et al. 1978; Al-Dulaimi 1986; Geden et al. 1988, 1990; Geden and Axtell 1988; Almeida 1994; Perotti 2001; Guanilo and Cordero 2003), little housefly (*Fannia canicularis* L.) (Singh et al. 1966; O'Donnell and Nelson 1967), face fly (*Musca autumnalis* De Geer) (Singh et al. 1966), black blow fly (*Phormia regina* Meigen) (Kinn 1966), stable fly (Kinn 1966) and horn fly (*Haematobia irritans* L.) (Perotti 2001). Predation rates were also reported for *Macrocheles peregrinus* Krantz on buffalo fly (*Haematobia irritans exigua* de Meijere) (Doube et al. 1986; Roth et al. 1988) and the African buffalo fly (*Haematobia thirouxi potans* Bezzi) (Doube et al. 1986), as well as for *Macrocheles glaber* (Müller) on housefly (Al-Dulaimi 2002).

Adult of *M. muscaedomesticae* have a marked preference for housefly eggs over first instar larvae (Filipponi 1955; Ito 1977; Almeida 1994; Perotti 2001; Al-Dulaimi 2002), while protonymphs and deutonymphs have a preference for nematodes (Rodriguez et al. 1962; Filipponi and Delupis 1963; Ito 1977).

Development and reproduction on housefly were reported for *Macrocheles insignitus* Berlese (Filipponi and Ilardi 1959); *M. muscaedomesticae* (Wade and Rodriguez 1961; Filipponi and Francaviglia 1964; Filipponi and Petrelli 1967; Filipponi et al. 1971; Soliman et al. 1978; Al-Dulaimi 1986; Ho 1989; Geden et al. 1990; Almeida 1994); *Macrocheles merdarius* (Berlese), *Macrocheles peniculatus* Berlese and *M. subbadius* (Filipponi and Francaviglia 1964); *Macrocheles perglaber* Filipponi and Pegazzano (Filipponi and Francaviglia 1964; Filipponi and Petrelli 1969), *Macrocheles matrius* Hull (Cicolani 1978) and *M. glaber* (Al-Dulaimi 2002).

Development and reproduction of *M. muscaedomesticae* on a lepidopteran species (*Phthorimaea operculella* Zeller, Gelechiidae, the potato tuber moth) was also reported by Hassan et al. (2002), who found a considerable number of these mites associated with *P. operculella* on stored potatoes.

Macrochelid predation on mites has been reported less frequently, and always involving species of Astigmatina, which usually require high levels of environmental humidity to develop. The life cycle of *Macrocheles rodriguezi* Oliver and Krantz on an unidentified astigmatine of the family Acaridae was studied by Oliver and Krantz (1963). The predator was able to develop and oviposit on that prey, but not on an unidentified astigmatine of the family Histiostomatidae (cited as Anoetidae). Predation of *Macrocheles robustulus* (Berlese) on *Carpoglyphus lactis* (L.) was reported by Grosman et al. (2011), while Saito and Takaku (2013) investigated the predation capacity of *Macrocheles similis* Krantz and Filipponi on *Tyrophagus similis* Volgin.

Consumption of nematodes by *Macrocheles* species has also been widely observed. It was reported for *M. muscaedomesticae* on an unidentified nematode (Filipponi and Delupis 1963), on *Rhabditis elongata* Schneider (Ito 1971, 1973, 1977), on *Rhabditella axei* (Cobbold) (Filipponi and Petrelli 1967), and on *Rhabditis teres* (Schneider), *Panagrellus redivivus* L., *Aphelenchus avenae* Bastian and *Aphelenchoides composticora* Franklin (Ito 1971). Predation has also been reported for *Macrocheles glaber*, *M. peniculatus*, *M. perglaber*, *M. robustulus* and *M. scutatus* (Berlese) on an unidentified nematode (Filipponi and Delupis 1963).

In addition, nematodes have been shown to allow development and reproduction of *Macrocheles*. Available reports include *M. merdarius*, *M. muscaedomesticae*, *M. peniculatus*, *M. perglaber* and *M. subbadius* on unidentified free living nematodes (Filipponi and Francaviglia 1964); *M. robustulus* and *Macrocheles parapisentii* Costa on *Panagrellus* sp. (Costa 1966, 1967, respectively); *M. muscaedomesticae*, *M. boudreauxi* Krantz and *M. matrius* on rhabditoid nematodes (Filipponi et al. 1971; Kinn and Witcoskjy 1977 and Cicolani 1978, respectively); *M. matrius* on *Rhabditella axei* (Filipponi and Cicolani 1974); *Macrocheles mycotrupetes* Krantz and Mellot on *Rhitis inermiformis* (Osche) (Krantz and Royce 1994); and *M. glaber*, *M. merdarius*, *M. penicilinger* (Berlese), *M. peniculatus*, *M. subbadius* and *M. virgo* Halliday on *Panagrellus* sp. (Manning and Halliday 1994).

The available information has demonstrated the ability of *Macrocheles* species to prey and reproduce on a range of other food items, demonstrating their polyphagous behavior. A few studies have reported several collembolan (Arthropoda: Entognatha) and *Enchytraeus* (Annelida: Oligochaeta) species to allow development and reproduction of different species of *Macrocheles*. Predation on collembolans of the order Poduromorpha and on *Enchytraeus albidus* Henle was reported by Filipponi and Delupis (1963) and Filipponi and Petrelli (1967) for *M. muscaedomesticae*, and by Filipponi and Petrelli (1967) for *M. peniculatus*, *M. robustulus*, *M. perglaber*, *M. glaber* and *M. scutatus*. Feeding on wounded or dead enchytraeid, larvae of house cricket (*Acheta domesticus* (L.)) and housefly by *M. glaber* was reported by Ignatowicz (1974).

Aspects of food ingestion have been studied by different authors. Filipponi and Delupis (1963) observed that the food preferred by a species was not always the same as that ensuring its greatest reproductive rate. This has also been reported by other authors for other groups of Mesostigmata (Cavalcante 2014). Filipponi and Delupis (1963) also observed that the food preferred by immatures did not always coincide with that preferred by adults. Studies have also been conducted on the interference of a food item on the ingestion of other food items. Ito (1973) noted that although the availability of nematodes to *M. muscaedomesticae* reduced its predation rate on little housefly eggs, its reproduction rate was remarkably enhanced when nematodes were also available as food.

Several studies have been conducted on the effect of temperature on the biology of different *Macrocheles* species, including *M. glaber* (Filipponi 1959), *M. muscaedomesticae* (Wallwork and Rodriguez 1963; Filipponi 1971; Filipponi et al. 1971; Geden and Axtell 1988; Almeida 1994; Hassan et al. 2002), *M. matrius* (Filipponi and Cicolani 1974; Cicolani 1978), *M. penicillinger*, *M. peniculatus* and *M. subbadius* (Filipponi 1971). Biological parameters obtained by different authors under different temperature regimes were summarized by Cicolani (1979). She reported development of *Macrocheles* species from egg to adult to be completed in periods ranging from 1.4 (*M. robustulus*) to 4.6 (*M. penicilliger*) days; the fecundity to range from 53.3 (*M. penicilliger*) to 163.2 (*M. matrius*) adult offspring per female; and the intrinsic rate of population increase to range from about 0.323 (*M. penicilliger*) to 0.906 (*M. muscaedomesticae*). These values are significantly higher than those reported for other groups of Mesostigmata. For instance, the highest values of intrinsic rate of population increase reported for Phytoseiidae mites correspond to the lowest levels determined for those *Macrocheles* species (Sabelis 1985). The duration of immature development at temperatures ranging from 15 to 35 °C has been observed to vary between 2 and 5 days, while fecundity ranged between 0.2 and 8 eggs per female per day.

A simulation model for *M. muscaedomesticae* population dynamics and predation rates on immature housefly, taking into account temperature as well as other environmental factors, was first proposed by Geden et al. (1990). The authors reported close agreement between mite populations predicted by the model with actual populations from validation experiments. Simulations models may be helpful in describing interactions of biotic factors and thus aid decision making in fly management.

4.6.2 Glyptholaspis

Predation rates were reported for *G. confusa* on housefly (Axtell 1961; Geden et al. 1988), horn fly (Halliday and Holm 1987; Perotti 2001), and Australian bush fly (*Musca vetustissima* Walker) (Halliday and Holm 1987). Only one study has been published on the development and reproduction of *Glyptholaspis*. Afifi (1989) reported that *Glyptholaspis americana* (Berlese) can complete immature development on housefly eggs and larvae. Adult female longevity and fecundity of *G. americana* were determined to be higher than those observed for *M. robustulus* and *M. carinatus*. Immature development of *G. americana* was completed in 6.5 days, females lived for an average of 71.4 days, while average fecundity was 14.3 eggs per female when feeding on housefly eggs. The average total number of housefly eggs consumed during the adult female stage by this species was 247.

Studies on the feeding behavior of some manure-inhabiting mesostigmatids mites on free-living nematodes concluded that *G. confusa* was not attracted to nematodes (Ito 1971). However, Afifi (1989) reported successful development and reproduction of *G. americana* feeding on free-living nematodes, although oviposition rates and longevities were lower on mites fed upon nematodes.

4.7 Reproduction

Apparently one of the first reports concerning the modes of reproduction of macrochelids was published by Kramer (1876), who referred to the scarcity of males in *M. muscaedomesticae*. Canestrini (1881) interpreted this as an indication of the occurrence of parthenogenesis in this species. Almost 70 years later, Pereira and Castro (1947) demonstrated experimentally the occurrence of arrhenotoky in *M. muscaedomesticae*.

Norton et al. (1993) summarized the published information regarding the modes of reproduction of macrochelids and presented a list of species undergoing thelytoky (species of *Geholaspis* and *Macrocheles*) and arrhenotoky (species of

Glyptholaspis, *Holostaspella* and *Macrocheles*). Thelytoky was also reported by Halliday (1993) and Manning and Halliday (1994) (*Macrocheles*), while arrhenotoky was also reported by Cicolani (1983) and Manning and Halliday (1994) (*Glyptholaspis* and *Macrocheles*).

The role of pheromones has been investigated in macrochelid reproduction. Krantz and Redmond (1989) suggested the opening of the cribral glands of *M. perglaber* (localized in the posterior portion of female ventrianal shield) to be sites of sex pheromone release, the cribrum (on the posterior end of the ventrianal or anal shield) functioning as a dispersal platform for these substances. Adult males of *M. muscaedomesticae* were observed to mount on pharate female deutonymphs, preventing other males from approaching them, mating soon after emergence of the adult (Yasui 1988, 1992). According to Yasui (1996) males are able to assess the insemination status of other males, adjusting duration of copulation according to the probability of fertilizing eggs with their own sperm. Males were able to assess the value of their mates according to the age and reproductive status of the latter (Yasui 1996).

Krantz and Wernz (1979) studied the mechanism of sperm transfer and various aspects of pre-mating and mating behavior in *G. americana*. They observed that the oozing sperm droplet is taken up by one of the male chelicerae, which then remain closely appressed to each other throughout the transfer process, allowing uninterrupted sperm flow from the droplet to the spermadactyl of the opposing chelicera, which ultimately does the actual insemination.

The relationship between distribution of macrochelid species and type of reproduction was examined by Manning and Halliday (1994). Not needing to mate before reproduction, thelytokous species should have an advantage in colonising patchy and temporary habitats over sexually reproducing species. However, Manning and Halliday (1994) concluded that thelytokous species (*M. penicilliger*, *M. peniculatus* and *M. virgo*) were usually found in temporarily stable, physically continuous habitats such as leaf litter and compost. They also observed that arrhenotokous species (*M. glaber*, *M. merdarius*, *M. robustulus*, *M. subbadius* and *G. americana*) were found in a variety of habitats, but especially in ephemeral scattered habitats. This occurs because arrhenotokous macrochelid females can mate with their own sons (Manning and Halliday 1994), producing fertilized eggs to establish a new population even when males are not available beforehand in the newly reached habitat.

4.8 Behaviour

Laboratory and field observations have confirmed the importance of semiochemicals in the relation between macrochelids and other organisms serving as their carriers or their prey (Krantz 1998). Strong positive responses of *Macrocheles pisentii* (Berlese), *Macrocheles saceri* Costa and *M. mycotrupetes* were reported to be elicited by surface extracts of their scarab carriers (Krantz 1991, 1998). The effect of those chemicals may be remarkably strong in some cases. Krantz and Royce (1994) reported the ability of *M. mycotrupetes* to commute between underground galleries of its beetle hosts by moving through a sandy substrate for up to 80 mm, stimulated by chemicals produced by beetles in a different gallery.

Specialized receptors present on tarsus I are presumed to play a major role in the perception of volatile compounds. Farish and Axtell (1966) demonstrated that the removal of tarsi I of *M. muscaedomesticae* resulted in lack of attraction of the mite to housefly eggs, pupae or adults. This result led the authors to concluded that receptors of olfactory stimuli are present on tarsi I. Jalil and Rodriguez (1970) concluded that among the several setae of tarsi I, the longer, terminal setae were the most important in odor perception. This was later corrected by Coons and Axtell (1973), who reported that the setae involved in odor perception were the neighboring (also terminal) peg-like setae. Jalil and Rodriguez (1970) and Wicht et al. (1971) demonstrated the attraction of *M. muscaedomesticae* to water-soluble compound(s) released by adult houseflies. Wicht et al. (1971) pointed out that water extracts from adult houseflies could contain an amino sugar or a polysaccharide present on both males and females.

Jalil and Rodriguez (1970) showed preference of *M. muscaedomesticae* for odors of housefly adults over housefly eggs. Borden (1989) reported higher olfactory attraction of the same predator to the little housefly adults, followed by the attraction to eggs, larvae and pupae. This author concluded that the predators would benefit from choosing adults, to assure dispersal to new breeding sites.

Adults of *M. muscaedomesticae* have been shown to congregate around the anterior end of stable fly pupae, attaching to the adults as they emerge (Greenberg and Carpenter 1960). The predator often attaches to abdomen of adult housefly (Jalil and Rodriguez 1970), especially on the ventral surface (Ho 1990; Rodrigueiro and Prado 2004).

Perception of non-volatile compounds was assumed to be determined by different sensors located on the palpi of *M. muscaedomesticae* (Farish and Axtell 1966; Coons and Axtell 1973).

Other chemicals seem to affect other types of behavior of macrochelids. In laboratory experiments, addition of ammonia to the substrate was observed to increase the predation rate of *M. muscaedomesticae* (Rodriguez and Wade 1961; Wallwork and Rodriguez 1963). This is not surprising, given that its main prey, the housefly, is attracted to ammonia, naturally released from the nitrogenous compounds in bovine manure (Wallwork and Rodriguez 1963). Attraction of *M. muscaedomesticae* to ammonia and to skatole was also demonstrated (Jalil and Rodriguez 1970). These observations might be important in the implementation of applied biological control programs.

4.9 Potential for the Practical Use of Macrochelids in Biological Control

Soil mesostigmatid mites have the potential to control plant pests that spend part of their lives in the soil or litter, or that attack underground plant parts (Eickwort 1983; Lesna et al. 1995; Gerson et al. 2003). Some of them, including the macrochelids, also have the potential to reduce population levels of noxious flies associated with animal excreta or decomposing animal cadavers (Wallace et al. 1979; Krantz 1983a). Given the ability of macrochelids to feed on a variety of soil invertebrates, it is certainly conceivable that they could be used for applied biological control (Walter and Ikonen 1989).

Pereira and Castro (1945) were the first to claim that a macrochelid species had a definitive role as a biological control agent, referring to the predation of the housefly by *M. muscaedomesticae*. The first thorough appraisal of macrochelids as biological control agents was done by Axtell (1969), centering his discussion on the potential of augmentative release of these mites to control synanthropic flies. He highlighted positive and negative aspects of macrochelid biology and ecology in relation to their potential as control agents of the housefly and other manure inhabiting synanthropic flies. He emphasized the need for the development of means to mass produce macrochelids at low cost, making their field releases possible early in the fly breeding season. In addition, Axtell (1969) suggested that better control might be achieved by the concurrent use of one or more macrochelid species in association with other types of natural enemies. He was quite positive about the potential use of macrochelids for fly control, especially within the scope of integrated management programs.

The second comprehensive appraisal of the potential use of macrochelids as control agents was by Krantz (1983a), who summarized the differences between the mite fauna found in what he termed domestic and pastoral settings, referring to dung accumulations and to individual dung pads, respectively. He placed special importance on understanding these differences when proposing biological control programs, due to the lower diversity of insect groups that could serve as carriers for macrochelids in domestic rather than in pastoral settings. According to this author, dung beetles tend to favor isolated dung pads, so that synanthropic flies are the only common potential carriers for the macrochelids of domestic settings (where synanthropic flies predominate). He listed a series of factors considered necessary to allow the widespread use of macrochelids as biological control agents in both settings, within the concept of integrated pest management programs. In relation to pastoral settings, Krantz (1983a) emphasized the importance of considering the introduction of prospective macrochelids and their carriers from other regions in a classical biological control approach.

Beaulieu and Weeks (2007) reviewed the role of free-living mesostigmatid mites as biological control agents in Australia. They mentioned the following traits as making some macrochelids suitable as fly control agents: (1) aggressive behaviour and feeding preference for fly eggs and larvae; (2) high rate of population increase; (3) efficient dispersal on a variety of dung-inhabiting beetles; and (4) the ease with which they can be reared on flies or nematodes (Krantz 1983a).

A discussion is presented below on the potential of macrochelid species as biological control agents of specific groups of harmful organisms.

4.9.1 Housefly

Applied biological control activities are usually classified in three general tactics, conservation, augmentation and classical biological control. While considering the conclusion of Pereira and Castro (1945) on the potential for the use of *M. muscaedomesticae* for housefly control, Axtell (1969) listed 11 attributes that in his view made this species a prospective control agent of that prey. These essentially referred to the preference of the predator to inhabit sites also preferred by the fly, to its phoretic behavior on the fly, and to the biological characteristics that enable it to survive under unfavorable environmental conditions.

Practical use of *M. muscaedomesticae* could benefit from the results of the work of Wallwork and Rodriguez (1963), which suggested that the addition of ammonia to the substrate may increase predation. Gerson et al. (2003) concluded that the ability of *M. muscaedomesticae* to bring about satisfactory housefly control also depended on the presence of alternate prey. According to these authors, the alternative prey (e.g. acarid mites, nematodes and larvae of other flies) helped to maintain predator fecundity and longevity when housefly numbers declined. From the previous sections of this chapter, it is clear that this predator has the ability to feed on several other organisms in addition to housefly. The release of nematodes to increase the population of macrochelids has also been proposed by different authors (Axtell 1969; Wicht and Rodriguez 1970; Ito 1973).

Effective control of housefly by *M. muscaedomesticae* and *G. confusa* was also reported by Axtell (1963a, b) in calf and poultry manure. Significant reductions in fly populations by these predators were observed under natural conditions, but the reduction was much more significant, reaching up to 85.5 %, when the predators were augmented (Axtell 1963a; Singh et al. 1966). Effective control of houseflies in poultry houses, with up to 99 % reduction in population level, was obtained by Rodriguez et al. (1970). Krantz (1983a) commented that despite this positive evidence, widespread use of the macrochelids for fly control in calf and poultry manure had not occurred due to several factors, including the total removal of manure, lack of understanding about the interactions between and among biotic and abiotic components of the system, and lack of efficient rearing methods.

4.9.2 Other Flies

Given the demonstrated ability of macrochelids to feed on the housefly, an obvious question to be asked is what is their potential use as control agents of other important flies.

Works conducted so far do not show a clear potential of macrochelid species as biocontrol agents, suggesting that further studies are necessary. Several studies highlight the importance of the early arrival of the mites, on their dung beetle carriers, for maximum predation to take place. Wallace et al. (1979) suggested that the native Australian macrochelids, *M. glaber* and *M. perglaber* (according to Beaulieu and Weeks 2007, both referred to as *M. glaber*) could control the Australian bush fly, provided the predators could be brought to the dung pad soon after the flies laid their eggs. They concluded that significant improvement in Australian bush fly control could be achieved in southeastern Australia by introducing a dung beetle that could carry the mites. The dung beetles should have high flight activity in the afternoon and/or evening, coinciding as close as possible with the flight activity of the fly.

Macrocheles peregrinus was introduced from South Africa to Australia for control of the Australian bush fly and the buffalo fly (Wallace and Holm 1983). Roth et al. (1988) considered it had a significant effect on the Australian bush fly, which oviposites in parts of the dung pad accessible to the mite, i.e. on the surface. However, they considered the predator to be ineffective in the control of the buffalo fly, providing an estimated suppression of 33 % of the population under field conditions.

Another promising result was obtained by Singh et al. (1966). The authors reported 92 % control of little housefly in an experiment conducted under semi-field condition, with the release of *M. muscaedomesticae* at a ratio of 1 mite to 5 muscid eggs. Further studies to confirm that efficiency are necessary.

4.9.3 Other Arthropods

Regardless of the most common occurrence of macrochelids in association with patchy habitats, some inhabit more uniform environments. *Macrocheles robustulus* has often been found in greenhouse soils (Messelink and van Holstein-Saj 2008), where it arrives with the introduction of compost (Messelink and De Kogel 2005). This species performed better as predator of the thrips (Thripidae) *Frankliniella occidentalis* (Pergande) (Messelink and van Holstein-Saj 2008) and of fungus gnats *Bradysia* spp. (family Sciaridae) (Grosman et al. 2011), than the predatory mite *Gaeolaelaps aculeifer* (Canestrini) (Laelapidae). In the case of the sciarids, the predator's population increased with the addition of mulch containing an astigmatid mite that served as an alternative prey, on which the predators were previously reared. *Macrocheles robustulus* is presently commercially available for the control of prepupae and pupae of thrips as well as larvae of *Lyprauta* sp. (Keroplatidae) sciarids.

4.10 Rearing Techniques

Efforts to develop rearing techniques for macrochelid mites under laboratory conditions started long ago. Pereira and Castro (1946, 1947) were apparently the first to rear *M. muscaedomesticae* for experimental purposes, using porous clay substrates to prevent water condensation, and allowing humidity retention and air exchange.

The effect of environmental factors that could interfere with the mass production of *M. muscaedomesticae* was discussed by Rodriguez and Wade (1961), based on laboratory and semi-field observations. The type of substrate, livestock feed additives, pH, temperature and relative humidity had an effect on the reproduction of *M. muscaedomesticae*. They proposed a substrate for the production of macrochelids, which included housefly eggs and nematodes as food. Filipponi (1964) presented a series of recommendations for rearing macrochelid mites for experimental taxonomic studies, including a substrate, food items (housefly eggs and nematodes), and the containers to be used. He also mentioned the procedure used for the production of the nematodes.

Singh and Rodriguez (1966) reported successful results in their efforts to mass rear *M. merdarius*, *M. muscaedomesticae* and *M. subbadius* on the nematode *Rhabditella leptura* (Cobb). They also standardized a method for rearing the nematode species in the laboratory. Ho et al. (1990) developed a mass-production method for *M. muscaedomesticae*, involving the use of spent housefly media, frozen housefly eggs, and the nematode *Protorhabditis* sp. At about the same time, Royce and Krantz (1991) developed a rearing method that was successfully used for the production of several macrochelid species including, *M. muscaedomesticae*, *M. mycotrupetes*, *M. perglaber*, *M. pisentii*, *M. saceri*, *M. subbadius*, *M. vernalis* and *G. confusa*. It consisted of the production of nematodes in a fly rearing medium, which was then periodically supplied to the mite colonies in jars lined with plaster-of-paris/charcoal.

An important aspect to consider is the common cannibalistic behavior of the macrochelids (Wade and Rodriguez 1961; Ignatowicz 1974). Almeida (1994) proposed the use of pieces of ceramics at the bottom of the rearing units to reduce the rate of cannibalism.

More recently, Grosman et al. (2011) reported the use of the astigmatid mite *C. lactis* for the mass production of *M. robustulus*. Details about the rearing process were not provided, except that the prey was reared on bran. Astigmatine mites have long been widely used for the mass production of other predatory mites (Ramakers and van Lieburg 1982). However, the use of astigmatines for the production of macrochelids is interesting for the fact that, as mentioned previously in this chapter, relatively few studies have reported predation by macrochelids on astigmatids.

4.11 Integrated Management

The natural occurrence of macrochelids in animal excreta under different ecological conditions was known before the widespread use of chemical insecticides became common for fly control. The deleterious effect of insecticides on those predators soon became evident to researchers (Axtell 1963a, 1966, 1968, 1970; Axtell and Rutz 1986; Rodriguez et al. 1970; Anderson 1983; Axtell and Arends 1990; Mattos 1992). It was then concluded that conservation of predators could improve fly control, especially implementing techniques to reduce the humidity in dung piles, judicious removal of manure, and use of selective pesticides (Axtell 1986, 1999).

Axtell (1991) constructed simulation models to evaluate the effect of different densities of *M. muscaedomesticae*, application of different insecticides, presence or absence of alternative prey and manure-removal on housefly population. Maintaining the manure as dry as possible through proper air ventilation, drainage and phased manure removal, were considered essential for the success of fly control. Stafford and Bay (1987) determined that *M. muscaedomesticae* preferred slightly drier areas than the third instar larvae of the housefly; most of the mites were found in the manure cones, where fly eggs were also more abundant.

Deleterious effect of bi-weekly and monthly manure removal on predatory mites, mainly macrochelids, was shown by Peck and Anderson (1970), Erõss and Mahunka (1971), Wills et al. (1990) and Mullens et al. (1996; 2001). The first authors noticed that at least 4 weeks were required before *M. muscaedomesticae* population could reach the densities found in older, unremoved manure. The authors suggested different manure management schedules to allow more effective predation of the housefly by macrochelids.

4.12 Possible Limitations

Axtell (1969) listed six aspects he considered as possible limitations in relation to the use of macrochelids for fly control. Among these were humidity requirements, population increase and predation rates, limited prey stages attacked (mostly eggs and first instar larvae) and susceptibility of macrochelid immatures to predation by other organisms.

Some studies suggest that egg structure may have an important bearing on the predation potential of macrochelids. The limited ability of *M. muscaedomesticae* to feed on stable fly eggs was attributed to their thicker chorion in comparison with housefly eggs (Kinn 1966).

All nine macrochelid species studied by Halliday and Holm (1987) were observed to consume eggs of the Australian bush fly, but only the larger species were able to feed on the eggs of the buffalo fly, probably due to the thicker chorion of the latter. In a subsequent experiment, the ineffective control of the buffalo fly in Australia by the introduced *M. pereginus* was attributed mostly to its preference for fly species that had softer egg chorion (Roth et al. 1988). Before any effort is initiated for the mass production of macrochelids for practical use, adequate selection of predators is essencial.

Success in biological control of a particular species of unwanted organism cannot be immediately deduced from laboratory tests, most often conducted under artificial conditions. This is exemplified by the observations by Kinn (1966), who reported effective predation of eggs and first instar larvae of black blow fly by *M. muscaedomesticae* under laboratory condition, but concluded that it was unlikely that it played an important role in the control of that insect, as *M. muscaedomesticae* is seldom found on carrion, the preferred habitat of the black blow fly.

Still another limitation to predation of fly eggs relates to macrochelid oviposition behavior and fly biological traits. *Musca sorbens* Wiedemann deposits clusters of several 100 eggs into fresh dung. These hatch quickly enough to escape predation by *M. muscaedomesticae*, which cannot colonise the droppings in time to attack most eggs. The most exposed eggs of the cluster can be attacked, but these become hard and dry, forming a barrier that protects the inner eggs of the cluster from predation (Toyama and Ikeda 1976).

In relation to the use of macrochelids for nematode control, possible limitations refer mainly to the relatively large size of these mites, physically preventing their movement deep into the soil. Most plant parasitic nematodes are found in association with roots, at least several centimeters below the soil surface. Parasitic nematodes are found on aerial parts of some low growing plants, but macrochelids have rarely been found on aerial plant parts (a *Macrocheles* species was recently found by T. Ranielle in the state of Pernambuco, in northeast Brazil, in flowers of *Heliconia bihai* L.), and their use in this type of habitat seems improbable.

Predation of macrochelids on annelids of the family Enchytraeidae could be seen as a limitation in the use of these predators, given that enchytraieds are considered beneficials for their natural role in the decomposition process of organic matter. However, the alternative control of pest species with agrochemicals may cause considerable mortality of enchytraeids. In addition, most macrochelids considered for practical use are found in patchy habitats, implying that they do not seem to have a close association with enchytraeids, of seemingly more uniform distribution in the environment.

The introduction of a macrochelid from a different region of the world warrants a careful evaluation of benefits and potential problems that could result from the introduction. However, the use of macrochelid species naturally occurring in the area where the pest is present, is not expected to cause major impact on enchytraeid population, especially in areas of traditional agriculture, where other factors (as the use of fertilizer) is already expected to reduce enchytraeid populations.

4.13 Suggested Complementary Studies

It is widely known that the success in biological control efforts of unwanted organisms is dependent upon adequate taxonomic and ecological knowledge of the species to be controlled as well as of the prospective biological control agent. Macrochelids have received limited attention in most countries, and their potential should still be considered to be limited (Krantz 1983a). However, researchers involved in the biological control of organisms that spend part of their life in discontinuous habitats, where these mites are most commonly found (especially fly species), might consider macrochelids to deserve special attention. The initial effort in these countries should concentrate on faunistic studies to enable correct determination of the species present, their population dynamics, and their relationships with other organisms.

A series of other more applied topics for research could then be established, involving the evaluation of groups of organisms to be targeted for control. One of these special cases refers to nematodes that are intestinal parasites of farm animals. These are periodically eliminated in the faeces of their hosts. The potential role of macrochelids in the control of these organisms has not been adequately explored. Given the reported suitability of many soil nematodes as food for macrochelids, it seems likely that these predators could have some effect on those nematodes, which have free-living stages that are taken up by new hosts as part of their biological cycle.

Another seemingly important topic for research refers to a possible limitation to the use of macrochelids due to intra-guild predation. Intra-guild predation involving organisms of the same genus has been mentioned in the literature. In the laboratory, competition between *M. robustulus* (Berlese) and *M. glaber* resulted in reduced densities of the latter when both were placed together in cow dung (Wallace and Holm 1984), despite its larger size. Such competition however, is not expected to occur in the field, given that while *M. robustulus* favors domestic settings, *M. glaber* favors pastoral settings. There are also cases of intra-guild predation involving macrochelids and other species of Mesostigmata. Axtell (1969) mentioned the predation of macrochelid larvae and protonymphs by species of Parasitidae. Preliminary experimental evidence supporting that statement was presented by Ignatowicz (1974), who observed predation of *M. glaber* immatures by all mobile stages of *Parasitus coleoptratorum* (L.). Thus, the presence of parasitids in the substrate might inhibit colonization by macrochelids.

However, it has also been suggested (Axtell 1962, 1963a) that the usual pattern of succession of mite groups in aging manure is characterized by the pioneer colonization by parasitids, followed by macrochelids and then other mite groups. While this assumed pattern seems to be the opposite of what was observed by Axtell (1969) and Ignatowicz (1974), additional studies on this subject are warranted, to determine the possible limitation of macrochelids by high abundance of parasitids.

Acknowledgements We are especially thankful to the following researchers who provided copies of the papers referred to in this publication: C.H.W. Flechtmann, C.C. Ho, G. Takaku, G.W. Krantz, M. Knapp and R. Abo-Shnaf.

References

Adis J, Krantz GW (1985) Notes on the natural history of *Macrocheles* (Acari: Gamasida: Macrochelidae) associated with 3-toed sloths *Bradypus* spp. (Edentata: Bradypodidae) in the Central Amazon. Zool Anz 214:22–224

Afifi AM (1989) Laboratory studies on the biology and feeding habitats of three macrochelid mites (Acari: Gamasida: Macrochelidae). Bull Entomol Soc Egypt 68:169–174

Al-Dulaimi SI (1986) Biological studies on the mite *Macrocheles muscaedomesticae* Acarina Macrochelida. J Biol Sci Res 1:329–338

Al-Dulaimi SI (2002) Predation by the mite *Macrocheles glaber* (Müller) (Acarina: Macrochelidae) on the house fly *Musca domestica* L. with some notes on its biology. Bull Iraq Nat Hist Mus 9:7–11

Almeida NW (1994) Caracterização de alguns parâmetros biológicos de *Macrocheles muscaedomesticae* (Scopoli, 1772) (Acari: Gamasida) associada às moscas sinantrópicas em granja de aves poedeiras de Monte-Mor, S.P. (Acarina: Mesostigmata: Macrochelidae). MS dissertation, Universidade Estadual de Campinas, Campinas, p 115

Anderson JR (1983) Mites as biological control agentes of dung-breeding pests: practical considerations and selection for pesticide resistance. In: Hoy MA, Cunningham GL, Knutson L (eds) Biological control of pests by mites, vol 3304. University of California, Berkeley, pp 99–102, Special publication

Athias-Henriot C (1968) L'appareil d'insémination laelapoïde (Acariens anactinotriches: Laelapoidea, ♀♀). Premières observations. Possibilité d'emploi à des fins taxonomiques. Bull Sci Bourgogne 25:229–274

Axtell RC (1961) New records of North American Macrochelidae (Acarina: Mesostigmata) and their predation rates on the house fly. Ann Entomol Soc Am 54:748

Axtell RC (1962) Macrochelidae (Acarina: Mesostigmata) inhabiting manure and their effect on house fly production. PhD thesis, Cornell University, Ithaca, p 151

Axtell RC (1963a) Effect of Macrochelidae (Acarina: Mesostigmata) on house fly production from dairy cattle manure. J Econ Entomol 56:317–321

Axtell RC (1963b) Acarina occurring in domestic animal manure. Ann Entomol Soc Am 56:628–633

Axtell RC (1964) Phoretic relationship of some common manure-inhabiting Macrochelidae (Acarina: Mesostigmata) to the house fly. Ann Entomol Soc Am 57:584–587

Axtell RC (1966) Comparative toxicities of insecticides to house fly larvae and *Macrocheles muscaedomesticae*, a mite predator of the house fly. J Econ Entomol 59:1128–1130

Axtell RC (1968) Integrated house fly control: populations of fly larvae and predaceous mites, *Macrocheles muscaedomesticae,* in poultry manure after larvicide treatment. J Econ Entomol 61:245–249

Axtell RC (1969) Macrochelidae (Acarina: Mesostigmata) as biological control agents for synanthropic flies. Proceedings of the 2nd international congress of acarology, Sutton Bonington. Akadamia Kiado Sutton Bonington, Budapest, 1967, pp 401–416

Axtell RC (1970) Fly control in caged-poultry houses: comparison of larviciding and integrated control programs. J Econ Entomol 63:1734–1737

Axtell RC (1986) Fly management in poultry production: cultural, biological, and chemical. Poult Sci 65:657–667

Axtell RC (1991) Role of mesostigmatid mites in integrated fly control. In: Dusbábek F, Bukva V (eds) Modern acarology, vol 2. Academia, Prague and SPB Academic Publishing bv, The Hague, pp 639–646

Axtell RC (1999) Poultry integrated pest management: status and future. Integr Pest Manag Rev 4:53–73

Axtell RC, Arends JJ (1990) Ecology and management of arthropod pests of poultry. Annu Rev Entomol 35:101–126

Axtell RC, Rultz DA (1986) Role of parasites and predators as biological fly control agents in poultry production facilities. In: Biological control of muscoid flies, vol 61. Entomological Society of America, Miscellaneous Publications, College Park, pp 88–100

Balogh J (1958) Macrocheliden aus Bulgarien (Acarina: Mesostigmata). Acta Entomol Mus Natl Prague 32:247–256

Beaulieu F, Weeks AR (2007) Free-living mesostigmatic mites in Australia: their roles in biological control and bioindication. Aust J Exp Agr 47:460–478

Behan-Pelletier VM (2003) Acari and Collembola biodiversity in Canadian agricultural soils. Can J Soils Sci 83:279–288

Blaszak C, Ehrnsberger R, Schuster R (1990) Beiträge zur Kenntnis der Lebensweise der Litoralmilbe *Macrocheles superbus* Hull,1918 (Acarina: Gamasina). Osnabrücker Naturwissenschaftliche Mitt 16:51–62

Borden EER (1989) The phoretic behavior and olfactory preference of *Macrocheles muscaedomesticae* (Scopoli) (Acarina: Macrochelidae) in its relationship with *Fannia canicularis* (L.) (Diptera: Muscidae). Pan Pac Entomol 65:89–96

Braig HR, Perotti MA (2009) Carcases and mites. Exp Appl Acarol 49:45–84

Bregetova NG (1977) Sem. Macrochelidae. In: Bregetova NG, Vaînshteîn BA, Kadite BA, Koroleva EV, Petrova AD, Tikhomirov SI, Shcherbak GI (eds) Opredelitel obitayushchikh v pochve kleshchei, Mesostigmata. Izdatelstvo Nauka, Leningrad, pp 621–632

Bregetova NG, Koroleva EV (1960) The macrochelid mites (Gamasoidea: Macrochelidae) in the USSR. Akademii Nauk SSSR Zoolog Inst Parazitologicheskii Sbornik 29:32–154

Canestrini G (1881) Osservazioni intorno al genere Gamasus. Atti del Reale Istituto Veneto di Scienze, Lettere ed Arti, 7, serie 5. Apud G. and R. Canestrini (1882)

Cavalcante AC (2014) Ácaros predadores da família Phytoseiidae (Acari: Mesostigmata) para o controle do biótipo B da mosca-branca *Bemisia tabaci* (Gennadius) (Hemiptera: Aleyrodidae) no Brasil. PhD thesis, Universidade de São Paulo/USP, Piracicaba, p 93

Cicolani B (1978) The influence of temperature on the speed of development, from egg to adult, of *Macrocheles matrius* (Acarina: Mesostigmata). Acarologia 19:384–394

Cicolani B (1979) The intrinsic rate of natural increase in dung macrochelid mites, predators of *Musca domestica* eggs. Boll Zool 46:171–178

Cicolani B (1983) Foresia tra nematodi rabditidi, acari macrochelidi e scarabei coprofagi: una originale strategia ecologica. Atti XIII Congresso Nazionale Italiano di Entomologia, Torino, pp 629–634

Cicolani B, Passariello S, Petrelli G (1977) Influenza dela temperatura su l'incremento di popolazione in *Macrocheles subbadius* (Acarina: Mesostigmata). Acarologia 19:563–578

Coons LB, Axtell RC (1973) Sensory setae of the first tarsi and palps of the mite *Macrocheles muscaedomesticae*. Ann Entomol Soc Am 66:539–544

Costa M (1966) Notes on macrochelids associated with manure and coprid beetles in Israel. I. *Macrocheles robustulus* (Berlese, 1904), development and biology. Acarologia 8:532–548

Costa M (1967) Notes on macrochelids associated with manure and coprid beetles in Israel. II. Three new species of the *Macrocheles pisentii* complex, with notes on their biology. Acarologia 9:304–329

Doube BM, Macqueen A, Huxham KA (1986) Aspects of predatory activity of *Macrocheles peregrinus* (Acarini: Acrochelidae [sic]) on two species of *Haematobia* fly (Diptera: Muscidae). In: Biological control of muscoid flies, vol 61. Entomological Society of America, Miscellaneous Publications, College Park, pp 132–141

Eickwort GC (1983) Potential use of mites as biological control agents of leaf-feeding insects. In: Hoy MA, Cunningham GL, Knutson L (eds) Biological control of pests by mites, vol 3304. University of California, Berkeley, pp 41–52, Special publication

Emberson RM (2010) A reappraisal of some basal lineages of the family Macrochelidae, with the description of a new genus (Acarina: Mesostigmata). Zootaxa 2501:37–53

Erõss J, Mahunka S (1971) Investigations on coprophilous and stercoricolous macrochelids (Acari, Gamasina) in Hungary, as possible agents in the control of synanthropous flies. Parasitol Hung 4:215–226

Estébanes-González ML, Cervantes FA (2009) Mites and ticks associated with some small mammals in Mexico. Int J Acarol 31:23–37

Evans GO (1956) On the classification of the family Macrochelidae with particular reference to the subfamily Parholaspinae (Acarina: Mesostigmata). Proc Zool Soc 127:345–377

Evans GO (1963) Observations on the chaetotaxy of the legs in the free-living Gamasina (Acari: Mesostigmata). Bull Br Mus (Nat Hist) Zool 10:277–303

Evans GO, Browning E (1956) British mites of the subfamily Macrochelinae Trägårdh (Gamasina-Macrochelidae). Bull Br Mus (Nat Hist) Zool 4:1–55

Evans GO, Hyatt KH (1963) Mites of the genus *Macrocheles* Latr. (Mesostigmata) associated with coprid beetles in the collections of the British Museum (Natural History). Bull Br Mus (Nat Hist) Zool 9:327–401

Farish DJ, Axtell RC (1966) Sensory functions of the palps and first tarsi of *Macrocheles muscaedomesticae* (Acarina: Macrochelidae), a predator of the house fly. Ann Entomol Soc Am 59:165–170

Farish DJ, Axtell RC (1971) Phoresy redefined and examined in *Macrocheles muscaedomesticae* (Acarina: Macrochelidae). Acarologia 13:16–29

Filipponi A (1955) Sulla natura dell'associazione tra *Macrocheles muscaedomesticae* e *Musca domestica*. Riv Parassitol 16:84–102

Filipponi A (1959) Due specie riproduttivamente isolate sotto il nome di *Macrocheles gabler* (Mueller) (Acarina: Mesostigmata). Riv Parassitol 20:199–212

Filipponi A (1964) The feasibility of mass producing macrochelid mites for field trials against houseflies. Bull World Health Organ 31:499–501

Filipponi A (1971) Influence of temperature on population increase of four dung-living macrochelid mites within optimum range. In: Daniel M, Rosický B (eds) Proceedings of the 3rd international congress of acarology, 31 Aug–6 Sept 1971. S. Junk, The Hague and Academia, Prague, pp 775–779

Filipponi A, Cicolani B (1974) Influenza della temperatura sulla fecondita' longevita' e capacita' moltiplicativa nello intervallo ottimale di *Macrocheles matrius* (Acarina: Mesostigmata). Riv Parassitol 35:291–306

Filipponi A, Delupis GD (1963) Sul regime dietetico di alcuni Macrochelidae (Acari: Mesostigmata), associati in natura a muscidi di interesse sanitario. Riv Parassitol 24:277–287

Filipponi A, Francaviglia G (1964) Larviparità facoltativa in alcuni macrochelidi (Acari: Mesostigmata) associati a muscidi di interesse sanitario. Parassitologia 6:99–113

Filipponi A, Ilardi PA (1959) Alcuni dati sulla biologia di *Macrocheles insignitus*, Berl. (Acarina: Mesostigmata). Riv Parassitol 20:79–90

Filipponi A, Mosna B (1968) Influenza di fattori ecologici e genetici sulla natalitá e mortalitá di *Macrocheles robustulus* (Berlese, 1904). Ann Is Super Sanita (Roma) 4:551–571

Filipponi A, Petrelli MG (1967) Autoecologia e capacita' moltiplicativa di *Macrocheles muscaedomesticae* (Scopoli) (Acari: Mesostigmata). Riv Parassitol 28:129–156

Filipponi A, Petrelli G (1969) Sulla capacità moltiplicativa di *Macrocheles perglaber* (Acarina: Mesostigmata) in condizioni di laboratorio. Parassitologia 11:1–4

Filipponi A, Mosna B, Petrelli G (1971) L'ottimo di temperatura di *Macrocheles muscaedomesticae* (Scopoli), come attributo di popolazione. Riv Parassitol 32:193–218

Geden CJ, Axtell RC (1988) Predation by *Carcinops pumilio* (Coleoptera: Histeridae) and *Macrocheles muscaedomesticae* (Acarina: Macrochelidae) on the house fly (Diptera: Muscidae): functional response, effects of temperature, and availability of alternative prey. Environ Entomol 17:739–744

Geden CJ, Stinner RE, Axtell RC (1988) Predation by predators of the house fly in poultry manure: effects of predator density, feeding history, interspecific interference, and field conditions. Environ Entomol 17:320–329

Geden CJ, Stinner RE, Kramer DA, Axtell RC (1990) MACMOD: a simulation model for *Macrocheles muscaedomesticae* (Acari: Macrochelidae) population dynamics and rates of predation on immature house flies (Diptera: Muscidae). Environ Entomol 19:578–586

Gerson U, Smiley RL, Ochoa R (2003) Mites (Acari) for pest control. Blackwell Science Ltd, London, p 539

Glida H, Bertrand M, Peyrusse V (2003) A limiting factor in the abundance of predatory phoretic mites (Acari: Macrochelidae): the seasonal abundance of their phorionts (dung beetles) in Southern France. Can J Zool 81:2066–2072

Greenberg B, Carpenter PD (1960) Factors in phoretic association of a mite and a fly. Science 132:738–739

Grosman A, Messelink G, Groot E (2011) Combined use of a mulch layer and the soil-dwelling predatory mite *Macrocheles robustulus* (Berlese) enhance the biological control of sciarids in potted plants. IOBC/WPRS Bull 68:51–54, Integrated control in protected crops, temperate climate

Guanilo A, Cordero J (2003) Comportamiento depredador de *Macrocheles muscaedomesticae* (Scopoli) (Acarina: Macrochelidae) sobre huevos de *Musca domestica* Linnaeus (Diptera: Muscidae). Rev Peruana Entomol 43:137–142

Halliday RB (1986) Mites of the genus *Glyptholaspis* Filipponi and Pegazzano (Acarina: Macrochelidae) in Australia. J Aust Entomol Soc 25:71–74

Halliday RB (1987) Further observations on the dorsal idiosomal chaetotaxy in the Macrochelidae (Acarina). Int J Acarol 13:51–53

Halliday RB (1993) Two new species of *Macrocheles* from Australia (Acarina: Mesostigmata: Macrochelidae). Aust Entomol 20:99–106

Halliday RB (2000) The Australian species of *Macrocheles* (Acarina: Macrochelidae). Invert Taxon 14:273–326

Halliday RB, Holm E (1987) Mites of the family Macrochelidae as predators of two species of dung-breeding pest flies. Entomophaga 32:333–338

Hassan MF, Ali FS, Hussein AM, Mahgoub MH (2002) Biological studies on *Macrocheles muscaedomesticae* Scopoli fed on different stages of potato tuber moth, *Phthorimaea operculella* Zeller. Egypt J Biol Pest Control 12:43–46

Ho CC (1989) Studies on the biology of *Macrocheles muscaedomesticae* (Scopoli) (Acarina: Macrochelidae). Chin J Entomol 3:181–187, Special publication

Ho TM (1990) Phoretic association between *Macrocheles muscaedomesticae* (Acari: Macrochelidae) and flies inhabiting poultry manure in peninsular Malaysia. Exp Appl Acarol 10:61–68

Ho CC, Cromroy HL, Patterson RS (1990) Mass production of the predaceous mite, *Macrocheles muscaedomesticae* (Scopoli) (Acarina: Macrochelidae), a predator of the house fly. In: Rutz DA, Patterson RS (eds) Biocontrol of arthropods affecting livestock and poultry. Westview Press, Colorado, pp 201–213

Hunter PE, Rosario RMT (1988) Associations of Mesostigmata with other arthropods. Annu Rev Entomol 33:393–417

Hyatt KH, Emberson RM (1988) A review of the Macrochelidae (Acari: Mesostigmata) of the British Isles. Bull Br Mus (Nat Hist) Zool 54:63–125

Ignatowicz S (1974) Inter and intraspecific competition in mites: *Parasitus coleoptratorum* (L.) and *Macrocheles glaber* (Müller). In: Edward P (ed) Proceedings of 4th international congress acarol saalfelden (Austria). Akademiai Kiado, Budapest, 1979, pp 623–625

Ito Y (1971) Predation by manure-inhabiting mesostigmatids (Acarina: Mesostigmata) on some free-living nematodes. Appl Entomol Zool 6:51–56

Ito Y (1973) The effects of nematode feeding on the predatory efficiency for house fly eggs and reproduction rate of *Macrocheles muscaedomesticae* (Acarina: Mesostigmata). Japan Soc Med Entomol Zool 23:209–213

Ito Y (1977) Predatory activity of mesostigmatid mites (Acarina: Mesostigmata) for house fly eggs and larvae under feeding of nematodes. Japan Soc Med Entomol Zool 28:167–173

Jalil M, Rodriguez JG (1970) Studies of behaviour of *Macrocheles muscaedomesticae* (Acarina: Macrochelidae) with emphasis on its attraction to the house fly. Ann Entomol Soc Am 63:738–744

Karg W (1993) Einzelbehandlung der Familien mit Bestimmungsschlüsseln der Gattungen und Arten sowie Autökologie der Arten. In: Karg W (ed) Acari (Acarina), Milben Parasitiformes (Anactinochaeta) Cohors Gamasina Leach. Gustav Fischer Verlag Jena, New York

Kinn DN (1966) Predation by the mite, *Macrocheles muscaedomesticae* (Acarina: Macrochelidae), on three species of flies. J Med Entomol 3(2):155–158

Kinn DN, Witcoskjy JJ (1977) The life cycle and behaviour of *Macrocheles boudreauxi* Krantz. Z Angew Entomol 84:136–144

Koehler HH (1999) Predatory mites (Gamasina: Mesostigmata). Agr Ecosyst Environ 74:395–410

Kramer P (1876) Zur Naturgeschichte Einiger Gattungen aus der Famillie der Gamasiden. Arch Naturgeschichte 42:46–105

Krantz GW (1960) A re-evaluation of the Parholaspinae Evans, 1956 (Acarina: Mesostigmata: Macrochelidae). Acarologia 2:293–433

Krantz GW (1962) A review of the genera of the family Macrochelidae Vitzthum, 1930 (Acarina: Mesostigmata). Acarologia 4:143–173

Krantz GW (1967) A review of the genus *Holocelaeno* Berlese, 1910 (Acarina: Macrochelidae). Acarologia Suppl 9:1–89

Krantz GW (1983a) Mites as biological control agents of dung-breeding flies, with special reference to the Macrochelidae. In: Hoy MA, Cunningham GL, Knutson L (eds) Biological control of pests by mites, vol 3304. University of California, Berkeley, pp 91–98, Special publication

Krantz GW (1983b) Three new species of *Macrocheles* (Acari: Gamasida: Macrochelidae) associated with 3-toed sloths, *Bradypus* spp. (Edentata: Bradypodidae), in Brazil and Surinam. Acarologia 24:3–12

Krantz GW (1991) Nature oh the association between *Pisentii*-group mites (Acari: Macrochelidae: *Macrocheles*) and dung beetles of the genus *Scarabaeus* (Coleoptera: Scarabaeidae) in southern France. Acarologia 32:3–11

Krantz GW (1998) Reflections on the biology, morphology and ecology of the Macrochelidae. Exp Appl Acarol 22:125–137

Krantz GW, Ainscough BD (1990) Acarina: Mesostigmata (Gamasida). In: Dindal DL (ed) Soil biology guide. Wiley, New York, pp 583–666

Krantz GW, Redmond BL (1987) Identification of glandular and poroidal idionotal systems in *Macrocheles perglaber* F. and P. (Acari: Macrochelidae). Exp Appl Acarol 3:243–253

Krantz GW, Redmond BL (1989) On the structure and function of the cribrum, with special reference to *Macrocheles perglaber* (Gamasida: Macrochelidae). In: Channabosavanna GE, Viraktamath CA (eds) Progress in acarology, vol 1. Oxford/IBH Publishing, New Delhi, pp 179–185

Krantz GW, Royce LA (1994) Observations on the biology and behavior of *Macrocheles mycotrupetes* Krantz and Mellott (Acari: Macrochelidae). Int J Acarol 20:115–121

Krantz GW, Wernz JG (1979) Sperm transfer in *Glyptholaspis americana*. In: Rodriguez JG (ed) Recent advances in acarology, vol 2. Academic, New York, pp 441–446

Krantz GW, Whitaker JO Jr (1988) Mites of the genus *Macrocheles* (Acari: Macrochelidae) associated with small mammals in North America. Acarologia 29:225–259

Krauss W (1970) Die europäischen Arten der Gattungen *Macrocheles* Latreille 1829 und *Geholaspis* Berlese 1918. (Eine systematische Studie aus dem Jahr 1960). Acarol Schriftenr Vergleichende Milbenkunde 14:1–43

Lesna I, Sabelis MW, Bolland HR, Conijn CGM (1995) Candidate natural enemies for control of *Rhizoglyphus robini* Claparede (Acari: Astigmata) in lily bulbs: exploration in the field and pre-selection in the laboratory. Exp Appl Acarol 19:655–669

Lindquist EE, Evans GO (1965) Taxonomic concepts in the Ascidae, with a modified setal nomenclature for the idiosome of the Gamasina (Acarina: Mesostigmata). Mem Entomol Soc Can 47:1–47

Lindquist EE, Krantz GW, Walter DE (2009) Order Mesostigmata. In: Krantz GW, Walter DE (eds) A manual of acarology. Texas Tech University Press, Lubbock, pp 124–232

Manning MJ, Halliday RB (1994) Biology and reproduction of some Australian species of Macrochelidae (Acarina). Aust Entomol 21:89–94

Mašán P (2003) Macrochelid mites of Slovakia (Acari: Mesostigmata: Macrochelidae). Institute of Zoology, Slovak Academy of Sciences, Bratislava, p 149

Mattos MR (1992) Macroquelídeos associados a fezes acumuladas em granja de aves poedeiras do município de montemor, Estado de São Paulo: levantamento, taxonomia e estudos populacionais (Acari, Gamasida: Macrochelidae). PhD thesis, Universidade Estadual de Campinas/Unicamp, Campinas, p 78

Messelink GJ, De Kogel WJ (2005) Impact of chrysanthemum cultivar, fertilization and soil-dwelling predatory mites on *Frankliniella occidentalis*. Proc Neth Entomol Soc Meet 16:101–107

Messelink G, van Holstein-Saj R (2008) Improving thrips control by the soil-dwelling predatory mite *Macrocheles robustulus* (Berlese). IOBC/WPRS Bull 32:135–138, Integrated control in protected crops, temperate climate

Moraza ML (2004) The phoretic genus *Neopodocinum* (Oudemans, 1902) in the Iberian Peninsula (Acari: Mesostigmata: Macrochelidae). Rev Iberica Arachnol 10:261–269

Mullens BA, Hinkle NC, Szijj CE (1996) Impact of alternating manure removal schedules on pest flies (Diptera: Muscidae) and associated predators (Coleoptera: Histeridae, Staphylinidae; Acarina: Macrochelidae) in caged-layer poultry manure in Southern California. Biol Microb Control 89:1406–1417
Mullens BA, Hinkle NC, Szijj CE, Kuney DR (2001) Managing manure and conserving predators helps control flies in caged-layer poultry systems. Calif Agric 55:26–30
Niogret J, Lumaret JP, Bertrand M (2006) Review of the phoretic association between coprophilous insects and macrochelid mites (Acari: Mesostigmata) in France. Elytron 20:99–121
Norton RA, Kethley JB, Johnston DE, O'Connor BM (1993) Phylogenetic perspectives on genetic systems and reproductive modes of mites. In: Wrensch DL, Ebbert MA (eds) Evolution and diversity of sex ratio in insects and mites. Chapman and Hall, New York, pp 8–99
O'Donnell AE, Axtell RC (1964) Predation by *Fuscuropoda vegetans* (Acarina: Uropodidae) on the house fly (*Musca domestica*). Ann Entomol Soc Am 58:403–404
O'Donnell AE, Nelson EL (1967) Predation by *Fuscuropoda vegetans* (Acarina: Uropodidae) and *Macrocheles muscaedomesticae* (Acarina: Macrochelidae) on the eggs of the little house fly, *Fannia canicularis*. J Kansas Entomol Soc 40:441–443
Oliver JH Jr, Krantz GW (1963) *Macrocheles rodriguezi*, a new species of mite from Kansas (Acarina: Macrochelidae) with notes on its life cycle and behavior. Acarologia 5:519–525
Peck JH, Anderson JR (1970) Influence of poultry-manure-removal schedules on various diptera larvae and selected arthropod predators. J Econ Entomol 63:82–90
Pereira C, Castro MP (1945) Contribuição para o conhecimento da espécie tipo de "Macrocheles Latr". ("Acarina"): "M. muscaedomesticae (Scopoli, 1772)" Emend. Arq Inst Biol 16:153–186
Pereira C, Castro MP (1946) Cultura de ácaros e de alguns outros aracnídeos predadores. In: Imprensa oficial do estado de São Paulo, Livro comemorativo de Romulo Ferreira de Almeida, vol 33. pp 285–288. São Paulo, SP
Pereira C, Castro MP (1947) Forese e partenogênese arrenótoca em "*Macrocheles muscaedomesticae*" (Scopoli) ("Acarina: Macrochelidae") e sua significação ecológica. Arq Inst Biol 18:71–89
Perotti MA (2001) Prey location and predation rates of predatory mites (Acari: Macrochelidae) on immature stages of pest flies (Diptera: Muscidae). Syst Appl Acarol 6:27–33
Perotti MA, Braig HR (2009) Phoretic mites associated with animal and human decomposition. Exp Appl Acarol 49:85–124
Petrova AD (1960) Materialien über den Bau des inneren Sackchens des Receptaculum seminis der gamasoiden Milben Macrochelidae Vitzt. Zool Anz 165:393–400
Polak M (1996) Ectoparasitic effects on host survival and reproduction: the *Drosophila-Macrocheles* association. Ecology 77:1379–1389
Ramakers PMJ, van Lieburg MJ (1982) Start of commercial production and introduction of *Amblyseius mckenzie* Sch. and Pr. (Acarina: Phytoseiidae) for the control of *Thrips tabaci* Lind. (Thysanoptera: Thripidae) in glasshouses. Med Van de Faculteit Landbouwwetenschappen 47:541–545
Rettenmeyer CW (1962) Notes on host specificity and behavior of myrmecophilous macrochelid mites. J Kansas Entomol Soc 35:358–360
Rodrigueiro TSC, Prado AP (2004) *Macrocheles muscaedomesticae* (Acari: Macrochelidae) and a species of *Uroseius* (Acari: Polyaspididae) phoretic on *Musca domestica* (Diptera: Muscidae): effects on dispersal and colonization of poultry manure. Iheringia Sér Zool 94:181–185
Rodriguez JG, Wade CF (1961) The nutrition of *Macrocheles muscaedomesticae* in relation to its predatory action on the house fly egg. Ann Entomol Soc Am 54:782–788
Rodriguez JG, Wade CF, Wells CN (1962) Nematodes as a natural food for *Macrocheles muscaedomesticae* (Acarina: Macrochelidae), a predator of the house fly egg. Ann Entomol Soc Am 55:507–511
Rodriguez JG, Singh P, Taylor B (1970) Manure mites and their role in fly control. J Med Entomol 7:335–341
Roth JP, MacQueen A, Bay DE (1988) Predation by the introduced phoretic mite, *Macrocheles peregrinus* (Acari: Macrochelidae), on the buffalo fly, *Haematobia irritans exigua* (Diptera: Muscidae), in Australia. Environ Entomol 17:603–607

Royce LA, Krantz GW (1991) A new rearing method for nematophagous mites. In: Dusbábek F, Bukva V (eds) Modern acarology, vol 2. Academia, Prague and SPB Academic Publishing, The Hague, pp 619–622

Sabelis MW (1985) Reproduction. In: Helle W, Sabelis MW (eds) Spider mites. Their biology, natural enemies and control, vol 1B, Natural enemies of the Tetranychidae. World Crop Pests/ Elsevier, Amsterdam, pp 73–82

Saito M, Takaku G (2013) Predation of *Tyrophagus similis* Volgin (Acari: Acaridae) by indigenous predatory mites (Acari: Gamasina) found in spinach fields. J Acarol Soc Japan 22:37–43

Scopoli JA (1772) Annus V. Historico naturalis. I. Emendationes et Additamenta ad Ann. I. II. III. IV. II. Tentamen Mineralogicum I. De Minera Argenti alba. III. Tentamen Mineralogicum II. De Sulphure. IV. Tentamen Mineralogicum III. De Pseudogalena, Auripigmento, aliisque. V. Observationes Zoologicae. Lipsiae Chr G Hilscher 5:128

Singh P, Rodriguez JG (1966) Food for Macrochelid mites (Acarina) by an improved method for mass rearing of a nematode, *Rhabditella leptura*. Acarologia 8:549–550

Singh P, King WE, Rodriguez JG (1966) Biological control of muscids as influenced by host preference of *Macrocheles muscaedomesticae* (Acarina: Macrochelidae). J Med Entomol 3:78–81

Singh P, McEllistrem MT, Rodriguez JG (1967) The response of some macrochelids to temperature and humidity (Acarina: Macrochelidae). Acarologia 9:1–20

Soliman ZR, Zaher MA, Mohamed MI (1978) Biology and predaceous efficiency of *Macrocheles matrius* Hull Acari, Mesostigmata. Z Angew Entomol 85:225–230

Stafford KC, Bay DE (1987) Dispersion pattern and association of house fly, *Musca domestica* (Diptera: Muscidae), larvae and both sexes of *Macrocheles muscaedomesticae* (Acari: Macrochelidae) in response to poultry manure moisture, temperature, and accumulation. Environ Entomol 16:159–164

Toyama GM, Ikeda JK (1976) An evaluation of fly predators at animal farms on Leeward and central Oahu. Proc Hawaiian Entomol Soc 22:369–379

Valle A (1953) Revisione di generi e sottogeneri Berlesiani di Acari. (Primo contributo). Redia 38:316–360

Van der Hammen L (1964) The morphology of *Glyptholaspis confusa* (Foà, 1900) (Acarida: Gamasina). Zool Verh 71:1–56

Vitzthum HG (1930) Acarologische Beobachtungen (14. Reihe). Zool Jahrb Abt Syst 59:281–350

Wade CF, Rodriguez JG (1961) Life history of *Macrocheles muscaedomesticae* (Acarina: Macrochelidae), a predator of the house fly. Ann Entomol Soc Am 54:776–781

Wallace MMH, Holm E (1983) Establishment and dispersal of the introduced predatory mite, *Macrocheles peregrinus* Krantz, in Australia. J Aust Entomol Soc 22:345–348

Wallace MMH, Holm E (1984) Competition between species of predatory macrochelid mites in the cow-dung pad. In: Griffiths DA, Bowman CE (eds) Acarology VI, vol 1. Ellis Horwood, Chichester, pp 581–584

Wallace MMH, Tyndale-Biscoe M, Holm E (1979) The influence of *Macrocheles glaber* on the breeding of the Australian bushfly, *Musca vetustissima* in cow dungs. In: Rodriguez JG (ed) Recent advances in acarology, vol 2. Academic, New York, pp 217–222

Wallwork JH, Rodriguez JC (1963) The effect of ammonia oh the predation rate of *Macrocheles muscaedomesticae* on house fly eggs. In: Naegele JA (ed) Advances in acarology, vol 1. Cornell University Press, Ithaca, pp 60–69

Walter DE, Ikonen EK (1989) Species, guilds, and functional groups: taxonomy and behavior in nematophagous arthropods. J Nematol 21:315–327

Whitaker JO (1977) Food and external parasites of the Norway rat, *Rattus norvegicus*, in Indiana. Proc Indiana Acad Sci 86:193–198

Whitaker JO, Spicka EJ (1976) Ectoparasites of squirrels of the genus *Sciurus* from Indiana. Proc Indiana Acad Sci 85:431–436

Wicht MC, Rodriguez JG (1970) Integrated control of muscid flies in poultry houses using predator mites, selected pesticides and microbial agentes. J Med Entomol 7:687–692

Wicht MC, Rodriguez JG, Smith WT, Jalil M (1971) Attractant to *Macrocheles muscaedomesticae* (Acarina) present in the housefly, *Musca domestica*. J Insect Physiol 17:63–67

Willis RR, Axtell RC (1968) Mite predators of the house fly: a comparison of *Fuscuropoda vegetans* and *Macrocheles muscaedomesticae*. J Econ Entomol 61:1669–1674

Wills LE, Mullens BA, Mandeville JD (1990) Effects of pesticides on filth fly predators (Coleoptera: Histeridae, Staphylinidae; Acarina: Macrochelidae, Uropodidae) in caged layer poultry manure. J Econ Entomol 83:451–457

Womersley H (1955) The acarina fauna of mutton birds' nests on a Bass Strait island. Aust J Zool 3:412–414

Yasui Y (1988) Sperm competition of *Macrocheles muscaedomesticae* (Scopoli) (Acarina: Mesostigmata: Macrochelidae), with special reference to precopulatory mate guarding behavior. J Ethol 6:83–90

Yasui Y (1992) The existence of sex pheromone eliciting precopulatory mate guarding behavior in a male predatory mite, *Macrocheles muscaedomesticae* (Scopoli). J Ethol 10:15–19

Yasui Y (1996) Males of a mite, *Macrocheles muscaedomesticae*, estimate a female's value on the basis of her age and reproductive status. J Insect Behav 9:517–524

Chapter 5
The Phytoseiidae (Acari: Mesostigmata) as Biological Control Agents

James A. McMurtry, Nazer Famah Sourassou, and Peterson Rodrigo Demite

5.1 Introduction

The Phytoseiidae have been intensely studied for nearly six decades because of their recognized potential as biological control agents of mites as well as of thrips and whiteflies. In most ecosystems the family is first represented among all families of acarine predators, as the prime exploiter of the foliage habitat (McMurtry 2010).

Although much less frequently, the phytoseiids are also found in the soil. It is believed that the "protophytoseiids" lived under tree bark or in the soil/humus habitat (McMurtry 2010).

J.A. McMurtry (✉)
Department of Entomology, University of California, Riverside,
P.O. Box 4487, Sunriver, Oregon 97707, USA
e-mail: mcmurtry921@gmail.com

N.F. Sourassou
Departamento de Entomologia e Acarologia, Escola Superior de Agricultura "Luiz de Queiroz", Universidade de São Paulo, 13418-900 Piracicaba, SP, Brazil
e-mail: sfamah@yahoo.com

P.R. Demite
Departamento de Entomologia e Acarologia, Escola Superior de Agricultura "Luiz de Queiroz", Universidade de São Paulo, 13418-900 Piracicaba, SP, Brazil

Instituto Federal Goiano, câmpus Urutaí, Urutaí, GO 75790-000, Brazil
e-mail: peterson_demite@yahoo.com.br

D. Carrillo et al. (eds.), *Prospects for Biological Control of Plant Feeding Mites and Other Harmful Organisms*, Progress in Biological Control 19,
DOI 10.1007/978-3-319-15042-0_5

5.2 Taxonomy

Taxonomically, the phytoseiids are placed in the order Mesostigmata, cohort Gamasina, and superfamily Phytoseioidea (Lindquist et al. 2009), together with their closest relatives the Blattisociidae, Otopheidomenidae and Podocinidae, all having a "phytoseiid-type" spermathecal apparatus (Krantz and Walter 2009).

The first phytoseiid species to be described were *Amblyseius obtusus* (Koch) and *Transeius similis* (Koch) (Koch 1839). Other species described by Koch (1839) and mentioned in the literature at some point in time as phytoseiids, such as *Gamasus vepallidus*, *Zercon ovalis* and *Zercon pallens*, are currently considered *incertae sedis*. Up to the first half of the twentieth century, only 45 phytoseiid species had been described (Fig. 5.1). The number of described species has increased tremendously since then, given the initial reports about their possible use as biological control agents (Huffaker et al. 1969, 1970; Flaherty and Huffaker 1970). These reports generated increased interest in the study of the biology and ecology of the phytoseiids. Thus, the number of taxonomic studies of this family has increased considerably from the 1950s (Fig. 5.2).

An update of the taxonomic catalogs of the phytoseiids published by Moraes et al. (1986) and Moraes et al. (2004) was recently made available on the internet by the taxonomy team of ESALQ-USP and collaborators (Demite et al. 2014b; http://www.lea.esalq.usp.br/phytoseiidae). About 1850 papers were reported in that work, dealing with phytoseiid taxonomy and distribution. By April 2014, a total of 2,709 species had been included in that database, of which 2,436 are valid.

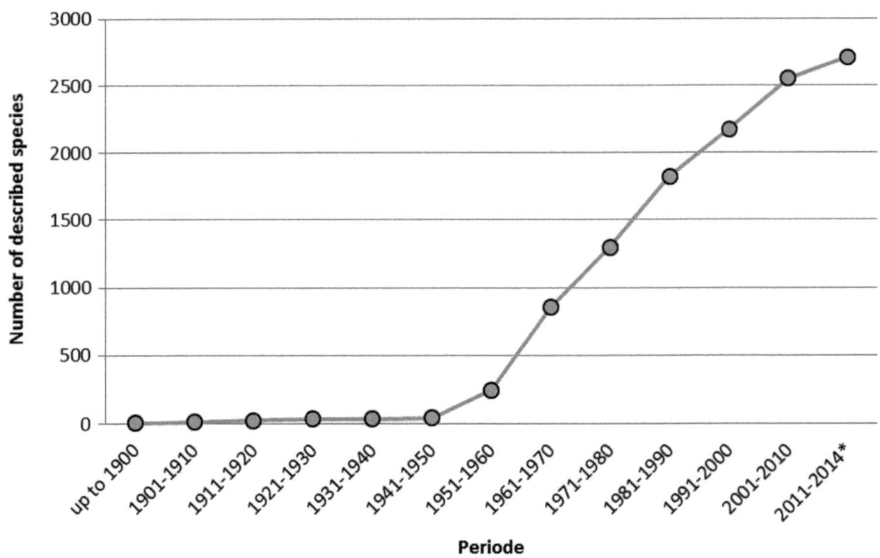

Fig. 5.1 Cumulative number of scientific papers on phytoseiid taxonomy (Based on Demite et al. 2014b). *Up to April 2014

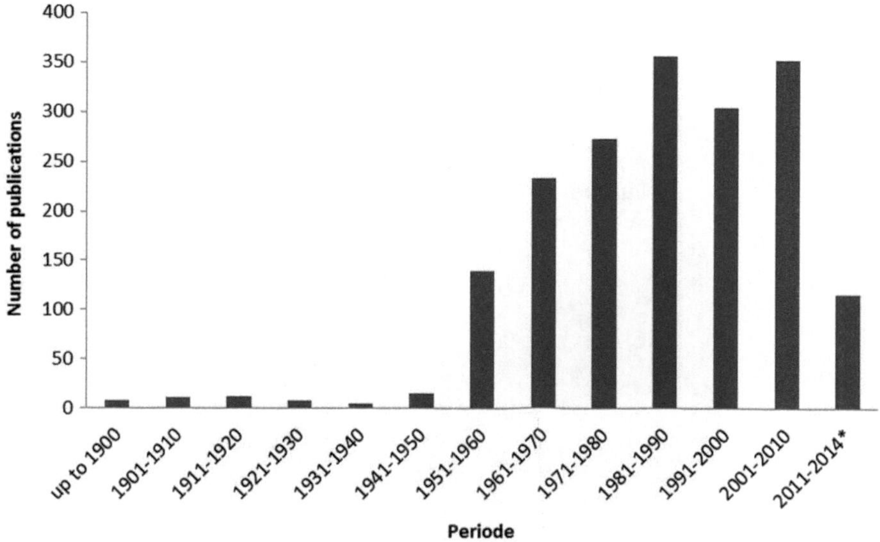

Fig. 5.2 Cumulative number of described phytoseiid species (Based on Demite et al. 2014b). *Up to April 2014

Phytoseiidae is organized in three subfamilies containing altogether 91 genera: Amblyseiinae (1,748 species in 65 genera), Phytoseiinae (229 species in 3 genera) and Typhlodrominae (732 species in 23 genera) (Demite et al. 2014a, b). The family is divided in three subfamilies (Amblyseiinae, Phytoseiinae and Typhlodrominae). Amblyseiinae is divided in nine tribes (Afroseiulini, Amblyseiini, Euseiini, Indoseiulini, Kampimodromini, Macroseiini, Neoseiulini, Phytoseiulini and Typhlodromipsini) and Typhlodrominae in six tribes (Chanteiini, Galendromimini, Metaseiulini, Paraseiulini, Typhlodromini and Typhloseiopsini) (Chant and McMurtry 2007). The Phytoseiinae are not divided in tribes. In descending order, *Typhlodromus*, *Amblyseius*, *Neoseiulus*, *Phytoseius*, *Euseius* and *Proprioseiopsis* are the most speciose genera. Fifty-one other genera contain between 2 and 93 species, whereas the remaining 34 genera are monotypic, i.e. each contains a single described species (Demite et al. 2014a, b).

Phytoseiid species have been recorded from all continents except Antarctica. The countries with the highest number of reported species are USA, China, India, Brazil and Pakistan (see Demite et al. 2014a). However, most countries have less than 15 reported species and for many countries there are no records.

The presently high rate of descriptions of new species and genera in countries of Central and South America, Africa and Asia (Tixier et al. 2012) indicate that the diversity of this family is much greater than presently known. Asia is the continent from which the largest number of species have been described so far (857 = 35 %), followed by the Americas (720 = 30 %) and Africa (420 = 17 %) (Fig. 5.3). At the country level, the USA ranks highest in number of species described, followed by China, Pakistan and India (Table 5.1).

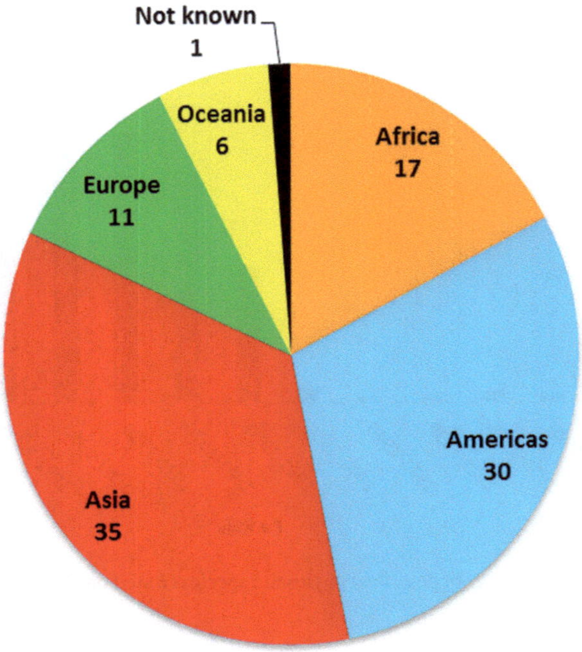

Fig. 5.3 Proportion (%) of valid phytoseiid species described per continent (Based on Demite et al. 2014b)

Phytoseiids are primarily plant-inhabiting mites, with more than 2,077 species (representing about 85 % of species) described from diverse plant species. The other 150 species (6 %) have been described from soil, humus, litter, and moss, and 25 species (1 %) in association with vertebrates and invertebrates (Table 5.2). There was no information regarding host or habitat of the remaining 184 species (7 %). Although a small proportion of phytoseiid species have been described from humus, litter, moss and soil, for some genera the proportions of species described from these habitats are much higher: *Proprioseiulus* (75 %), *Chelaseius* (72 %), *Proprioseiopsis* (25 %), *Arrenoseius* (21 %), *Amblyseius* (13 %) and *Graminaseius* (12 %).

5.3 Some Biological and Behavioural Aspects

The phytoseiids have a typical Gamasina life history, usually with five life stages: egg, larva, protonymph, deutonymph and adult. The entire life cycle may be completed in only 5 days in *Phytoseiulus* spp., faster than their spider mite prey, but most commonly 7–10 days. Eggs are laid at a maximum rate of about four eggs per day, again mainly for *Phytoseiulus* spp., but the usual rate is around two eggs per

Table 5.1 Number of phytoseiid species described from each country or region

Type locality	Species[a]	Type locality	Species[a]
USA	225	Armenia, Chile, Trinidad	22
China	210	Algeria, Georgia, Israel	21
Pakistan	167	Honduras, Puerto Rico	19
India	147	Hawaii, Madagascar	18
Australia	121	Argentina, Colombia	17
South Africa	111	New Zealand	15
Brazil	109	Guyana, Nigeria, Peru, Spain, Zambia	14
Kenya	71	South Korea	13
Philippines	65	France, Indonesia, Sudan	12
Japan	63	Cameroon, Hong Kong	11
Mexico	59	Azerbaijan, Kazakhstan, Kyrgyzstan	10
Canada	54	Ghana, Norway, Tanzania	9
Russia	42	Canary Islands, Ecuador, England, Malaysia, Papua New Guinea, Uganda	8
Ukraine	38	Benin, Iran, Tajikistan, Thailand	7
Costa Rica, Greece	37	Finland, Guatemala, Reunion Islands, Switzerland	6
Germany	36	Moldova, Mozambique, Zimbabwe	5
Egypt	35	Belize, Hungary, Lebanon, Namibia, Netherlands, Rwanda, Turkmenistan, Vanuatu, Yemen	4
Not known	30	Angola, Jamaica, Morocco, New Caledonia, Nicaragua, Saint Lucia, Sri Lanka, Uruguay, Venezuela	3
DR Congo	28	Alaska, Belfium, Burundi, Dominican Republic, El Salvador, Guadeloupe, Ivory Coast, Malawi, Paraguay, Portugal, Sierra Leone	2
Taiwan	26	Antigua, Austria, Bulgaria, Comores, Croatia, Cuba, Cyprus, Fiji, Guinea, Israel/Lebanon/Syria, La Desirade, Lesotho, Madeira Islands, Martinique, Motenegro, Panama, Poland, Saudi Arabia, Scotland, Seychelles, Sweden, Tahiti, Tortola Island, Tunisia, Turkey	1
Galapagos	25		

Based on Demite et al. (2014b)
[a]Only valid species included

day (Sabelis 1985). With few exceptions, phytoseiids are parahaploid (pseudoarrhenotokous) – females must mate in order to produce progeny; female progeny arise from fertilized eggs and are diploid, while male progeny arise from fertilized eggs that become haploid after losing the set of chromosomes received from the father (Hoy and Cave 1979; Nelson-Rees et al. 1980). Development and reproduction can occur year-round in warm regions, while in temperate regions, phytoseiids undergo facultative winter diapause in the mated adult female stage. Diapause is induced under short-day conditions (Overmeer 1985).

Some phytoseiids have been shown to be attracted to herbivore-induced plant chemical cues (see Sabelis and Dicke 1985). These volatiles are reliable indicators

Table 5.2 Number of valid phytoseiid species of each genus described from each type of habitat

Genus	P	S	A	O	Genus	P	S	A	O
Amblyseiinae					*Evansoseius*	1			
Amblyseius	267	47	4	41	*Honduriella*	1			
Neoseiulus	292	28	4	31	*Macrocaudus*	1			
Euseius	183			8	*Macroseius*	1			
Proprioseiopsis	82	35	3	18	*Maunaseius*				1
Typhlodromips	83	3		2	*Metadromips*	1			
Scapulaseius	59			3	*Moraeseius*	1			
Transeius	41	5	3	5	*Neoparaphytoseius*	1			
Arrenoseius	13	5	1	5	*Paraamblyseiulella*	1			
Typhlodromalus	20			2	*Parachelaseius*			1	
Ueckermannseius	21				*Parakampimodromus*	1			
Amblydromalus	19			1	*Pholaseius*	1			
Paraphytoseius	16			3	*Prasadromalus*	1			
Okiseius	15			2	*Quadromalus*	1			
Graminaseius	13	2		1	*Ragusaseius*	1			
Kampimodromus	15				*Ricoseius*	1			
Amblyseiulella	14			1	*Rubuseius*	1			
Phytoscutus	14				*Serraseius*	1			
Gynaeseius	11				*Swirskiseius*	1			
Chelaseius	1	8	1	1	*Tenuisternum*	1			
Proprioseius	10				**Phytoseiinae**				
Iphiseiodes	10				*Phytoseius*	188			15
Paragigaghathus	9				*Platyseiella*	5			1
Paraamblyseius	9				*Chantia*	1			
Aristadromips	7			1	**Typhlodrominae**				
Eharius	6				*Typhlodromus*	378	6	3	28
Asperoseius	3		2		*Kuzinellus*	46			
Phytoseiulus	4				*Galendromus*	18			
Tenorioseuis		4			*Meyerius*	16			2
Knopkirie	4				*Typhloseiulus*	10			
Proprioseiulus	1	3			*Chanteius*	8			1
Olpiseius	3				*Typhloseiopsis*	7			
Phytodromips	3				*Typhlodromina*	6			
Amblyseiella	2				*Australiseiulus*	5			1
Chileseius	2				*Galendromimus*	6			
Kampimoseiulella	2				*Cydnoseius*	3			1
Pararrenoseius		2			*Cocoseius*	3			
Phyllodromus	2				*Africoseiulella*	1			
Typhloseiella	2				*Africoseiulus*	1			

(continued)

Table 5.2 (continued)

Genus	P	S	A	O	Genus	P	S	A	O
Flagroseius	1				*Breviseius*	1			
Iphiseius	1				*Gigagnathus*	1			
Afrodromips			1		*Indiraseiulus*	1			
Afroseiulus			1		*Leonseius*	1			
Amblymexica			1		*Papuaseius*	1			
Archeosetus	1				*Silvaseius*	1			
Diaphoroseius				1	**Total**	**2,077**	**150**	**25**	**184**

Based on Demite et al. (2014b)
P plant, *S* humus, litter, moss, soil, *A* associated with other animals, *O* others/not informed

of prey presence and prey identity, and are used by them in long-range prey-habitat location and extermination. The composition of these volatiles depends on plant species and plant cultivar, and may vary with prey species infesting a plant (see Dicke et al. 1999). From a pragmatic point of view, the extensive works conducted in this area are certainly important for the practice of biological control, as for the selection of natural enemies to be used to control particular pest species, both for classical biological control and augmentation projects.

Cannibalism and intraguild predation have been shown to occur in Phytoseiidae (Schausberger and Croft 2000; Schausberger 2003; Zannou et al. 2005). These studies have centered mainly on the ability and frequency of predation of immatures by adults of its own (cannibalism) or of different (intraguild predation) species. Recently, intraguild predation involving adults of different phytoseiid species has also been studied (Famah Sourassou et al. 2013). Although found among generalist and specialist phytoseiid predators, it has been shown that specialist phytoseiids act less often as cannibals or intraguild predators than are generalist predators (Schausberger and Croft 2000). These processes may have negative effects on biological control of pests, particularly when the intraguild predator is less effective than intraguild prey. We know much about these two processes from studied conducted under laboratory conditions, but very little is known about the occurrence and intensity of cannibalism and intraguild predation under field conditions, especially between predators naturally found together. This can have an important bearing on classical biological control efforts, given that quite often the most dominant predators are considered more promising than others.

Flechtmann and McMurtry (1992) described how phytoseiids feed on spider mites and pollen. The phytoseiids studied cut into the cuticle of the prey with their chelicerae. Evidence suggested that proteolytic enzymes are injected into the prey to liquefy its content, which is then taken up by the predator (Evans 1992). *Euseius stipulatus* Athias-Henriot and *Amblyseius similoides* Buchellos & Pritchard pick up pollen grains individually with one of the chelicerae. The exine of the pollen grain is then ruptured during alternate movements of the chelicerae and the content of the pollen grain is withdrawn.

Some species are able to counteract the protective function of spider mite (Tetranychidae) webbing. In one type of behaviour reported by Shimoda et al. (2009) the phytoseiids were observed to grab a silken thread of *Tetranychus urticae* Koch and then cut it with the chelicerae. This behaviour allowed the predators *Neoseiulus fallacis* (Athias-Henriot), *N. womersleyi* (Schicha) and *Phytoseiulus persimilis* to enter the prey colony. In another type of behaviour *Neoseiulus californicus* (McGregor) used the front legs to rip into the walls of webbed nests of *Oligonychus perseae* Tuttle, Baker and Abbatiello and thus enter the nests to exploit the prey (Montserrat et al. 2008).

Phytoseiids apparently have a relatively high rate of dispersal. A mechanism for *Neoseiulus fallacis* (Garman) newly emerged females is to assume a posture of a raised body, waving the outstretched front legs, to be carried off the plant in air currents (Johnson and Croft 1976). From numbers of *Galendromus* (*Galendromus*) *occidentalis* (Nesbitt) caught on sticky panels, it was estimated that millions were dispersing from an almond orchard following mass dispersal of *T. urticae* (Hoy et al. 1985). *Euseius stipulatus* colonized in a single season an entire block of lemon trees in California, USA from a single release point of less than 1,000 mites (McMurtry et al. 1992). Many decades after termination of releases, this species was collected on plants many kilometres from the closest release sites (McMurtry and Show 2012). Similarly, *Typhlodromalus aripo* DeLeon has spread to an area of about 1,500 km^2 during a single year (Yaninek and Hanna 2003). This species resides in cassava tips, and may be transported by the usual transferral of cassava stems used as planting material, from one location to another.

5.4 Life Styles

McMurtry and Croft (1997) established a life-style concept as follows: Type I lifestyle – specialized predators of *Tetranychus* spp. that produces copious webbing, composed of *Phytoseiulus* spp. with all of the predator's life functions taking place in the prey colonies; Type II lifestyle – selective predators of tetranychid mites, represented by *Galendromus*, a few *Neoseiulus* and a few *Typhlodromus* species, having a wider range of acceptable mite prey than Type I; Type III lifestyle – generalist predators with broad feeding habits, represented by many *Neoseiulus*, most *Amblyseius* and *Typhlodromus* as well as species in most other genera; and Type IV lifestyle – specialized pollen feeders/generalist predators, represented mainly by *Euseius* spp. Alternate prey foods include mites of various families, as well as small insects such as thrips and whiteflies, and also nematodes. In addition to pollen, alternate non-prey foods include honeydew, fungus spores and plant exudates.

A revision of the McMurtry and Croft (1997) system was recently published by McMurtry et al. (2013), the main differences being the subdivisions of Types I and III. In Type I the specialized predators were subdivided into three categories based on the group to which their special prey mite belonged: Ia –*Tetranychus* spp., producing copious webbing, composed of *Phytoseiulus* spp.; I-b –web-nest producing

Table 5.3 Phytoseiid life styles according to previous (McMurtry and Croft 1997) and most recent (McMurtry et al. 2013) classification systems

McMurtry and Croft (1997)	McMurtry et al. (2013)
Type I – Highly specific on *Tetrnychus* spp.	Type I – Subdivided according to mite prey specificity: I-a, b, c
Type II – Broadly specific tetranychids most favoured	Type II – Nearly the same
Type III – Generalists – wide range of foods acceptable	Type III – Subdivided according to habitat: III-a, b, c, d, e
Type IV – Primarily pollen feeders Generalist predators	Type IV – Same, but with new information

prey; and I-c –tydeoid mites. In Type III the generalist predators were subdivided into five categories based on their preferred habitat: III-a –hairy leaves; III-b –glabrous leaves; III-c –confined spaces on dicotyledonous plants; III-d –confined spaces on monocotyledonous plants; and III-e –soil/humus habitats. There are no definite lines between the different types, so that some species could be placed in more than one type (see Table 5.3).

5.5 Biological Control Efforts

Initially, biological control programs were centered on *Tetranychus* spp., using *Phytoseiulus persimilis* Type I to control *T. urticae* on greenhouse cucumbers and outdoor strawberries, and *G. (G.) occidentalis* (Type II) for control of *Tetranychus mcdanieli* McGregor on apple (Hoyt 1969) and *Tetranychus pacificus* (McGregor) on grape (Flaherty and Huffaker 1970). Later, a program was successfully implemented on almond, utilizing an O-P-resistant *G.* (*G.*) *occidentalis* strain to control *T. pacificus* (Hoy et al. 1985). These projects demonstrated that Type I and Type II lifestyles of phytoseiids could be effective biological control agents of *Tetranychus* species in outbreak situations. These predators might be replaced by Type III generalists when *Tetranychus* species were at low (endemic) levels. Important Type III lifestyle predators included *Typhlodromus* (*Typhlodromus*) *pyri* Scheuten on apple, *Typhlodromus* (*Anthoseius*) *caudiglans* Shuster on apple and other deciduous fruits (Croft and MacRae 1992; Dunley and Croft 1990).

In the early stage of the development of biological control of pest mites with predatory phytoseiids, Chant (1961) demonstrated the ability of *P. persimilis* to control *T. urticae* when periodically released in a greenhouse experiment, as did Hussey et al. (1965) and Bravenboer and Dosse (1962) on greenhouse cucumbers and Oatman and McMurtry (1966) as well as Oatman et al. (1977) in field-grown strawberry plants. Other examples involving periodic releases of predatory mites refer to the use of phytoseiids on cotton, hops and mint (McMurtry and Croft 1997). More recently, a project involved finding a prospective predator for the biological control of *Tetranychus evansi* (Baker and Pritchard). To date, a single Type I-a predator

Phytoseiulus longipes (Evans) was shown to be potentially useful for control of this pest (Furtado et al. 2006; Ferrero et al. 2007; Silva et al. 2010; Navajas et al. 2012). Among Type I-c predators, *Paraseiulus talbii* (Athias-Henriot) is the most extensively studied. Its impact on tydeoid populations on grape was determined to be moderate (Camporese and Duso 1995). It was later realized that Types III and IV lifestyle predators could be effective at keeping pest mites and small insects at low densities. *Amblyseius swirskii* Athias-Henriot, *Transeius montdorensis* (Schicha), *Neoseiulus cucumeris* (Oudemans) and *Neoseiulus barkeri* Hughes were used primarily for thrips control and it was shown that provisioning the plants with pollen before the pest appeared could maintain the predators at higher numbers for subsequent control of thrips (see McMurtry and Croft 1997). Alternative hosts and other foods such as pollen are probably the very reason Type III and IV predators can be effective in controlling the target pests. *Typhlodromus* (*T.*) *pyri*, a cosmopolitan species on apple, is apparently able to keep the European red mite, *Panonychus ulmi* Koch, in balance at low densities, despite showing no preference for preying on that species (e.g. Nyrop 1988). Although it is more difficult to conclusively demonstrate the effects of Type III or Type IV predators, some studies demonstrated these abilities: Type III *Amblydromalus limonicus* (Garman and McGregor) on *Oligonychus punicae* (Hirst) by McMurtry and Scriven (1971); *T.* (*T.*) *pyri* on *P. ulmi* by various workers (see McMurtry and Croft 1997); and Type IV *Euseius* spp. by McMurtry and Scriven (1968) and McMurtry et al. (1992).

The coconut mite *Aceria guerreronis* Keifer is a serious pest of coconut in the tropics. In the last decade efforts have been dedicated to define control strategies against that pest. A search for prospective natural enemies conducted in three continents – Africa, Americas and Asia – along with laboratory works identified the Type IIIb predators *Neoseiulus baraki* (Athias-Henriot) and *Neoseiulus paspalivorus* (De Leon) as promising predators in the control of this pest (Fernando et al. 2010; Negloh et al. 2008; Lawson-Balagbo et al. 2007a; Navia et al. 2013). However, despite a significant number of papers published regarding biological, ecology and behavioural aspects of these two predatory mites, controlling that pest has been a difficult task, particularly because of its refuge (Lawson-Balagbo et al. 2007b).

The question which remains to be answered after so many years of that project is whether there is any prospect for the biological control of the coconut mite. The coconut mite, tiny in size, is well protected under highly appressed bracts of the fruit, and difficult to reach by its predators (Lawson-Balagbo et al. 2007b). Detailed data about dispersal of coconut mites are still scanty. Recently, efforts have been dedicated to evaluate the ability of these mites to disperse from infested to non-infested fruits. Dispersal between fruits within bunches or between bunches can occur by short-distance walking, however, the walking capacity of these mites is not sufficient to allow dispersal between plants (Galvão et al. 2012; Melo et al. 2014). Precise information about this respect would be helpful to define a control strategy.

Despite the extensive effort to find prospective natural enemies of the coconut mite, the results obtained until now are not very promising. Only two predators, *N. paspalivorus* and *N. baraki*, have been consistently found in association with it, even in areas where the pest causes major damage. Thus, one could wonder whether it should be concluded that there is no biological solution to this case, or whether other strategies should be adopted in the search of promising predators. It has been reported that when searching for host or prey, a natural enemy is often first attracted to their habitat, and only then they actually search for their host or prey (Vinson 1976). Thus, Navia et al. (2013) suggested that prospective predators should be searched for in parts of Asia where this pest is not present, but where its counterpart, *Colomerus novahebridensis* Keifer, is found, usually causing little or no damage to coconut. They argue that the little damage caused by *C. novahebridensis* could be due to the effect of natural enemies, and that these could also be effective against the coconut mite in other parts of the world, given that even the more specialized predatory mites usually feed on various prey species (McMurtry et al. 2013). Thus, there is good reason to conduct searches for prospective species to be used against the coconut mite in regions where *C. novahebridensis* is present, as done by Da Silva et al. (2014).

The work recently published by Lesna et al. (2014) seems to fit the same reasoning. In that work, *N. paspalivorus* was shown to be a good predator of the dry bulb mite, *Aceria tulipae* Keifer, in stored tulip bulbs. The authors highlighted the ability of this predator to control a different species from the one it usually is associated with, in a completely different micro- and macro-habitat. They concluded that the result of their work confirms the new association hypothesis of Hokkanen and Pimentel (1989).

The red palm mite, *Raoiella indica* (Hirst), has become a global pest in the tropics. Unlike the coconut mite (*A. guerreronis*) which seems to be host-specific, *R. indica* attacks a wide range of plant species, mostly the palm species. Type III life-style predator *Amblyseius largoensis* (Muma) has so far been the predator most frequently found on coconut, the main plant species attacked by that pest (Carrillo et al. 2012; Moraes et al. 2012). This predator is cosmopolitan in the tropics, and it seems to prefer glabrous leaves (Type III-b), as coconut. Thus far, no phytoseiids have been shown to be specialists of tenuipaplpid mites (Gerson et al. 2003).

5.6 Mass-Rearing and Releases

Early rearing methods were described by Scriven and McMurtry (1971) and McMurtry et al. (1991). Pollen is an effective and cheap food source for rearing some phytoseiids. Sometimes a relatively low number of spider mite eggs and larvae supplemented the pollen diet (largely ice plant – *Malephora crocea* and cattail – *Typha latifolia* pollens). Commercial insectaries now produce and sell various species of phytoseiids, as seen by a search in the internet. Life-style types I and II still require spider mites for efficient mass-rearing. Different factitious hosts were tried for a cheaper rearing method of Type III life-style predators, and acarid mites were found to be satisfactory for rearing some species, like *A. swirskii, T. montdorensis, N. cucumeris* and *N. barkeri* (McMurtry and Croft 1997).

Periodic release/delivery of phytoseiids in small areas are still most often done manually. However, large scale release methods have gone through great development. Pickett et al. (1987) showed that releases of phytoseiid mites by aircraft might be feasible for controlling *Oligonychus pratensis* (Banks) on corn, while Takano-Lee and Hoddle (2001) used a modified mistblower on a backpack for distributing *N. californicus* on avocado to control *O. perseae*. Another example of release system is the sophisticated equipment developed by Koppert®, allowing quick delivery over wide areas, and allowing the concurrent delivery of more than one species of predatory mite. The development of those methods makes the use of predatory mite more competitive with other control methods.

5.7 Classical Biological Control Programs

There are two examples of extensive long-term projects – the cassava green mite in Africa and the citrus and avocado mites in California – where classical biological control has proven to be a powerful tool for suppression of exotic mite pests.

Cassava Green Mite in Africa The cassava green mite, *Mononychellus tanajoa* (Bondar) is an exotic pest, accidentally introduced into Africa in the 1970s. The pest spread rapidly over many cassava growing areas, causing up to 50 % reduction in cassava production. A continent-wide classical biological control programme was initiated to control that pest. This program of many years' duration has involved personnel in South America, where the pest is thought to be native, and Africa, where it was introduced (see Yaninek and Hanna 2003). Although several phytoseiid species were introduced and released in Africa, only two Type III life-style species *Amblydromalus manihoti* (Moraes) and *T. aripo* De Leon were established. The high dispersal rate of *T. aripo* together with its high impact on pest populations (reduce the pest population by >50 %) and its ability to persist in cassava fields makes this species the best predator of the cassava green mite in Africa. *Typhlodromalus aripo* was first released in Benin in 1993 and by 2001 it was established in 16 countries (Yaninek 2007). Successful biological control of the cassava green mite produced direct economic benefits of US$1.7 billion for Nigeria, US$384 million for Ghana, and US$74 million for Benin (Alen et al. 2005).

Citrus and Avocado Mites in California These crops are often plagued by the citrus red mite *Panonychus citri* (McGregor) and the avocado brown mite *O. punicae* respectively. More than 25 species representing Types II, III and IV were introduced and released in California, but only the Type IV *E. stipulatus* has spread extensively from release sites. Four decades after releases, mostly in southern California, it has been collected in valleys and regions where it was never released. Two Type II predators showed limited establishment: *Galendromus* (*Galendromus*) *helveolus* (Chant) [synonymized as *G.* (*G.*) *floridanus* (Muma) by Chant and McMurtry (1994)], only after the establishment in California of *O. perseae*, a web nest making species on avocado (attempted establishment against *O. punicae* failed), and

Typhlodromus (*Anthoseius*) *rickeri* Chant, found in colonies of *Eotetranychus sexmaculatus* (Riley). As a consequence of the positive results in California, *E. stipulatus* was introduced to Peru in 2006 for the control of the citrus red mite. It is now produced commercially in Peru for the control of that pest (Aguirre-Gil et al. 2013).

Not a part of this program was the introduction and release of *P. persimilis* to control *T. urticae* on strawberry in California. This predator now occurs on various weeds (McMurtry et al. 1978) and it has replaced the native *Phytoseiulus macropilis* (Banks) in some stands of castor bean *Ricinus communis* L. (JAM pers. obs.). Another separate project was the introduction of OP-resistant *G.* (*G.*) *occidentalis* to Australia in 1972 and the subsequent control of *T. urticae* on apple (see McMurtry 1982).

5.8 Conclusions

Because of their important role in biological control of mite and insect pests, phytoseiid mites have been the subject of intensive studies around the world, resulting in a large number of species described, as well as an extensive knowledge about the distribution, biology, ecology, behaviour etc.…of mites of this family. However, despite the large number of species known in this family, few of them are effectively used by biological control practitioners. There is therefore a need to continue the investigations on the ability of other, unknown species to be used in biological control.

The progress in the area of biological control of organisms by predatory mites will benefit from adequate follow-ups of the results of each project, regardless of the positive or negative result. In the case of classical biological control, it would be important to accumulate data for a certain number of years, so that the actual establishment or lack of establishment can be adequately determined. Analyses of reasons leading to the observed results would be most wanted, so that in future projects mistakes could be corrected, increasing the chances of success of new endeavours.

A need often not considered is the deposit of voucher specimens of each project, which should be made available for loan to taxonomists. An example is *N. californicus*, commercialized in different countries but whose identity still remains in dispute. Given possible sub-specific variations within each species, voucher specimens should be kept not only on regular microscope slides, but also in absolute alcohol, so that they would be available for molecular comparisons.

Genetic selection has been taken into consideration mainly for resistance to pesticides and tolerance to high temperatures and/or low humidity (Kostiainen and Hoy 1994; Roush and Hoy 1981). Because desirable traits acquired through artificial selection could be lost or diluted, it would seem that searching for desirable traits in natural populations would be more productive. "Tailoring" the ideal predator through genetic engineering does not seem feasible at the present time.

Though seldom published, lack of positive results in biological control can also be instructive. For example, no encouraging results have yet occurred for the citrus leprosis mite, *Brevipalpus phoenicis* (Geijskes) (and indeed for no tenuipalpid species), for the coconut mite and for the red palm mite pests.

Acknowledgements Thanks are due to Prof. Dr. Gilberto J. de Moraes (ESALQ-USP, Piracicaba, São Paulo, Brazil) for his numerous comments and suggestions in the preparation of this chapter. Peterson R. Demite and N. Famah Sourassou are FAPESP Postdoctoral fellows (Process # 2011/08941-3 and 2012/21422-2, respectively).

References

Aguirre-Gil OJ, Jorge SJ, Busoli AC (2013) Control biologico classic na America Latina: o caso do Peru. In: Busoli AC, de Alencar JRCC, Fraga DF, de Souza LA, de Souza BHS, Grigolli JFJ (eds) Tópicos em entomologia agrícola, vol IV. Gráfica Multipress, Jaboticabal-SP, p 66–5

Alen AD, Neuenschwander P, Manyyong VM, Coulibaly O, Hanna R (2005) The impact of IITA-led biological control of major pests in Sub-Saharan African agriculture, Impact series. IITA, Ibadan, p 26

Bravenboer L, Dosse G (1962) *Phytoseiulus riegeli* Dosse als Prädator einiger Schadmilben aus der *Tetranychus urticae* Gruppe. Entomol Exp Appl 5:291–304

Camporese P, Duso C (1995) Life history and life table parameters of the predatory mite *Typhlodromus talbii*. Entomol Exp Appl 77:149–157

Carrillo D, Frank GH, Rodrigues JC, Peña JE (2012) A review of the natural enemies of the red palm mite, *Raoiella indica* (Acari: Tenuipalpidae). Exp Appl Acarol 57:347–360

Chant DA (1961) An experiment in biological control of *Tetranychus telarius* (L.) (Acarina: Tetranychidae) in a greenhouse using the predacious mite *Phytoseiulus persimilis* A (Phytoseiidae). Can Entomol 93:437–443

Chant DA, McMurtry JA (1994) A review of the subfamilies Phytoseiinae and Typhlodrominae (Acari: Phytoseiidae). Int J Acarol 20:223–310

Chant DA, McMurtry JA (2007) Illustrated keys and diagnoses for the genera and subgenera of the Phytoseiidae of the world (Acari: Mesostigmata). Indira Publishing House, West Bloomfield, p 219

Croft BA, MacRae IV (1992) Persistence of *Typhlodromus pyri* and *Metaseiulus occidentalis* (Acari: Phytoseiidae) on apple after inoculative release and competition with *Zetzellia mali* (Acari: Stigmaeidae). Environ Entomol 21:1168–1177

Da Silva RV, Narita JPZ, Vichitbandha P, Chandrapatya A, Konvipasruang P, Kongshuensin M, Moraes GJ (2014) Prospection for predatory mites to control coconut pest mite in Thailand, with taxonomic descriptions of collected Mesostigmata (Acari). J Nat Hist 48:699–719

Demite PR, McMurtry JA, de Moraes GJ (2014a) Phytoseiidae database: a website for taxonomic and distribution information on phytoseiid mites (Acari). Zootaxa 3795:571–577

Demite PR, Moraes GJ, McMurtry JA, Denmark HA, Castilho RC (2014b) Phytoseiidae Database. Available at: www.lea.esalq.usp.br/phytoseiidae. Accessed 8 Apr 2014

Dicke M, Takabayashi J, Posthumus MA, Schütte C, Krips OE (1999) Behavioural ecology of plant-phytoseiid interactions mediated by herbivore-induced plant volatiles. In: Bruin J, van der Geest LPS, Sabelis MW (eds) Ecology and evolution of the Acari. Kluwer Academic Publishers, Dordrecht, pp 251–268

Dunley JE, Croft BA (1990) Dispersal between and colonization of apple by *Metaseiulus occidentalis* and *Typhlodromus pyri* (Acarina: Phytoseiidae). Exp Appl Acarol 10:137–149

Evans GO (1992) Principles of acarology. CAB International, University Press, Cambridge, UK, Wallingford, p 563

Famah Sourassou N, Hanna R, Negloh K, Breeuwer JAJ, Sabelis MW (2013) Females as intraguild predators of males in cross-pairing experiments with phytoseiid mites. Exp Appl Acarol 61:173–182

Fernando LCP, Waidhyarathne KP, Pereira KFG, De Silva PHPR (2010) Evidence for suppressing coconut mite, *Aceria guerreronis* by inundative release of the predatory mites *Neoseiulus baraki*. Biol Control 53:108–111

Ferrero M, Moraes GJ, Kreiter S, Tixier M-S, Knap M (2007) Life tables of *Phytoseiulus persimilis* feeding on *Tetranychus evansi* at four temperatures (Acari: Phytoseiidae, Tetranychidae). Exp Appl Acarol 41:4–53

Flaherty DL, Huffaker CB (1970) Biological control of Pacific mites and Willamette mites in San Joaquin Valley vineyards. 1. Role of *Metaseiulus occidentalis*. II. Influence of dispersion patterns of *Metaseiulus occidentalis*. Hilgardia 40:267–330

Flechtmann CHW, McMurtry JA (1992) Studies on how phytoseiid mites feed on spider mites and pollen. Int J Acarol 18:157–162

Furtado IP, Moraes GJ, Kreiter S, Knapp M (2006) Search for effective natural enemies of *Tetranychus evansi* in south and southeast Brazil. Exp Appl Acarol 40:157–174

Galvão AS, Melo JWS, Monteiro VK, Lima DB, Moraes GJ, Gondim MGC (2012) Dispersal strategies of *Aceria guerreronis* (Acari: Eriophyidae), a coconut pest. Exp Appl Acarol 57:1–13

Gerson U, Smiley RL, Ochoa R (2003) Mites (Acari) for pest control. Blackwell Publishing, Oxford, p 539

Hokkanen HMT, Pimentel D (1989) New associations in biological control—theory and practice. Can Entomol 121:829–840

Hoy MA, Cave FE (1979) Parahaploidy of the arrhenotokous predator, *Metaseiulus occidentalis* (Acari: Phytoseiidae), demonstrated by X-irradiation of males. Entomol Exp Appl 26:97–104

Hoy MA, Groot JJR, Baan HE (1985) Influence of aerial dispersal on persistence and spread of pesticide-resistant *Metaseiulus occidentalis* in California almond orchards. Entomol Exp Appl 37:17–31

Hoyt SC (1969) Population studies of five mite species on apple in Washington. In: Evans GO (ed) Proceedings of 2nd international congress of acarology. Akademiai Kiado, Budapest, Canberra, pp 117–133

Huffaker CB, van de Vrie M, McMurtry JA (1969) The ecology of tetranychid mites and their natural control. Annu Rev Entomol 14:125–174

Huffaker CB, van de Vrie M, McMurtry JA (1970) Ecology of tetranychid mites and their natural enemies: a review. II. Tetranychid populations and their possible control by predators: an evaluation. Hilgardia 40:391–458

Hussey NW, Parr WJ, Gould HJ (1965) Observations on the control of *Tetranychus urticae* Koch on cucumbers by the predatory mite *Phytoseiulus riegeli* Dosse. Entomol Exp Appl 8:271–281

Johnson DT, Croft BA (1976) Laboratory study of the dispersal behavior of *Amblyseius fallacis* (Acarina: Phytoseiidae). Ann Entomol Soc Am 69:1019–1023

Koch CL (1839) Deutschlands Custaceen, Myriapode und Arachniden. Regensburg, 5–6(25), 22:5–6(27), 6, 13

Kostiainen T, Hoy MA (1994) Genetic improvement of Amblyseius finlandicus (Acari: Phytoseiidae): laboratory selection for resistance to azinphosmethyl and dimetoathe. Exp Appl Acarol 18:469–484

Krantz G, Walter DE (2009) A manual of acarology, 3rd edn. Texas Tech Univ Press, Texas, p 807

Lawson-Balagbo ML, Gondim MGC Jr, Moraes GJ, Hanna R, Schausberger P (2007a) Life history of the predatory mites *Neoseiulus paspalivorus* and *Proctolaelaps bickleyi*, candidates for biological control of *Aceria guerreronis*. Exp Appl Acarol 34:49–61

Lawson-Balagbo ML, Gondim MGC Jr, Moraes GJ, Hanna R, Schausberger P (2007b) Refuge used by the coconut mite *Aceria guerreronis*: fine scale distribution and association with other mites under the perianth. Biol Control 43:102–110

Lesna I, Da Silva FR, Sato Y, Sabelis MW, Lommen STE (2014) *Neoseiulus paspalivorus*, a predator from coconut, as a candidate for controlling dry bulb mites infesting stored tulip bulbs. Exp Appl Acarol 63:189–204

Lindquist EE, Krantz GW, Walter DE (2009) Order Mesostigmata. In: Krantz GW, Walter DE (eds) A manual of acarology, 3rd edn. Texas Tech University Press, Lubbock, pp 124–232

McMurtry JA (1982) The use of phytoseiids for biological control: progress and future prospects. In: Hoy MA (ed) Recent advances in knowledge of the Phytoseiidae. Division of Agricultural Science, University of California, Berkeley, pp 23–48

McMurtry JA (2010) Concepts of classification of the Phytoseiidae: relevance to biological control. In: Sabelis MW, Bruin J (eds) Trends in acarology: proceeding of 12th international congress. Springer, New York, pp 393–397

McMurtry JA, Croft BA (1997) Life-styles of phytoseiid mites and their roles in biological control. Annu Rev Entomol 42:291–321

McMurtry JA, Scriven GT (1968) Studies on predator-prey interactions between *Amblyseius hibisci* and *Oligonychus punicae*: effects of host-plant conditioning and limited quantities of an alternative food. Ann Entomol Soc Am 61:393–397

McMurtry JA, Scriven GT (1971) Predation by *Amblyseius limonicus* on *Oligonychus punicae*: effects of initial predator-prey ratios and prey distribution. Ann Entomol Soc Am 64:219–224

McMurtry JA, Show ED (2012) Phytoseiids on blackberry in Central California. Syst Appl Acarol 17:384–387

McMurtry JA, Oatman ER, Phillips PH, Wood CW (1978) Establishment of *Phytoseiulus persimilis* (Acari: Phytoseiidae) in southern California. Entomophaga 23:175–179

McMurtry JA, Scriven GT, Newbeger SN, Johnson HG (1991) Methodologies of rearing, introducing and establishing Phytoseiid mites. Proceedings of 1989 ADAP crop protection conference, University Hawaii, Honolulu, Hawaii, pp 104–110

McMurtry JA, Moraes GJ, Johnson HG (1992) Studies of the impact of *Euseius* species (Acari: Phytoseiidae) on citrus mites using predator exclusion and predator release experiments. Exp Appl Acarol 15:233–248

McMurtry JA, Moraes GJ, Famah Sourassou N (2013) Revision of lifestyles of phytoseiid mites (Acari: Phytoseiidae) and implications for biological control strategies. Syst Appl Acarol 18:297–320

Melo JWS, Lima DB, Sabelis MW, Pallini A, Gondim MGC (2014) Limits to ambulatory displacement of coconut mites in absence and presence of food related-cues. Exp Appl Acarol 62:449–462

Montserrat M, de la Peña F, Hormaza JI, González-Fernández JJ (2008) How do *Neoseiulus californicus* (Acari: Phytoseiidae) females penetrate densely webbed spider mite nests? Exp Appl Acarol 44:101–106

Moraes GJ, McMurtry JA, Denmark HA, Campos CB (1986) A catalog of the mite family Phytoseiidae: references to taxonomy, synonymy, distribution and habitat. EMBREPA-DDT, Brasilia, p 353

Moraes GJ, McMurtry JA, Denmark HA, Campos CB (2004) A revised catalog of the mite family Phytoseiidae. Zootaxa 434:1–494

Moraes GJ, de Castro TMMG, Kreiter S, Quilict S, Gondim MGC, de Sá LAN (2012) Search for natural enemies de *Raoiella indica* Hurst in Réunion Island (Indian Ocean). Acarologia 52:129–134

Navajas M, Moraes GJ, Pigeon PAA (2012) Revision of the invasion of *Tetranychus evansi*, biology, colonization pathways, potential expansion and prospective for biological control. Exp Appl Acarol 59:43–65

Navia D, Gondim MG Jr, Aratchige NS, Moraes GJ (2013) A review of the status of the coconut mite, *Aceria guerreronis* (Acari: Eriophyidae) a major tropical mite pest. Exp Appl Acarol 59:67–94

Negloh K, Hanna R, Schausberger P (2008) Comparative demography and diet breadth of Brazilian and African populations of the predatory mite *Neoseiulus baraki*, a candidate for biological control of coconut mite. Biol Control 46:523–531

Nelson-Rees WA, Hoy MA, Roush RT (1980) Heterochromatinization, chromatin elimination and haploidization in the parahaploid mite *Metaseiulus occidentalis* (Nesbitt) (Acarina: Phytoseiidae). Chromosoma 77:263–276

Nyrop JP (1988) Spatial dynamics of an acarine predatory-prey system: *Typhlodromus pyri* (Acari: Phytoseiidae) preying on *Panonychus ulmi* (Acari: Tetranychidae). Environ Entomol 17:1019–1031

Oatman ER, McMurtry JA (1966) Biological control of the two-spotted spider mite on strawberry in southern California. J Econ Entomol 59:433–439

Oatman ER, McMurtry JA, Gilstrap FE, Voth V (1977) Effect of *Amblyseius californicus*, *Phytoseiulus persimilis* and *Typhlodromus occidentalis* on the two spotted spider mite on strawberry in southern California. J Econ Entomol 70:45–47

Overmeer WPJ (1985) Diapause. In: Helle W, Sabelis MW (eds) Spider mites; their biology, natural enemies and control, vol IB. Elsevier, Amsterdam, pp 95–102

Pickett CH, Gilstrap FE, Morrison RK, Bouse LF (1987) Release of predatory mites (Acari: Phytoseiidae) by aircraft for the biological control of the spider mites (Acari: Tetranychidae) infesting corn. J Econ Entomol 80:906–910

Roush RT, Hoy MA (1981) Genetic improvement of *Metaseiulus occidentalis*: selection with methomyl, dimethoate, and carbaryl and genetic analysis of carbaryl resistance. J Econ Entomol 74:138–141

Sabelis MW (1985) Development. In: Helle W, Sabelis MW (eds) Spider mites; their biology, natural enemies and control, vol IB. Elsevier, Amsterdam, pp 43–53

Sabelis MW, Dicke M (1985) Long-range dispersal and searching behaviour. In: Helle W, Sabelis MW (eds) Spider mites: their biology, natural enemies and control, vol 1B, World crop pests. Elsevier, Amsterdam, pp 141–160

Schausberger P (2003) Cannibalism among phytoseiid mites: a review. Exp Appl Acarol 29:173–191

Schausberger P, Croft B (2000) Cannibalism and intraguild predation among phytoseiid mites: are aggressiveness and prey preference related to diet specialization? Exp Appl Acarol 24:709–725

Scriven GT, McMurtry JA (1971) Quantitative production and processing of tetranychid mites for large-scale testing of predator production. J Econ Entomol 64:1255–1257

Shimoda T, Kishimoto H, Takabayashi J, Amano H, Dicke M (2009) Comparison of thread-cutting behavior in three specialist predatory mites to cope with complex webs of *Tetranychus* spider mites. Exp Appl Acarol 47:111–120

Silva FR, Moraes GJ, Gondim MGC Jr, Knapp M, Rouam SL, Paes JLA, Oliveira GM (2010) Efficiency of *Phytoseiulus longipes* Evans as a control agent of *Tetranychus evansi* Baker and Pritchard (Acari: Phytoseiidae: Tetranychidae) on screenhouse tomatoes. Neotrop Entomol 39:991–995

Takano-Lee M, Hoddle MS (2001) Predatory biological control of *Oligonychus perseae* (Acari: Tetranychidae) on avocado: IV. Evaluating the efficacy of a modified mistblower to mechanically dispense *Neoseiulus californicus* (Acari: Phytoseiidae). Int J Acarol 27:157–169

Tixier MS, Kreiter S, Douin M, Moraes GJ (2012) Rates of description of Phytoseiidae mite species (Acari: Mesostigmata): space, time and body size variations. Biodivers Conserv 21:993–1013

Vinson SB (1976) Host selection by insect parasitoids. Annu Rev Entomol 21:109–131

Yaninek S (2007) Biological control of the cassava green mite in Africa: overcoming challenges to implementations. In: Vincent C, Goethel MS, Lazaorvits G (eds) Biological control: a global perspective. CABI Publishing, Wallingford, pp 28–37

Yaninek JS, Hanna R (2003) Cassava green mite in Africa: a unique example of successful classical biological control of a mite pest on a continental scale. In: Neuenschwander P, Borgemeister C, Langewald L (eds) Biological control in IPM systems in Africa. CABI Publishing, Wallingford, pp 61–75

Zannou I, Hanna R, Moraes JG, Kreiter S (2005) Cannibalism and interspecific predation in a phytoseiid predator guild from cassava field in Africa: evidence from the laboratory. Exp Appl Acarol 37:27–42

Chapter 6
Prostigmata (Acari: Trombidiformes) as Biological Control Agents

Fabio A. Hernandes, Tatiane Marie M.G. de Castro, and Renan Venancio

6.1 Introduction

Prostigmata is the larger of the two suborders of the order Trombidiformes within the mite superorder Acariformes (Walter et al. 2009). The other suborder, Sphaerolichida, is much less diverse, consisting of small edaphic mites that usually have no tracheal system and whose chelicerae are chelate and dentate. Little is known about their biology.

The prostigmatid mites are a large assemblage of morphologically, biologically and ecologically diverse groups comprising 36 superfamilies, some of which consist largely of predatory or parasitic species. These mites occupy a wide range of habitats, many being edaphic, others living in the water, on plants or on animals. Feeding habits include predation, parasitism, fungivory, pollenophagy, etc. Several major groups contain pest species of agricultural importance. The terrestrial forms of Prostigmata are distinguished from the Sphaerolichida by being usually larger, having developed tracheal systems (absent in some important groups, as the Eriophyoidea, most Bdelloidea, Dolichocyboidea and Stigmaeidae), and chelicerae

F.A. Hernandes
Departamento de Zoologia, Universidade Estadual Paulista (UNESP), 13506-900 Rio Claro, SP, Brazil
e-mail: abakashi@gmail.com

T.M.M.G. de Castro
Universidade Estadual de Roraima (UERR), 69373-000 Rorainópolis, RR, Brazil
e-mail: tatianemariecastro@gmail.com

R. Venancio (✉)
Departamento de Entomologia e Acarologia, Escola Superior de Agricultura "Luiz de Queiroz" (ESALQ), Universidade de São Paulo (USP), 13418-900 Piracicaba, SP, Brazil
e-mail: renan453@yahoo.com.br

D. Carrillo et al. (eds.), *Prospects for Biological Control of Plant Feeding Mites and Other Harmful Organisms*, Progress in Biological Control 19,
DOI 10.1007/978-3-319-15042-0_6

usually not chelate and dentate (except in Labidostomatidae, Rhagidiidae and some Bdellidae) (Walter et al. 2009).

6.2 Prostigmata as Pest Natural Enemies

A comprehensive taxonomic treatment about the suprageneric groups of Prostigmata was presented by Walter et al. (2009). The same authors also summarized biology and ecology information of genera within those groups. Taxonomic details presented in this text are based mostly on that publication.

6.2.1 *Supercohort Labidostomatides*

This is a small group consisting of a single family (Labidostomatidae), which differs morphologically from most other Prostigmata by the extensive sclerotization of the dorsal and ventral idiosomal plates. They are edaphic and seemingly predators. Sidorchuk and Bertrand (2013) reported two morphological types within this family, one consisting of large species with long legs and large eyes, including species which live on the soil surface and litter, and the other consisting of smaller species, short-legged, with reduced eyes or eyeless, including species that colonize the deeper soil layer.

6.2.2 *Supercohort Eupodides*

This group consists of five superfamilies, one of the most important being Eriophyoidea, for containing three families of strictly phytophagous mites (Diptilomiopidae, Eriophyidae and Phytoptidae), which include many pest species (Jeppson et al. 1975; Amrine et al. 2003). The Halacaroidea consists of two families of fresh and salt water mites that are mainly predatory, algivorous and parasitic species. The Bdelloidea consists of two families (Bdellidae and Cunaxidae), which will be discussed later in this section.

The Eupodoidea is divided into seven families. Khaustov (2014) discussed the constitution of this superfamily and proposed a rearrangement of different genera between families. Within this superfamily, members of Pentapalpidae and Strandtmanniidae could be predators, judging from their chelate chelicerae (Halliday 1991; Olivier and Theron 2000; Walter et al. 2009). At least some species of Penthaleidae are phytophagous, of which a few are considered major pests in certain countries, conversely, phytophagy has not been reported for Penthalodidae or Rhagidiidae, which are believed to be predators. A revision of the world rhagidiids was published by Zacharda (1980). Since that work, several other taxa were described, mainly by Zacharda and co-workers (Walter et al. 2009). Adults of some rhagidiid have been observed to produce silk to catch prey and for protection (Ehrnsberger 1979). Eupodidae species seem to be predators or algivorous. A revision of the family, with a key to genera was recently published by Khaustov (2014). Details about the Tydeoidea are presented later in this section.

6.2.2.1 Bdelloidea

Mites of this superfamily have elongate chelicerae. The dorsum of the anterior part of the body, the propodosoma, bears two pairs of trichobothria (long setae inserted in cup-like bases); some of these mites have ocelli. The posterior part of the body, the hysterosoma, contains 8–10 pairs of dorsal setae. Legs with or without trichobothria. They are predominantly predators, although Australian species of *Rubrocirus* (Cunaxidae) have also been observed to ingest honeydew (Walter and Proctor 2013).

6.2.2.1.1 Bdellidae

Bdellids occupy many different habitats, ranging from the sea level to altitudes over 4,000 m (Thor 1931; Willmann 1951) in terrestrial and marine ecosystems, including plants, soil, intertidal rocks, caves (Hernandes et al. 2011) and rodent and bird nests (Oudemans 1914; Banks 1916; Allred and Beck 1953; Allred and Roscoe 1957; Costa 1961; Gupta and Paul 1985). They are fast moving and medium sized mites, most species range from 0.4 to 0.7 mm in length, but some reach up to 3.5 mm. Atyeo (1960) observed that within the same genus or species, bdellids from colder regions tend to be larger.

Most bdellids have soft, striated integument, except species of *Trachymolgus*, which have strongly sclerotized integument (Fisher et al. 2011). The body of bdellids is pear-shaped with the anterior part conical; they are usually red, green, purple or yellow. The chelicerae usually have fixed digit reduced and movable digit hook-like; some species have both digits well developed. Palps elbowed, each bearing a pair of long setae employed as sensory exploratory structures (only one seta in *Monotrichobdella* Baker and Balock). Their nickname "snout mites" derives from the long and conical rostrum formed by the chelicerae and hypostome. Bdellid eggs are often elaborately ornamented (Wallace and Mahon 1972), each being protected with webbing produced by the female.

6.2.2.1.1.1 Trends in Efforts to Understand the Group

Bdellids were among the first mites to be described. Two species were reported by Linnaeus (1758), namely *Acarus longicornis* L. and *Acarus littoralis* L.; these are now placed respectively in the genera *Bdella* and *Neomolgus*. Still, the total diversity of the group has never been compiled into a taxonomic catalog. Until 1900,

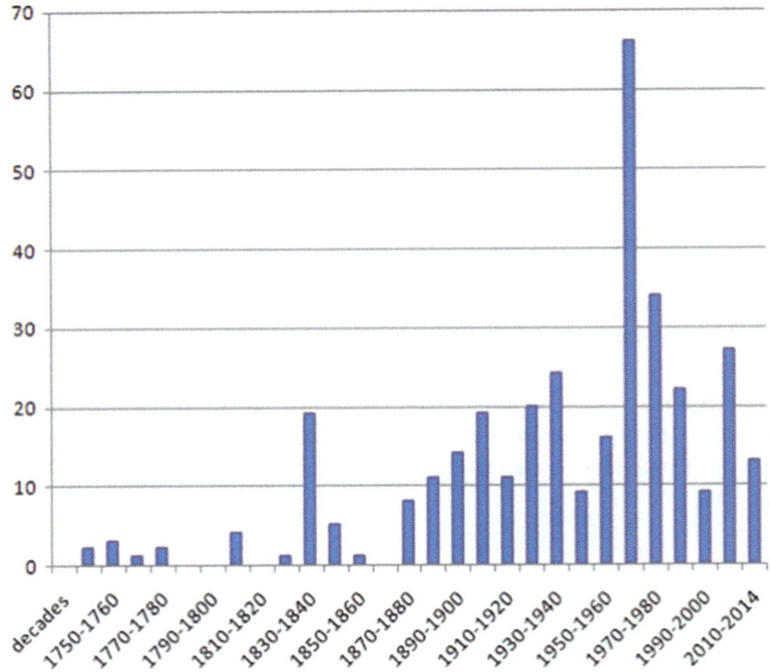

Fig. 6.1 Number of descriptions of species of Bdellidae since 1750

nearly 30 species had been described; in the 1960s and 1970s, knowledge about the taxonomy of the group flourished substantially (Fig. 6.1), due mostly to the efforts of Atyeo (1960, 1963a, b, 1964) and Wallace and Mahon (1972, 1976), who produced important monographic works on the fauna of North and Central America, and Australia, respectively. By the end of the 1970s, about 200 species were known; since then, about 70 new species were described.

The first biological studies of this group were conducted in Australia (Womersley 1933; Currie 1934). A few other studies were conducted later, to evaluate the influence of diets and temperature (Snetsinger 1956; Soliman and Mohamed 1972a, b; Foster et al. 1979; Sorensen et al. 1983; Pugh and King 1985, 1986; Ji et al. 2007a, b, 2012; Chen et al. 2007, 2008, 2011).

6.2.2.1.1.2 Taxonomy

Bdellids have been described from all continents, including Antarctica. The family is currently organized in 12 genera of five subfamilies (Atyeo 1960; Wallace and Mahon 1976; van der Schyff et al. 2003; Hernandes and Feres 2006b). Most species were described from the Australasian and West Palearctic biogeographic regions (Fig. 6.2), but that could be related to the higher sampling efforts in those regions.

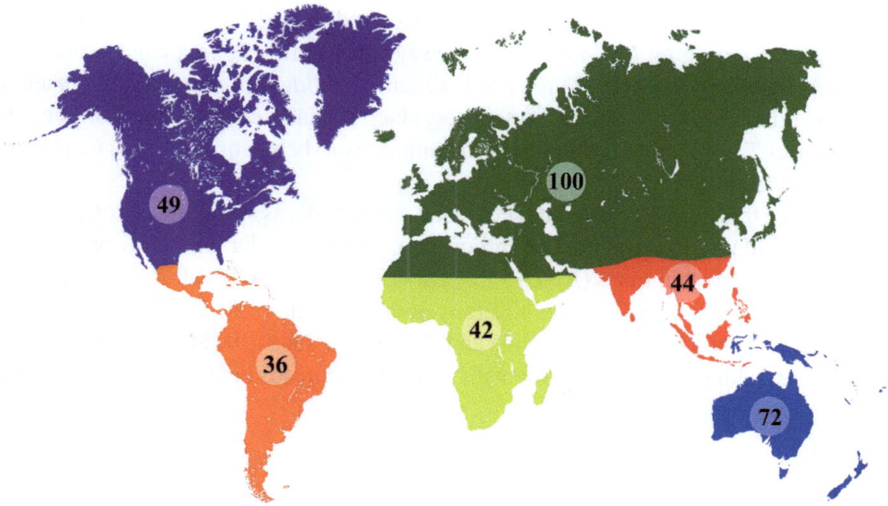

Fig. 6.2 Numbers of described bdellid species. Nearctic (*purple*), Neotropical (*orange*), Palearctic (*green*), Ethiopian (*yellow*), Oriental (*red*) and Australasian (*blue*) (Color figure online)

6.2.2.1.1.3 Bioecology

The fecundity of different species has been estimated to vary from 8 to 21 eggs, which are usually protected with webbing produced by the female (Soliman and Mohamed 1972b). The developmental stages are the egg, pre-larva (or deutovum), larva, protonymph, deutonymph and tritonymph, in addition to adults of both sexes (Atyeo 1960; Alberti 1975), each active stage being preceded by a pharate quiescent stage. When molting, bdellids often enter into a natural cavity or crevice on the plant or in the soil and then build a protective chamber, sealing the entrance with silk (Wallace and Mahon 1972; Hernandes and Feres 2006a).

Bdellids reproduce by indirect sperm transfer (Alberti 1974), males gluing an elaborate spermatophore to the substrate. Wrensch and Ebbert (1993) reported that nothing was known about bdellid genetic systems, although thelytoky was reported as suspected by Norton et al. (1993) for *Bdella captiosa* Atyeo and *Bdella tropica* Atyeo, and by Soliman and Mohamed (1972a) for *Spinibdella bifurcata* Atyeo.

Snetsinger (1956) observed that *Spinibdella depressa* (Ewing) required 21–30 days to develop from larva to adult at 15.5 °C, and from 14 to 21 days for development at 21.1 °C, in both cases feeding on the spider mites *Tetranychus urticae* Koch and *Bryobia praetiosa* Koch. *Spinibdella depressa* was also reported (Snetsinger 1956) to feed on the spider mites *Petrobia harti* (Ewing), *Petrobia latens* (Müller) and *Schizotetranychus* sp., in addition to the springtails *Bourletiella hortensis* (Fitch) and *Willowsia platani* (Nicolet). *Spinibdella bifurcata* Atyeo required 24–164 days to develop from larva to adult, depending on the season in which the experiment was carried out (Soliman and Mohamed 1972a). *Neomolgus aegyptiacus*

Soliman and Mohamed developed from larva to adult in 22 days at 17.4 °C when fed with the springtail *Lepidocyrtus incertus* (Handschin) (Soliman and Mohamed 1972b). Sorensen et al. (1983) reported predation of *Bdella longicornis* (Linnaeus) on springtails of the families Entomobryidae, Poduridae and Sminthuridae, in addition to *T. urticae*. He also observed cannibalistic behavior of *B. longicornis* in laboratory conditions.

Mites of this family are predators of small arthropods, as springtails (Collembola: Entomobryidae, Poduridae, Sminthuridae), dipteran larvae (Chironomidae, Borboridae) and other mites (Atyeo 1960; Alberti 1973; Gerson et al. 2003). Reported mite prey included spider mites (Tetranychidae) (Chen et al. 2007, 2008, 2011), Eupodidae, Nanorchestidae, Oribatida and Tydeidae (Alberti 1973). They immobilize the prey attaching it to the substrate with their silken threads before consuming them.

6.2.2.1.1.4 Practical Use in Biological Control Programs

The studies of Womersley (1933) and Currie (1934) are some of the oldest examples of pest control by predatory mites. They reported the effectiveness of *Bdellodes lapidarius* (Kramer) as a predator of the lucerne flea, *Sminthurus viridis* (Linnaeus) (Collembola: Sminthuridae), an important pest of the subterranean clover, *Trifolium subterraneum* Linnaeus (Fabaceae). *Bdellodes lapidarius* was reported to cause reductions of up to 70 % in the prey population in Western Australia (Currie 1934). This bdellid was later introduced to South Africa (Wallace and Walters 1974) to control *S. viridis* as well as the collembolan *Bourletiella arvalis* (Fitch) (Bourletiellidae). It was demonstrated to be an effective biological control agent, with the ability to feed on other food sources in the absence of the aforementioned pests (Wallace and Walters 1974). Conversely, other authors concluded that the same predator had little impact as a biological control agent of the lucerne flea in Southeastern Australia (Swan 1940; Ireson 1984; Roberts et al. 2011). Ireson and Webb (1995) reported reductions of up to 95 % in the population of *S. viridis* with the introduction of the bdellid *Neomolgus capillatus* (Kramer) from France to Northwestern Tasmania.

Ji et al. (2007a, 2012) concluded that *B. tropica* could be an effective predator of *Xenylla longauda* (Collembola: Hypogastruridae), an important pest of edible fungi. Another collembolan pest of edible fungi and vegetables, *Hypogastrura communis* (Folsom) (Hypogastruridae), has been reported to be effectively controlled by *Bdellodes japonicus* (Ehara), if the latter was released before the peak of the oviposition of the pest (Ji et al. 2007b).

6.2.2.1.2 Cunaxidae

Cunaxidae have been found on every continent except Antarctica (Skvarla et al. 2014). They are generalist predators of arthropods and nematodes (Moore et al. 1988; Walter and Kaplan 1991; Walter and Proctor 2013; Gerson et al. 2003).

Cunaxids are small to moderately large (0.3 to 1.1 mm), soft-bodied, as *Allocunaxa heveae* Den Heyer and Castro, to well sclerotized, as *Scutascirus polyscutosus* Den Heyer, whitish, red, orange, orange-brown or yellow. The chelicerae usually have fixed digit reduced and movable digit hook-like. Palps are often raptorial, with palp tarsus distally claw-like and often with internal spine-like seta that helps in grasping prey. Trichobothrium is often present on tibia IV (except in *Cunaxatricha* Castro and Den Heyer), smooth, as in Coleoscirinae and Cunaxiinae, or setose, as in Bonziinae. Cunaxid eggs may be highly ornamented with reticulate sheaths or spines (Krantz and Walter 2009) and often covered by a webbing, that may be absent as reported by Castro and Moraes (2010) for *Cunaxatricha tarsospinosa* Castro and Den Heyer for eggs laid on tree branches.

6.2.2.1.2.1 Trends in Efforts to Understand the Group

Until 1900, about 13 species of this group had been described; in the 1970s and 1980s the number of cunaxid species doubled, mostly due to the efforts of J. Den Heyer, M. Shiba and N. N. Kuznetzov, who worked on the fauna in Afrotropical, Oriental and Palearctic biogeographic regions, respectively. The first list of the world genera and species of cunaxids was published by Sepasgosarian (1984), including 124 valid species.

In the 1990s and 2000s, the number of species described doubled again, mainly by the efforts of L. A. Corpuz-Raros and R. L. Smiley, who worked on the fauna of Oriental and Nearctic biogeographic regions, respectively. By the end of the 2000s, 277 species were known, and 119 new species were described since then. Den Heyer (2011a) reported 329 valid species, while by mid 2014 the number of valid species increased to 375 (Skvarla et al. 2014).

Statistical data on the taxonomy of this group was reported by Den Heyer (2011b). According to that author, about 57.7 % of all cunaxid species were described by five authors and their co-authors (J. Den Heyer; L. A. Corpuz-Raros, R. L. Smiley, W. M. Chaudhri and M. H. Bashir). Still according to that author, the largest numbers of cunaxid species were described from the Oriental zoogeographic region (Fig. 6.3).

6.2.2.1.2.2 Taxonomy

The present classification system of the cunaxid was mainly established by Den Heyer (1980, 1981). Referring to those publications, Sepasgosarian (1984) stated “This work (a classification system for the family Cunaxidae) is so important that no taxonomist should describe a new genus or a new species without consulting Den Heyer’s work”.

Very recently, Skvarla et al. (2014) published a review of the Cunaxidae, providing keys to the world subfamilies, genera and species. That work is an essential reference for cunaxid identification today. According to those authors, the known cunaxid species are presently arranged in five subfamilies, six tribes and 27 genera (Table 6.1).

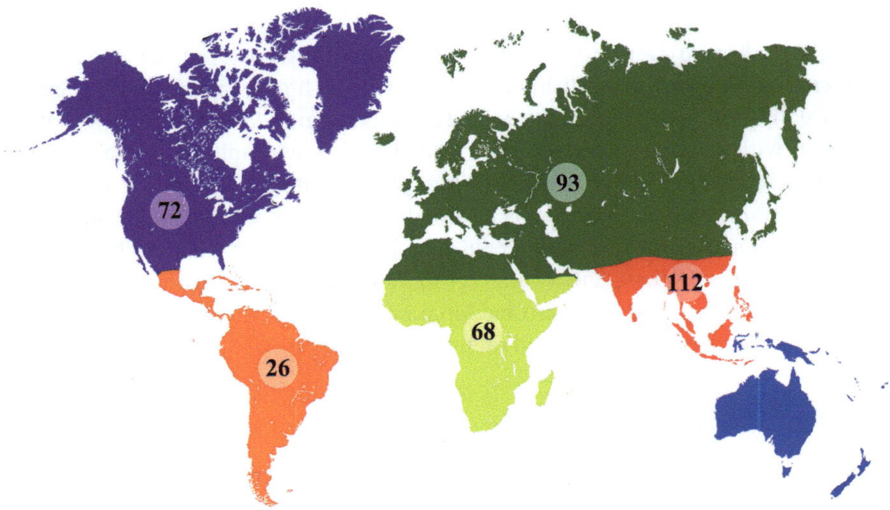

Fig. 6.3 Numbers of described cunaxid species. Nearctic (*purple*), Neotropical (*orange*), Palearctic (*green*), Ethiopian (*yellow*), Oriental (*red*) and Australasian (*blue*) (Adapted from Den Heyer (2011b)) (Color figure online)

6.2.2.1.2.3 Diversity and Habitats

Cunaxids are largely distributed and have the ability to exploit arboreal and terrestrial habitats. Species of Cunaxinae and Coleoscirinae are most common in the soil (respectively about 50 and 70 % of the species of these groups were described from soil) than on plans (about 30 and 15 %, respectively); other species of these groups were described based on specimens collected in the soil as well as on plants (Castro 2008). Conversely, Cunaxoidinae species have been described mostly from plants (about 40 % of the species) compared to those described from soil (about 30 %); the remaining species were described from specimens collected in the soil and on plants (Castro 2008). The remaining subfamilies, Bonziinae and Scirulinae, were described mostly from soil.

Comprehensive surveys in natural ecosystem have been reported by Michocka (1987) and Mejía-Recamier and Castaño-Meneses (2007), at specific level, and by Arruda Filho and Moraes (2002), Zacarias and Moraes (2002) and Castro and Moraes (2007), at the generic level only. Cunaxids have also been reported, at the specific level, from other ecosystems by different authors, namely from orchards (Muma 1960; Den Heyer and Ryke 1970; Forest et al. 1982; Mallikarjunappa and Nageshchandra 1990; Grout and Ueckermann 1999), stored food (Corpuz-Raros et al. 1988), ornamentals (Tagore and Putatunda 2003) and medicinal plants (Lahiri et al. 2004).

6.2.2.1.2.4 Bioecology

Sepasgosarian (1984) listed 110 papers on general biological aspects of this group, including few studies on life cycle. The first report on the feeding habits of the cunaxids was published by Ewing and Webster (1912), referring to a species then

Table 6.1 World diversity of Cunaxidae

Subfamily	Tribe	Genus	Number of species per genus	Number of species per subfamily
Bonziinae		*Parabonzia*	7	13
		Bonzia	6	
Coleoscirinae	Neoscirulini	*Neoscirula*	26	77
		Pseudobonzia	8	
		Neobonzia	22	
	Coleoscirini	*Orangescirula*	3	
		Coleoscirus	32	
		Scutascirus	8	
Cunaxinae	Cunaxini	*Cunaxa*	59	161
		Rubroscirus	24	
		Riscus	4	
		Cunaxatricha	1	
		Allocunaxa	1	
	Armascirini	*Armascirus*	39	
		Dactyloscirus	33	
Scirulinae		*Scirula*	2	2
Cunaxoidinae	Cunaxoidini	*Cunaxoides*	24	126
		Bunaxella	3	
		Funaxopsis	3	
		Dunaxeus	3	
		Qunaxella	1	
		Scutopalus	13	
		Denheyernaxoides	2	
		Paracunaxoides	1	
	Pulaeini	*Pulaeus*	40	
		Lupaeus	19	
		Neocunaxoides	17	
Total species				401

Adapted from Den Heyer (2011b) and updated according to the cunaxid database available at www.catalogueoflife.org. (Species 2000); including valid species and synonymies

identified as *Eupalus* sp., later described as *Eupalus parvus* Ewing (1917), and now placed in the genus *Cunaxoides*. The author referred to that as a "real enemy" of the oyster shell scale [*Lepidosaphes ulmi* (L.)].

Published life cycle studies are summarized in Table 6.2. An analysis of this table shows that studies conducted until now involve cunaxids feeding on a few species of plant damaging organisms, including root knot nematodes (*Meloidogyne* spp.) [*Coleoscirus simplex* (Ewing) and *Pulaeus* sp.], and the phytophagous mites *Eutetranychus orientalis* (Klein) [*Cunaxa capreolus* Berlese], *Calepitrimerus vitis* (Nalepa) [*Cunaxoides oliveri* (Schruft)], *Oligonychus iseilemae* (Hirst) (*Cunaxa*

Table 6.2 Studies on the life cycle and food consumption by cunaxid species

Species	Prey	Reference
Coleoscirus simplex	Nematodes	Walter and Kaplan (1991)
Cunaxa capreolus	*Eutetranychus orientalis*, Psocoptera	Zaher et al. (1975)
C. setirostris	*Oligonychus iseilemae*, *Tetranychus ludeni*	Sathiamma (1995), Arbabi and Singh (2000)
Cunaxatricha tarsospinosa	*Tenuipalpus heveae*	Castro and Moraes (2010)
Neoscirula sp.	Nematodes	Walter and Kaplan (1991)
Neocunaxoides andrei	*Caloglyphus rhizoglyphoides* *Panagrolaimus rigidus*	Taha et al. (1988)
Pulaeus sp.	Nematodes	Walter and Kaplan (1991)

setirostris Hermann) and *Tenuipalpus heveae* Baker (*C. tarsospinosa*). Compared to many other prostigmatid predators, the biology of cunaxids is poorly known. Consequently, little is known about their role as predators in natural and agricultural ecosystems.

The developmental stages of the cunaxids are the same as those of bdellids, except for the absence of pre-larvae. In most cunaxids, sperm transfer is indirect, but species of some genera (*Armascirus* Den Heyer, *Dactyloscirus* Berlese and *Rubroscirus* Den Heyer) have a variously sclerotized aedeagus. Studies about the mode of reproduction of cunaxids are scarce. Helle et al. (1984) reported the occurrence of haplo-diploidy in *C. capreolus*, what suggests that it could reproduce by arrhenotokous parthenogenesis. Castro and Moraes (2010) reported the reproduction of *C. tarsospinosa* by thelytokous parthenogenesis.

Bioecological information about cunaxids was published by the following authors: Schruft (1971), Soliman et al. (1975), Zaher et al. (1975), Taha et al. (1988), Walter and Kaplan (1991), Walter et al. (1993), Sathiamma (1995), Arbabi and Singh (2000) and Castro and Moraes (2010). Those studies have shown the ability of cunaxids to prey upon Eriophyoidea and Tetranychoidea, as well as upon other small arthropods and nematodes.

Coleoscirinae comprises species that feed upon plant parasitic nematodes, whereas Cunaxinae contains species that feed upon phytophagous mites and other arthropods, while Cunaxoidinae comprises species that consume arthropods and/or nematodes (Zaher et al. 1975; Moore et al. 1988; Taha et al. 1988; Walter et al. 1988, 1993; Walter and Kaplan 1991; Arbabi and Singh 2000). Nothing is known about the feeding habits of Bonziinae and Scirulinae.

6.2.2.1.2.5 Practical Use in Biological Control Programs

Gerson et al. (2003) stated that the potential of cunaxids as control agents of plant pests had not been adequately investigated. Eleven years later, the same is still true. However, cannibalism seems to be one of the main constraints in the practical use of those predators (Walter and Kaplan 1991), as it may hamper mass production and affect negatively their performance in the field.

In evaluating *Coleoscirus simplex* (Ewing), as predator of root knot nematodes, Walter and Kaplan (1991) concluded that it was not a promising candidate because eggs and adults were not attacked and population growth was slow, as well as because of its cannibalistic behavior and the fact that it also attacked other nematophagous mites. The results of the study by Castro and Moraes (2010) also could not show a significant role of *C. tarsospinosa* as a predator of *T. heveae*, a potential prey with which *C. tarsospinosa* was commonly found on rubber trees. Castro and Moraes (2010) observed that the total juvenile development of *C. tarsospinosa* required as long as 33.2 days to be completed, while fecundity was only 12.0 eggs per female at about 25 °C. Duration of the immature phase determined for *C. capreolus* at 25 °C (Zaher et al. 1975) and for *C. setirostris* at 24 °C (Arbabi and Singh 2000) were similar to that reported by Castro and Moraes (2010), but that duration was about five times longer than determined for *C. setirostris* at 28 °C by Sathiamma (1995).

While evaluating the role of predatory mites in citrus ecosystems, Nucifora and Vacante (1986) reported 28 species in lemon orchards, among which two cunaxids, *C. capreoleus* and *C. setirostris*. The authors considered those species as being part of a group considered to be auxiliary to other species he considered as more important in integrated control efforts adopted in Sicily.

Despite some promising reports in the literature, information available until now for the family as a whole does not allow a conclusion about the role of cunaxids under different situation and the potential to use them as biological control agents (Den Heyer et al. 2011, 2013).

6.2.2.2 Tydeoidea

These mites have chelicerae fused or closely contiguous, with reduced fixed cheliceral digit and stylet-like movable digit. They have a pair of trichobothria on the prodursum, tarsus I with or without claws; tarsi II–IV generally with paired claws and a stout and setose empodium. Generally with the following post-embryonic stages: larva, protonymph, deutonymph, tritonymph and adults of both sexes.

According to André and Fain (2000), this superfamily comprises four families, which are distributed worldwide, occupying diverse habitats and presenting different feeding habits. The first, ereynetids seem to favor humid habitats, sometimes associated with arthropods or vertebrates. Little is known about the second, the Triophtydeidae. Brickhill (1958) evaluated the suitability of a few items as food sources for *Triophtydeus bakeri* (Brickhill), reporting the ability of this species to develop when fed with the two spotted spider mite, *T. urticae* and with hemipteran honeydew, but not with eriophyid mites (Eriophyidae) or with eggs of the European red mite [*Panonychus ulmi* (Koch)]. The author also observed no oviposition of *T. bakeri* when it was maintained on leaves of *Abelia* R.Br. (Caprifoliaceae) or *Eugenia* L. (Myrtaceae) in the absence of any other potential food source, which suggests that they might not feed on those plants. Details about the other two families are presented subsequently.

6.2.2.2.1 Iolinidae

Mites of this family are unsclerotized, striate and whitish to yellowish. They have the palp tarsus with three or four elongate setae, movable cheliceral digit stylet-like or rarely whip-like and leg I without claws. Males usually have an aedeagus.

Important publications about the taxonomy of this group include André (1984) and André and Fain (2000). The Iolinidae comprises about 125 species. A key to the species of *Neopronematus* Panou, Emmanouel and Kaźmierski was recently published (Sadeghi et al. 2012). Some iolinids have been found in close association with insects, as in the tympana of noctuid moths (Noctuidae), in nests of stingless bees (Apidae: Meliponinae), in bark beetle (Curculionidae: Scolytinae) galleries or on cockroaches (Blattaria); they are also found in the soil and are extensively found on plants (Walter et al. 2009). Most recent records include *Citrus* (Horn et al. 2011) and *Coffea* sp. (Mineiro et al. 2010) in Brazil.

6.2.2.2.1.1 Bioecology

Knop and Hoy (1983) studied the biology of *H. anconai* under laboratory conditions. The mean time from egg to adult was 37.9 days at 18 °C, 13.6 days at 24 °C, and 8.1 days at 30 °C. Fecundity at 30 °C was nearly three times that at 24 °C, 45.2 and 15.6 eggs, respectively.

Iolinids seem to be mostly omnivorous. A review about the bioecology of these mites (mentioned as part of the tydeids) in different habitats was presented by Laing and Knop (1983).

Carmona (1970) reported *Pronematus ubiquitus* (McGregor) to prey on *Aculops lycopersici* (Massee). Jeppson et al. (1975) reported that same iolinid to feed on honeydew, fungi, dead arthropods as well as on *Eriophyes ficus* Cotte (Eriophyidae). Further reports of iolinid species preying on eriophyids include Baker (1968), Schruft (1972), Hessein and Perring (1986), Abou-Awad et al. (1999), Ferragut et al. (2008) and (Horn et al. 2011).

Hessein and Perring (1986) actually mentioned *Homeopronemautus anconai* (Baker) to be an effective predator of *A. lycopersici*. An unidentified Iolinidae has also been frequently seen associated with *A. lycopersici* on tomatoes in southern Brazil (G.J. de Moraes, personal communication). However, Brodeur et al. (1997) could not obtain complete immature development of *H. anconai* when fed exclusively with *A. lycopersici*. Similarly, De Vis et al. (2006) obtained virtually no oviposition of an unidentified species of *Pronematus* fed with the eriophyid *Calacarus heveae* Feres.

Predation by iolinids on mites of the family Tenuipalpidae has also been reported. Laing and Knop (1983) reported *P. ubiquitus* to prey on *Tenuipalpus granati* Sayed. Hessein and Perring (1986) found that *H. anconai* fed on *Brevipalpus lewisi* McGregor, while De Vis et al. (2006) observed a relatively high oviposition rate of *Pronematus* sp. on *Brevipalpus phoenicis* (Geijskes), but a much lower rate on *T. heveae*.

Mites of this family have also been reported in association with spider mites (Tetranychidae). Reports of those associations were summarized by Laing and Knop (1983), including *P. ubiquitus* on *E. orientalis*, and *H. anconai* on *Eotetranychus willametti* (McGregor) and *Tetranychus pacificus* (McGregor).

The ability of iolinids to feed on pollen has also been reported repeatedly in the literature. Flaherty and Hoy (1971) concluded that *H. anconai* fed primarily on pollen. Knop and Hoy (1983) were able to rear that species on pollen of *Callistemon* sp. (Myrtaceae) and *Magnolia* sp. (Magnoliaceae), but not on pollen of *Cosmos* sp. and *Dahlia* sp. (both Asteraceae) or of pine (Pinaceae). De Vis et al. (2006) observed high oviposition rate of *Pronematus* sp. on pollen of *Typha angustifolia* L. (Typhaceae).

Fungivory by iolinids has been rarely reported. One of the few records seems to be by McCoy et al. (1969), who reported *Parapronematus acaciae* Baker to feed on *Penicillium digitatum* Sacc and, mainly, *Colletorichum gloeosporioides* Penzig.

Iolinids have also been reported to serve as prey for phytoseiid predators. The possibility of using them as alternative food for phytoseiids in order to promote better control of damaging mites was discussed by Lindquist (1983). The author referred to the possibility of enhancing the increase of *Pronematus* species (and possibly of tydeids) on crops by the addition of pollen, which in turn could lead to higher phytoseiid populations and better pest control.

6.2.2.2.2 Tydeidae

These mites are weakly to heavily sclerotized, striate or reticulate, usually greenish-yellow, white or pink, and fast moving. They have palp tarsus with no elongate setae, movable cheliceral digit stylet-like or needle-like, two pairs of genital papillae and leg I with claws. Males have not been reported to have aedeagus.

One of the major contributions to the taxonomy of this group was published by Kaźmierski (1998). André and Fain (2000) redefined the Tydeidae, so that it now contains about 340 species, found on insects, on plants and in the soil (Walter et al. 2009). However, still little is known about the interactions of tydeids with their habitats.

6.2.2.2.2.1 Bioecology

Silva et al. (2014) studied the biology of *Tydeus californicus* (Banks) on *Inga marginata* L. (Fabaceae) leaves with and without pollen of *T. angustifolia* at 25 °C and 75 % R.H. They observed that the females were larviparous; development from the larval to the adult stage was completed in 21.4 days. Silva et al. (2014) compared their results with two other similar studies conducted at the same temperature (Liguori et al. 2002; Hernandes et al. 2006c), in some of which the developmental time was considerably shorter.

A bioecological review of the tydeids, including species placed today in the family Iolinidae, was presented by Laing and Knop (1983).

A revision of the literature suggests that most tydeids are unspecialized feeders. They have been reported by different authors to prey on eriophyid mites, including *T. californicus* on the citrus bud mite, *Aceria sheldoni* (Ewing) (Baker and Wharton 1952; Rizk et al. 1979a); *Tydeus goetzi* Schruft on *Calepitremerus vitis* and *Colomerus vitis* (Pagenstecher) (Schruft 1972); an unidentified *Tydeus* species on the citrus rust mite, *Phyllocoptruta oleivora* (Ashmead) (Chen et al. 1988); *Tydeus caryae* Khanjani and Ueckermann on the walnut leaf gall mite, *Aceria tristriata* (Nalepa) (Khanjani and Ueckermann 2003); *Tydeus caudatus* (Dugès) on *Calepitrimerus vitis* (Camporese and Duso 1995; Ferragut et al. 2008). Predation of tydeids on spider mites has also been reported. Brickhill (1958) observed the suitability of *T. urticae* as prey for *L. ferulus*, whereas Nucifora and Inserra (1967) believed *T. californicus* to be a predator of *P. ulmi* and Pringle (1975) believed *Tydeus grabouwi* (Meyer and Ryke) to prey upon *Tetranychus cinnabarinus* (Boisduval) but not on *P. ulmi*.

Reports of tydeid attacking insects are rare. Samarasinghe and Leroux (1966) reported a species identified as *Lorryia* sp. to cause low mortality of the oyster shell scale. Predation on nematodes has also been reported. Santos et al. (1981) studied the relationships between mites identified as Tydeidae (could be Iolinidae) and bateriophagous nematodes in early stages of litter decomposition. They found that the mites affected the nematode numbers and colonization rate in litter decomposition; they showed that elimination of tydeids with the insecticide chlordane resulted in an drastic increase in the population of free living nematodes.

Several species have been reported to be phytophagous, as *T. californicus* on avocado, citrus and sweet potato (Fleschner and Arakawa 1953; Zaher and Shehata 1963) and *Lorryia formosa* Cooreman on avocado, citrus, gardenia, and rubber trees (Jeppson et al. 1975; Aguilar et al. 2001; Hernandes et al. 2006c). However, some studies have demonstrated the unsuitability of some plant species as hosts for tydeids, as *Abelia* and *Eugenia* species for *L. ferulus* (Brickhill 1958) and *I. marginata*. for *T. californicus* (Silva et al. 2014).

Some species have also been reported as pollenophagous, as *T. californicus* (Baker 1965; Liguori et al. 2002; Silva et al. 2014). Feeding on homopteran honeydew has also been reported for a *Lorryia* species (Smirnoff 1957), *L. ferulus* (Brickhill 1958), *T. californicus* (Bayan 1986) and *L. formosa* (Mendel and Gerson 1982).

Some tydeid species have been reported to feed on fungi, as *L. formosa* (Jeppson et al. 1975; Kaźmierski 1998). In fact, some fungivorous species have been considered beneficial for feeding on undesirable fungi, as *L. formosa* on sooty mold on citrus (Mendel and Gerson 1982), and *T. caudatus* and *Tydeus lambi* (Baker) on powdery and downy mildews on grapes (English-Loeb et al. 1999, 2007; Norton et al. 2000; Duso et al. 2005; Melidossian et al. 2005).

Tydeids seem important not only as biological control agents, but also for serving as alternative prey for phytoseiid predators, as mentioned for *T. californicus* on apple (Calis et al. 1988; Niemczyk and Kaźmierski 2002) and on citrus (Rizk et al. 1979b; Ueckermann and Grout 2007).

6.2.3 *Supercohort Anystides*

This is a large group, divided in two cohorts. In the first cohort (Anystina), Adamystoidea and the Caeculoidea are considered primarily predators. The Paratydeoidea consists of two families, of which Paratydeidae has been assumed to contain predatory species.

The Anystoidea consists of three families. Members of Teneriffiidae live in seashore habitats, where they probably act as predators (Walter et al. 2009; Bernardi et al. 2012). Pseudocheylids are mainly soil mites, whose feeding habits are unknown. An unidentified species of this family has recently been found in sandy soils of two widely separated regions in Brazil (A.F. Duarte and U.S. da Cunha, unpublished; Santos 2013). Species of Anystidae are mainly predators. *Anystis agilis* (Banks) has been reported to prey on the mites *Panonychus citri* (McGregor) and *T. urticae*, the thrips *Scirtothrips citri* Moultan, the grape leafhopper *Erythroneura elegantula* Osborn and the purple scale, *Lepidosaphes beckii* (Newman) (Laing and Knop 1983). Sorensen et al. (1976) studied the biology of *A. agilis*, concluding that immature development was completed in about 49 days and that each female laid 1–3 clusters of about 23.4 eggs each. The authors concluded that the predator was expected to complete two generations a year in the field and that the average density was about one predator per 100 leaves. They considered *A. agilis* to be an important predator of *E. elegantula*, but not of spider mites. Nesbitt (1947) had previously concluded that *A. agilis* was too erratic, not remaining long enough to control either clover mites, aphids or bud moths in orchards in Canada. Another species of this family, *Anystis wallacei* Otto [reported as *Anytis salicinus* (L.)] was introduced from France to Australia and South Africa for the control of *Halotydeus destructor* (Tucker) (Meyer and Ueckermann 1987; Otto and Halliday 1991). McDonald et al. (1999) stated that the introduction of that species had a limited impact, because of its slow rate of dispersal.

The second cohort (Parasitengonina) is divided in four subcohorts whose members have parasitic larvae and predatory nymphs and adults. The subcohorts Hydrachnidiae and Stygothrombiae consist respectively of seven and one superfamilies of aquatic mites, which will not be treated in this book. The subcohort Erythraiae consists of two superfamilies; larvae, nymphs and adult of one of these (Calyptostomatoidea) have been reported to attack Diptera (Walter et al. 2009). Details about the other superfamily (Erythraeoidea) are presented in Chap. 8.

The subcohort Trombidiae consists of four superfamilies. Larvae of all of those groups parasitize arthropods, but larvae of Trombiculoidea (Leeuwenhoekiidae and Trombiculidae) can also parasitize vertebrates (Walter et al. 2009).

6.2.4 *Supercohort Eleutherengonides*

This large group consists of two cohorts. The first cohort (Raphignathina) is divided in five superfamilies. Members of Myobioidea and Pterygosomatoidea are parasites of vertebrates, although some species of Pterygosomatidae have also been reported on insects and scorpions. Pterygosomatid species of the genus *Pimeliaphilus* Trägårdh have been reported to parasitize cockroaches and bugs of the subfamily Triatominae (Reduviidae) (Walter et al. 2009). The Tetranychoidea consists of five families, all of which are phytophagous. Several species of Tenuipalpidae and Tetranychidae, of this superfamily, are major pests (Jeppson et al. 1975). The Raphignathoidea consists mainly of predatory mites, although some of them have also been observed to feed on pollen (Fan and Zhang 2005). An extensive revision of this superfamily, including keys to family and genera as well as information about their biology and ecology were presented by Fan and Zhang (2005). Stigmaeidae is the largest and most common family of this superfamily on cultivated plants, being the most widely studied raphignathoid group. Further details about this family are presented in Chap. 7. Species of the families Camerobiidae and Eupallopsellidae, of this superfamily, are also often found on cultivated plants, both preying upon scale insects (Coccoidea), and eupallopsellids also on phytophagous mites of the families Eriophyidae and Tarsonemidae. Details about the Cheyletoidea will be presented in Chap. 9.

The second cohort (Heterostigmatina) contains eight superfamilies. Members of Microdispidae, of the superfamily Scutacaroidea, are suspected to be insect parasitoids, whereas at least one species has been mentioned as a mushroom pest (Clift and Toffolon 1981). Members of some of the other superfamilies (Dolichocyboidea, Heterocheyloidea, Pygmephoroidea, Tarsocheyloidea and Trochometridioidea) are commonly associated with insects, but many species are known to be fungivorous (Walter et al. 2009); some species of Pygmephoroidea are mushroom pests. Details about the Pyemotoidea and Tarsonemoidea will be subsequently presented.

6.2.4.1 Pyemotoidea

Members of this superfamily have palpi usually small or undifferentiated, movable cheliceral digits short, stylet-like. Tracheal system of adult female opening through a pair of stigmata on the anterolateral margin of the prodorsum; the latter with 2–3 pairs of tactile setae and with or without a pair of anterolateral trichobothria, which are usually capitate setae. Leg I of female similar to or thicker than other legs, which in turn are similar to each other. Male provided with a short aedeagus. Only the adults are active; larval and the single nymphal stages are inactive. This superfamily comprises four families. Two of them (Caraboacaridae and Resinacaridae) are insufficiently known. Details about the other two families are given below.

6.2.4.1.1 Acarophenacidae

Prodorsum of females and males with 2–3 pairs of tactile setae and without trichobothria. Gnathosoma hypognathous; palpi undifferentiated. Male genital capsule ventrocaudal and with leg IV similar in form to leg III, with pretarsus and symmetrically paired claws.

Acarophenacidae comprises approximately 35 species. Adult females are parasitoids of insect eggs (Lindquist 1983). They have been reported in association with Coleoptera, Diptera, Hemiptera and Thysanoptera (Magowski 1994; Goldarazena et al. 1997; Khaustov 1999; Oliveira et al. 2002a, b).

6.2.4.1.1.1 Biological and Ecological Information

Few studies on the biology of Acarophenacidae have been conducted (Rack 1959; Steinkraus and Cross 1993; Faroni et al. 2000, 2001). These mites have been reported in association with several families of Coleoptera (Bostrichidae, Cerambycidae, Cucujidae, Dermestidae, Curculionidae and Tenebrionidae) and one family each of Thysanoptera (Phlaeothripidae), Diptera (Sciaridae) and Hemiptera (Inkaidae) (Wadhi and Kishore 1975; Magowski 1994; Goldarazena et al. 1997; Khaustov 1999; Oliveira et al. 2002a, b).

One of the first studies about these mites was conducted by Newstead and Duvall (1918), who reported a very small proportion of males in the progeny of *Acarophenax tribolii* Newstead and Duvall parasitizing eggs of *Tribolium* spp. (Coleoptera: Tenebrionidae) in stored products, which is typical of physogastric groups. Species of *Acarophenax* Newstead and Duvall have been commonly reported in association with beetles infesting stored grains or in poultry houses (Steinkraus and Cross 1993; Oliveira et al. 2003a; Walter et al. 2009; Yu et al. 2010; Gordon 2011). *Acarophenax mahunkai* Steinkraus and Cross, parasitoid of the lesser mealworm, *Alphitobius diaperinus* (Panzer), was studied by Steinkraus and Cross (1993). They reported high host egg mortality under laboratory condition, suggesting the mite to be a promising parasitoid for practical use, because of its high host specificity, rapid development and other biological characteristics. *Acarophenax lacunatus* (Cross and Krantz) was reported by Faroni et al. (2000) to parasitize eggs of the beetle *Rhyzoperta dominica* (Fabricius) (Bostrichidae). They found that optimal temperature and humidity for the development of this mite was 30 °C and 60 % relative humidity. Several other papers about *A. lacunatus* have been published by the same research group, related to: host preference (Oliveira et al. 2002a), host range (Oliveira et al. 2002b, 2006c, 2007), interaction with insecticides (Gonçalves et al. 2002, 2006a), selectivity of insecticides (Gonçalves et al. 2004, 2007), interaction with another biological control agent (Gonçalves et al. 2006b), dispersion in storage (Oliveira et al. 2006b) and tolerance to starvation (Oliveira et al. 2006a). Thus, *A. lacunatus* has been quite well studied under laboratory condition. According to L.R.D'A. Faroni (personal communication), field studies require the previous development of a feasible mass rearing process.

Adactylidium Cross species have been found in association with phlaeothripid thrips (Thysanoptera: Phlaeothripidae), some of which are important pests. A revision of this genus was published by Goldarazena et al. (2001), summarizing the knowledge about the taxonomy, host association and geographic distribution of all species in the genus. The life cycle of *Adactylidium ficorum* Goldarazena and Ochoa (mentioned as *Adalactylidium* sp., according to Goldarazena et al. 2001) was described by ElBadry and Tawfik (1966). Other five genera of this family are much less specious and are more poorly known (Walter et al. 2009).

6.2.4.1.1.2 Practical Use in Biological Control Programs

Acarophenax lacunatus was evaluated, under laboratory conditions, as a natural enemy of pests in stored grains, such as *Dinoderus minutus* (Fabricius), *Oryzaephilus surinamensis* (L.), *R. dominica* and *Tribolium castaneum* (Herbst) (Oliveira et al. 2003a, b). Oliveira et al. (2003a) observed that in free-choice experiments females showed greater preference for eggs of *R. dominica*, followed by *D. minutus* and *T. castaneum*, but did not parasitize *O. suranimensis*. In laboratory tests conducted in Petri dishes, Oliveira et al. (2003b) showed that *A. lacunatus* caused a significant reduction of 61 % in the larval population of *R. dominica*, and reductions of 53 and 26 % in the larval populations of *T. castaneum* and *Cryptolestes ferrugineus* (Stephens) (Coleoptera: Laemophloeidae). A complementary study about the potential of *A. lacunatus* to control *T. castaneum* was published by Oliveira et al. (2006a, b, c). In addition to egg parasitism, *A. lacunatus* also affects the host in their phoretical dispersal. Rocha et al. (2009) showed that the ability of *R. dominica* to take-off was impaired when more than seven mites were on the host.

Adactylidium ficorum was mentioned by ElBadry and Tawfik (1966) as an efficient natural enemy of *Gynakothrips ficorum* (Marchal). Bennett (1965) reported what seems to be an *Acactylidium* species (which he reported as *Pyemotes* sp.; see Flechtmann and Arruda 1967 and Gerson et al. 2003) associated with *G. ficorum* in different parts of Brazil, in some of which apparently causing high pest mortality.

The reported results suggest that some acarophenacids are potentially useful for the control of beetles infesting stored products and thrips.

6.2.4.1.2 Pyemotidae

Prodorsum of females with three pairs of tactile setae and a pair of capitate trichobothria. Gnathosoma prognathous; palpi distinguishable. Male genital capsule caudal and with leg IV stout, with a single sessile claw.

Pyemotidae comprises about 20 species, all placed in the genus *Pyemotes* Amerling. They are parasites of insects. A historic review of the genus was provided by Yu et al. (2010).

6.2.4.1.2.1 Biological and Ecological Information

Pyemotes are divided in two groups, *Pyemotes scolyti* (Oudemans) and *Pyemotes ventricosus* (Newport) species groups, based on their toxicity to the hosts and their degree of host specificity (Cross and Moser 1975; Moser 1975; Moser et al. 1987). Members of the *scolyti* group are more restricted in relation to host range and habitat occupation, and apparently they do not inject toxins. Although the use of mites of this group would be safe as far as their possible effect on human beings, they seem to be restricted to attacking bark beetles. *Pyemotes boylei* Krczal was originally described (Krczal 1959 apud Vaivanijkul and Haramoto 1969) attacking termites and a beetle species of the family Anthribidae. This mite is known today to parasitize insects of several orders and to cause dermatitis to humans (Vaivanijkul and Haramoto 1969). *Pyemotes scololyti* and *Pyemotes herfsi* Oudemans were reported by Weiser and Hardy (1962) to be able to kill termites under laboratory condition. *Pyemotes johnmoseri* (Khaustov) was recently studied by Akşit et al. (2007) for the control of *Hypoborus ficus* Erichson (Scolytinae) on fig trees, *Ficus carica* L., in Turkey. The authors showed that *P. johnmoseri* fed mainly on larvae, occasionally on eggs and pupae, of the host. However, the rate of parasitism was relatively low along the season (maximum of 9.3 %).

Members of the *ventricosus* group have wide host ranges and occur in very diverse habitats. Mites of this group also inject toxins in their hosts, paralyzing them. According to Cross et al. (1975), mites identified as *P. ventricosus* or *P. tritici* (LaGreze-Fossat et Montané) had been recorded from 139 insect hosts of six orders, some of which are beneficial organisms. Ikegami and Ebihara (1980) reported the parasitism of the silkworm (*Bombyx mori* L.) by *P. ventricosus*. They are commonly found in stored products (Baker et al. 1956; Krczal 1959; Cross 1965; Moser 1975; Treat 1975; Southcott 1976; Le Fichoux et al. 1980), often causing problems in laboratories where insects are reared for different purposes (Hostounsky 1971). *Pyemotes tritici* is one of the most studied species of this group. It has been reported to parasitize more than one hundred insect species of different orders (Coleoptera, Diptera, Hemiptera, Hymenoptera, Lepidoptera, Neuroptera and Strepsiptera), as reported by Oliveira et al. (2007). Wrensch and Bruce (1991) said that the intrinsic rate of increase of this species exceeded the growth rate of any potential arthropod pest, each female producing 200–350 offspring (Tomalski et al. 1988; Hoschele and Tanigoshi 1993).

More recently, natural infestations of *P. ventricosus* were reported by Gordon (2011) to kill on average 27 % of larvae and pupae of *Dasineura strobila* Dorchin (Diptera: Cecidomyiidae), an introduced gall midge that attacks the invasive Australian myrtle tree, *Leptospermum laevigatum* (Gaertn.) F. Muell. (Myrtaceae), in South Africa. In that study, *P. ventricosus* was the most abundant natural enemy of *D. strobili*.

6.2.4.1.2.2 Practical Use in Biological Control Programs

Several cases of possible use of pyemotid species have been mentioned in the literature. De Santis (1945) mentioned *P. ventricosus* (mentioned as *Pediculoides ventricosus*) as the most important natural enemy of *Oiketicus kirbyi* Guild. (Lepidoptera), a serious pest of different trees in Argentina. Weiser and Hardy (1962) suggested the possible use of *P. scololyti* and *P. herfsi* for the control of termites, based on their promising results under laboratory observations, suggesting the conduction of field experiments for a final decision about their possible use for termite control.

Other species of this group have been considered potentially useful for the control of several pests, including several stored product insects, the pink bollworm [*Pectinophora gossypiella* (Saunders)], the fire ant (*Solenopsis invicta* Buren) and the tomato leafminer [*Tuta absoluta* (Meyerick)] (Bruce 1983; Cunha et al. 2006; Oliveira et al. 2007). Diets for the mass production of *P. tritici* were developed

(Bruce 1983), but the major constraints for the practical use of mites of this group in biological control projects are the lack of specificity of some of these species and the ability to cause dermatitis to humans (De Santis 1945; Vaivanijkul and Haramoto 1969; Moser 1975). For these reasons, not much has been done recently in terms of developing the practical use of the pyemotids as biological control agents.

6.2.4.2 Tarsonemoidea

Members of this superfamily have small palpi, movable cheliceral digits short, stylet-like. Tracheal system of adult female opening through a pair of stigmata on the anterolateral margin of the prodorsum; rarely present in males. Prodorsum with one or usually two pairs of tactile setae and usually with a pair of anterolateral trichobothria, which are usually capitate setae. Leg IV of female absent or distinctly smaller and narrower than other legs; without terminal claw and ending in 2–3 setae; leg IV of males absent or stout and ending usually with a claw. With the following post-embryonic stages: larva and adults of both sexes. This superfamily comprises two families (Podapolipidae and Tarsonemidae).

6.2.4.2.1 Podapolipidae

Females usually with 2–3 pairs of legs (with one pair in *Podapolipus* Rovelli and Grassi and four pairs in *Chrysomelobia* Regenfuss); males with 3–4 pairs of legs; in both sexes, when present, leg IV with five segments. Males with caudal or dorsocaudal genital capsule.

Podapolipidae comprises 251 species of 31 genera (Kurosa and Husband 2013), consisting of parasitic mites associated with insects of the orders Blattaria, Coleoptera, Hemiptera, Hymenoptera and Orthoptera (Regenfuss 1968; Husband 1991, 1998; Husband and Kurosa 2000; Kurosa and Husband 2013).

6.2.4.2.1.1 Bioecology

Podapolipids seem to vary widely in relation to host specialization. Some species appear to be restricted to one host family, while others parasitize hosts of different orders. Gerson et al. (2003) and Walter et al. (2009) summarized the biological information regarding the podapolipids. The data presented by Walter et al. (2009), and one additional record of Haitlinger (1989), show that these mites are more diverse on Coleoptera than on other groups. Species of 23 podapolipid genera have been found on Coleoptera (including Scarabaeidae with five; Carabidae and Curculionidae, four each; Tenebrionidae, three; Cerambycidae and Chrysomelidae, two each; Buprestidae, Coccinellidae and Silphidae, one each); species of five genera have been found on both Blattaria and Orthoptera, two on Hemiptera and one on Hymenoptera.

Results of biological works have indicated the potential of podapolipids as biological control agents. One of the first observations in this sense seems to have been published by Gauchat (1972). He observed *Podapolipoides alatus* Husband (reported as *Podapolipoides grassi* Berlese according to Gerson et al. 2003) to have an adverse effect on the locust *Chortoicetes terminifera* Walker (Orthoptera: Acrididae) under laboratory condition. Infested locusts were reported as less active, being often cannibalized by other locusts.

Large numbers of the podapolipid *Coccipolipus epilachnae* Smiley and a concurrent sudden decline of egg production of a laboratory colony of the Mexican bean beetle, *Epilachna varivestis* Mulsant (Coccinellidae), led Smiley (1974) to conclude that *C. epilachnae* could have an important impact on that host. A study was conducted by Schroder (1981) to evaluate the effect of temperature on *C. epilachne*, the author concluding that the optimal temperature for development of the species was 20 °C, at which each female laid a total of about 36 eggs. Schroder (1982) evaluated the effect of the parasite on *E. varivestis*. The results were promising, but different from those later obtained in a similar study conducted by Cantwell et al. (1985), probably because of the different methodologies adopted.

Chrysomelobia labidomerae Eickwort was reported on the chrysomelid beetles *Labidomera clivicolis* (Kirby) and *Leptinotarsa decemlineata* (Say) (Baker and Eickwort 1975), and considered as potentially useful for the applied biological control of those species. Drummond et al. (1989) found that the optimal development and survival of *C. labidomerae* occurred at 30 °C, each female laying about 30 eggs. That temperature is much higher than reported above as optimal for *C. epilachnae* by Schroder (1981), suggesting that podapolipid species of different genera may have very distinct temperature preferences.

The mode of dispersion of podapolipids from parasitized to unparasited hosts was studied by different authors, who referred to this process as a "sexual disease transmission". Gauchat (1972) found that *P. alatus* (reported as *P. grassi*) dispersed between specimens of *C. terminifera* during host copulation. Knell and Webberley (2004) and Rhule et al. (2010) found that specimens of *Coccipolipus hippodamiae* (McDaniel and Morrill) were also transferred from infested to uninfested *Harmonia axyridis* (Pallas) (Coccinellidae) during copulation.

Ecological studies conducted with different podapolipid species strongly suggest their potential as biological control agents. *Coccipolipus hippodamiae* was reported on *Hippodamia convergens* Guérin-Méneville (McDaniel and Morrill 1969), *Adalia bipunctata* L. (Husband 1981) and *H. axyridis* (Riddick 2010). Reduced fecundity of *A. bipunctata* was observed when each insect was parasitized by 10 specimens of *C. hippodamiae* (Ryder et al. 2007). Rhule et al. (2010) suggested that this parasite had a great potential to control that host. Riddick (2010) reported the coexistence of that parasite with the fungus *Hesperomyces virescens* Thaxer (Ascomycetes: Laboulbeniales) on *H. axyridis*, suggesting that parasitism by the mite and fungus combined could reduce winter survival of *H. axyridis*.

Coccipolipus synonychae Ramaraju and Poorani was described (Ramaraju and Poorani 2012) from a laboratory colony of the giant coccinellid beetle, *Synonycha grandis* (Coccinellidae). In the same paper in which that species was described,

Ramaraju and Poorani (2012) reported sharp decline in vigor and fecundity of severely infested beetles.

6.2.4.2.1.2 *Practical Use in Biological Control Programs*

Most of the works about the biology and ecology of podapolipids have referred to species parasitizing undesirable insects. However, some podapolipids have also been reported to affect beneficial insects. The genus *Locustacarus* Ewing contains three species, two of which, *Locustacarus trachealis* Ewing and *Locustacarus masoni* Husband, parasitize Acrididae (Orthoptera). The third species, *Locustacarus buchneri* (Stammer), parasitizes nearly 30 species of bumblebees (Apidae: Bombiliidae), all native to the Northern Hemisphere (Plischuk et al. 2013). Bumblebees are important pollinators. Taxonomical information about *Locustacarus* and biological information about *L. buchneri* were reviewed by Husband and Sinha (1970). It has been observed that high mite infestation can severely affect the bumblebees, some of them being unable to forage. With the increasing commercialization of bumblebees as pollinators, care should be taken to prevent the dispersal of *L. buchneri* (Goka et al. 2001).

However, the apparent specificity of some podapolipids renders them potentially useful as control agents, and some species have been used in different projects.

Coccipolipus epilachnae was introduced from El Salvador to the USA to control *E. varivestis* (Smiley 1974). Schroder (1979) evaluated the possible effect of this parasite on beneficial coccinellids, concluding that it could only attack phytophagous species, of the subfamily Epilachninae. More than 700 mite infested *E. varivestis* were released in 1979 in Maryland and West Virginia, USA, but the parasite did not establish (Schroder 1981). This negative result could either be due to intrinsic characteristics of the parasite, that made it unsuitable to survive in those regions, or to an insufficient effort spent in attempting the establishment.

A Mexican population of *C. labidomerae* was introduced to the USA for the biological control of the Colorado potato beetle, *L. decemlineata* (Drummond et al. 1985). Although this mite had been found to parasitize that beetle in Mexico, it had never been reported parasitizing it in the USA. Drummond et al. (1985) studied the host specificity of *C. labidomerae* before its introduction, concluding that it was limited to attacking chrysomelid beetles of the tribe Zygogrammini. The authors found many species to be unsuitable hosts for the Mexican populations, including the honeybee, *Apis melifera* (L.), and other Coleoptera of the families Carabidae, Cerambycidae, Chrysomelidae and Coccinellidae.

Rhule et al. (2010) suggested that *C. hippodamiae* had a great potential to control *H. axyridis*, a predatory ladybird that in the last decades was introduced to Europe and North America for the control of aphids and coccids, but that was later considered undesirable for different negative environmental impacts. The authors said that within 2–3 weeks of the initiation of the parasitism by the mite, reduction in egg hatch rate and ultimately complete sterility of the coccinellid occurred. They suggested that the introduction of *C. hippodamiae* into Great Britain for the control of *H. axyridis* should be considered, arguing that knowledge about the ecology of the mite suggested it to be unlikely to have negative side effects on British coccinellids. However, this type of project would be unthinkable in a number of other countries, where *H. axyridis* as well as other coccinellid beetles attacked by *C. hyppodamiae* (*A. bipunctata* and *H. convergens*) are considered beneficial.

Given the potential so far demonstrated by podapolipids as biological control agents, it seems that further efforts should be dedicated to demonstrate its possible practical use.

6.2.4.2.2 Tarsonemidae

Females and males with four pairs of legs; leg IV with three segments and usually much more slender than other legs in females and with three or four segments in males. Male genital capsule caudal.

Tarsonemidae comprises about 530 species of 40 genera (Lindquist 1986; Walter et al. 2009), which can be phytophagous, fungigovous, algivorous, predators or parasites of insects.

6.2.4.2.2.1 Bioecology

Some important plant pests belong to this family, especially the widespread broad mite, *Polyphagotarsonemus latus* (Banks), and the panicle rice mite, *Steneotar-sonemus spinki* Smiley (Jeppson et al. 1975; Hummel et al. 2009). Some *Tarsonemus* Canestrini and Fanzago species are also important as pests or as vectors of pathogenic fungi. Fungivorous tarsonemids include species of *Fungitarsonemus* Cromroy and *Tarsonemus*. Feeding on fungi or algae is suspected to be exerted by species of *Daidalotarsonemus* De Leon, *Ceratotarsonemus* De Leon and *Excelsotarsonemus* Ochoa and Naskrecki. Several genera are phoretic on insects (Lindquist 1986). Some, as *Flechtmannus* Moraes, Lindquist and Lofego, are seemingly parasitic on insects (Moraes et al. 2002). Others were shown to be natural enemies of insects and mites, including species of *Iponemus* Lindquist as parasitizing bark beetles (André 1980), *Acaronemus* Lindquist and Smiley preying spider mites and tenuipalpid mites, and *Dendroptus* Kramer either preying on eriophyid mites or occupying galls produced by the latter. Species of *Amcotarsonemus* Fain, *Asiocortarsonemus* Fain and *Coreitarsonemus* Fain are parasites of Coreidae (Hemiptera) (Lindquist 1986). Of major importance is *Acarapis woodi* (Rennie), an important parasite the lives in the tracheae of honeybees, being thus known as tracheal mite. Some cases of human allergies have been attributed to some *Daidalotarsonemus* and *Tarsonemus* (Walter et al. 2009).

6.3 Final Remarks

Many groups of Prostigmata are known or suspected to contain species that act as natural enemies of harmful organisms, including arthropods and nematodes. Presently, there is a great interest for the reduction of the use of pesticides for the control of harmful organisms. However, much still needs to be done in terms of using prostigmatid species to replace pesticides. In relation to the groups treated in this chapter, most studies concerning the potential as natural enemies of pest organisms refer to the Bdelloidea, Pyemotoidea and Tarsonemoidea. However, the studies are still limited, conducted by a reduced number of specialists, usually from restricted geographic areas. The few works that have been conducted usually involve mainly laboratory trials, with rare consistent attempts toward the expansion of the work to conditions more similar to the habitats in which those natural enemies are expected to be used. These rare exceptions involve mostly some bdellid, cunaxid and podapolip species. Considering the behavior and biological characteristics of the different groups discussed in more details in this chapter, several species of different groups have shown good chances for practical use.

It seems that the main obstacle to the expansion of the efforts dedicated to the discovery of promising prostigmatid natural enemies for practical use is the lack of trained personnel to conduct the research work. Yet, the conduction of biodiversity studies has been stimulated lately in some countries, especially in tropical and subtropical regions. This new emphasis, especially in the regions where biodiversity is assumed to be high, might facilitate the conduction of solid and comprehensive projects that could lead to an increase in the use of Prostigmata for the biological control of harmful organisms.

Acknowledgements To G.J. de Moraes for his invaluable help throughout the preparation of this chapter. To C.H.W. Flechtmann, for his valuable suggestions, for his help with the literature and for reviewing a previous version of this chapter.

References

Abou-Awad BA, El-Sawaf BM, Abdel Kader AA (1999) Life history and life table of *Pronematus ubiquitus* (McGregor) as a predator of eriophyoid mites in Egypt (Acari: Tydeidae). Acarologia 40:29–32

Aguilar H, Childers CC, Welbourn WC (2001) Relative abundance and seasonal occurrence of mites in the family Tydeidae on citrus in Florida. In: Halliday BR, Walter DE, Proctor HC, Norton RA, Collof MJ (eds) Proceedings of the 10th international congress. Acarol, CSIRO Publishing, Collingwood, Australia, pp 376–380

Akşit T, Çakmak I, Moser J (2007) Attack by *Pyemotes johnmoseri* (Acari: Pyemotidae) on *Hypoborus ficus* (Coleoptera: Scolytidae) in fig trees in Turkey. Exp Appl Acarol 41:251–254

Alberti G (1973) Ernährungsbiologie und Spinnvermögen der Schnabelmilben (Bdellidae, Trombidiformes). Z Morph Tiere 76:285–338

Alberti G (1974) Fortpflanzungsverhalten und Fortpflanzungsorgane der Schnabelmilben (Acarina: Bdellidae, Trombidiformes). Z Morph Tiere 78:111–157

Alberti G (1975) Prälarven und Phylogenie der Schnabelmilben (Bdellidae, Trombidiformes). J Zoolog Syst Evol Res 13:44–62

Allred DM, Beck AE (1953) Mite fauna of wood rat nests in Utah. Proc Utah Acad Sci Arts Lett 30:53–56

Allred DM, Roscoe EJ (1957) Parasitic mites in desert wood rat nests with notes on free-living forms. Trans Am Microsc Soc 76:389–403

Amrine JJW, Stasny TA, Flechtmann CHW (2003) Revised keys to world genera of Eriophyoidea (Acari: Prostigmata). Indira Publishing House, West Bloomfield, p 244

André H (1980) Acariens, Scolytes et lutte biologique. Z Angew Entomol 90:127–133

André HM (1984) Redefinition of the Iolinidae (Acari: Actinedida), with a discussion of their familial and superfamilial status. In: Griffiths DA, Bowman CE (eds) Acarology VI, vol 1. Ellis Horwool Publishers, Chichester, pp 180–185

André HM, Fain A (2000) Phylogeny, ontogeny and adaptive radiation in the superfamily Tydeoidea (Acari: Actinedida), with a reappraisal of morphological characters. Zoolog J Linnean Soc 130:405–448

Arbabi M, Singh J (2000) Studies on biological aspects of predacious mite *Cunaxa setirostris* on *Tetranychus ludeni* at laboratory condition in Varanasi, India. J Agric Rural Dev 20:13–23

Atyeo WT (1960) A revision of the family Bdellidae in North and Central America (Acarina: Prostigmata). Univ Kans Sci Bull 40:345–499

Atyeo WT (1963a) The Bdellidae (Acarina) of the Australian Realm. Part I: New Zealand, Campbell Island, and the Auckland Islands. pp 113–166; part II: Australia and Tasmania. pp 167–210. Bull Univ Nebr State Mus 4:113–210

Atyeo WT (1963b) New species and records of Bdellidae from MacQuarie and the Auckland Islands (Acarina). Pac Insects 5:445–450

Atyeo WT (1964) Insects of Campbell Island. Prostigmata: Bdellidae. Pac Insects Monogr 7:166–169

Baker EW (1965) A review of the genera of the family Tydeidae (Acarina). Advances in acarology. Cornell University Press, Ithaca, pp 96–133

Baker EW (1968) The genus *Lorryia*. Ann Entomol Soc Am 61:986–1008

Baker TC, Eickwort GC (1975) Development and bionomics of *Chrysomelobia labidomerae* (Acari: Tarsonemina; Podapolipidae), a parasite of the milkweed leaf beetle (Coleoptera: Chrysomelidae). Can Entomol 107:627–638

Baker EW, Wharton GW (1952) An introduction to acarology. Macmillan, New York

Baker EW, Evans TM, Gould DJ, Hull B, Keegan HL (1956) A manual of parasitic mites of medical or economic importance. Technical Publication of the National Pest Control Association, New York

Banks N (1916) Acariens from Australian and Tasmanian ants and ant nests. Trans R Soc S Aust 7:224–240

Bayan A (1986) Tydeid mites associated with apples in Lebanon (Acari: Actinedida: Tydeidae). Acarologia 27:311–316

Bennett FD (1965) Observations on the natural enemies of *Gynaikothrips ficorum* Marchal in Brazil. Commonw Inst Biol Control Tech Bull 5:117–125

Bernardi LFO, Pellegrini TG, Ferreira RL (2012) New species of Neoteneriffiola (Acari: Trombidiformes: Teneriffiidae) from Brazilian caves: geographical distribution and ecological traits. Int J Acarol 38:410–419

Brickhill CD (1958) Biological studies of two species of tydeid mites from California. Hilgardia 27:601–620

Brodeur J, Bouchard A, Turcotte G (1997) Potential of four species of predatory mites as biological control agents of the tomato russet mite, *Aculops lycopersici* (Massee) (Eriophyidae). Can Entomol 129:1–6

Bruce WA (1983) Mites as biological control agents of stored product pests. In: Hoy MA, Knutson L, Cunningham GL (eds) Biological control of pests by mites. University of California, Berkeley, p 185

Calis JNM, Overmeer WPJ, van der Geest LPS (1988) Tydeids as alternative prey for phytoseiid mites in apple orchards. Med Fac Landbouww Rijksuniv Gent 53:793–798

Camporese P, Duso C (1995) Life history and life table parameters of the predatory mite *Typhlodromus talbii*. Entomol Exp Appl 77:149–157

Cantwell GE, Cantelo WW, Cantwell MA (1985) Effect of a parasitic mite, *Coccipolipus epilachnae*, on fecundity, food consumption and longevity of the Mexican bean beetle. J Entomol Sci 20:199–203

Carmona MA (1970) Contribuição para o conhecimento dos ácaros de plantas cultivadas em Portugal. V Agronomia Lusit 31:137–183
Chen H, Xu Y, Fei D, Deng G (1988) A preliminary study on the predaceous mite *Tydeus* sp. of *Phyllocoptruta oleivora* Ashmead. Chin J Biol Cont 4:181–182, In Chinese
Chen X, Zhang Y, Lin J (2007) Predation of *Bdellodes japonicus* (Ehara) on four species of spider mites (Acari: Bdellidae and Tetranychidae). Syst Appl Acarol 12:1–4
Chen X, Zhang YX, Ji J, Lin JZ (2008) Predatory behavior of *Bdella tropica* Atyeo to *Tetranychus urticae* Koch. J Fujian Agric Forestry Univ (Nat Sci Ed) 37:341–343
Chen X, Zhang YX, Zhang YP, Ji J, Lin JZ (2011) Influence of temperature on development of *Bdella tropica* Atyeo (Acari: Bdellidae) with *Tetranychus urticae* Koch (Acari: Tetranychidae) as its prey. Int J Acarol 37:34–39
Clift AD, Toffolon RB (1981) Biology, fungal host preferences and economic significance of two pygmephorid mites (Acarina: Pygmephoridae) in cultivated mushrooms, NSW, Australia. Mushroom Sci 11:245–253
Corpuz-Raros LA, Sabio GC, Velasco-Soriano M (1988) Mites associated with stored products, poultry houses and house dust in the Philippines. Philippine Entomol 73:311–321
Costa M (1961) Mites recovered from the nests of the levant vole (*Microtus guentheri*) in Israel. Ann Mag Nat History 4:257–282
Cross EA (1965) The generic relationships of the family Pyemotidae (Acarina: Trombidiformes). Univ Kansas Sci Bull 45:29–275
Cross EA, Moser JC (1975) A new, dimorphic species of *Pyemotes* and a key to previously described forms (Acarina: Tarsonemoidea). Ann Entomol Soc Am 68:723–732
Cross WH, McGovern WL, Cross EA (1975) Insect hosts of the parasitic mites called *Pyemotes ventricosus* (Newport). J Georgia Entomol Soc 10:1–8
Cunha US, Silva ES, Moraes GJ, Vendramin JD (2006) Ocorrência do Ácaro *Pyemotes* sp. (Acari: Pyemotidae) em Criações de Insetos em Laboratório. Neotrop Entomol 35:563–565
Currie GA (1934) The bdellid mite *Biscirus lapidarius* Kramer, predatory on the lucerne flea *Sminthurus viridis* L. in western Australia. J Counc Sci Ind Res Austr 7:9–20
de Arruda Filho GP, de Moraes GJ (2002) Grupos de ácaros (Arthropoda: Acari) encontrados em Arecaceae da Mata Atlântica do Estado de São Paulo. Biota Neotrop 2:1–18
de Castro TMMG (2008) Estudos taxonômicos e biológicos de Cunaxidae (Acari: Prostigmata) do Brasil. Thesis, Jaboticabal, p 99
de Castro TMMG, de Moraes GJ (2007) Mite diversity on plants of different families found in the Brazilian Atlantic Forest. Neotrop Entomol 36:774–872
de Castro TMMG, de Moraes GJ (2010) Life cycle and behaviour of the predacious mite *Cunaxatricha tarsospinosa* (Acari: Prostigmata: Cunaxidae). Exp Appl Acarol 50:133–139
de Moraes GJ, Lindquist EE, Lofego AC (2002) A new genus and species of tasonemid mite (Acari: Tarsonemidae) associated with a neotropical curculionid beetle (Coleoptera). Invert Syst 16:687–695
De Santis L (1945) El bicho de cesto (*Oiketicus kirbyi* Guild). Apuentes de clases sobre el tema, dictadas por el autor, vol 8. Universidad Nacional de La Plata, Faculdad de Agronomia, Talleres Gráficos "El Sol", La Plata, Argentina, p 15
De Vis RMJ, Moraes GJ, Bellini MR (2006) Initial screening of little known predatory mites in Brazil as potential pest control agents. Exp Appl Acarol 39:115–125
Den Heyer J (1980) Systematics of the family Cunaxidae Thor, 1902 (Actinedida: Acarida). Publ Univ North Ser A 24:1–19
Den Heyer J (1981) A classification system for the family Cunaxidae (Actinedida: Acarida). Publ Univ North Ser A 23:1–12, Pietersburg
Den Heyer J (2011a) CunaxidBase: family Cunaxidae database (version Jan 2010). In: Bisby FA et al (eds) Species 2000 and IT IS catalogue of life: 2011 checklist. Species 2000: reading. Digital resources at http://www.catalogueoflife.org/col
Den Heyer J (2011b) Some statistics on the taxonomy of the family Cunaxidae (Acari: Prostigmata). In: Mores GJ, Proctor H (eds) Acarology XIII: proceedings of the international congress. Zoosymposia 6:34–38

Den Heyer J, Ryke PAJ (1970) The mite complex and associated insects on citrus trees at Zebediela Estates. Wetenskaplike bydraes van die PU vir CHO, Reeks B: Natuurwetenskappe 32:1–23

Den Heyer J, Ueckermann EA, Khanjani M (2011) Iranian Cunaxidae (Acari: Prostigmata: Bdelloidea): part 2. Subfamily Cunaxinae. J Nat Hist 45:1667–1678

Den Heyer J, Ueckermann EA, Khanjani M (2013) Iranian Cunaxidae (Acari: Prostigmata: Bdelloidea): part 3. Subfamily Cunaxoidinae. J Nat Hist 47:2049–2070

Drummond FA, Logan PA, Casagrande RA, Gregson FA (1985) Host specificity tests of *Chrysomelobia labidomerae*, a mite parasitic on the Colorado potato beetle. Int J Acarol 11:169–172

Drummond FA, Casagrande RA, Logan PA (1989) Population dynamics of *Chrysomelobia labidomerae* Eickwort, a parasite of the Colorado potato beetle. Int J Acarol 15:31–45

Duso C, Pozzebon A, Capuzzo C, Malagnini V, Otto S, Borgo M (2005) Grape downy mildew spread and mite seasonal abundance in vineyards: effects on *Tydeus caudatus* and its predators. Biol Control 32:143–154

Ehrnsberger R (1979) Spinnvermogen bei Rhagidiidae (Acari: Prostigmata). Osnabrücker Naturwissenschaftliche Mitt 5:95–107

Elbadry EA, Tawfik MSE (1966) Life cycle of the mite *Adactylidium* sp. (Acarina: Pyemotidae), a predator of thrips eggs in the United Arab Republic. Ann Entomol Soc Am 59:458–461

English-Loeb G, Norton AP, Gadoury DM, Seem RC, Wilcox WF (1999) Control of powdery mildew in wild and cultivated grapes by a tydeid mite. Biol Control 14:97–103

English-Loeb G, Norton AP, Gadoury DM, Seem RC, Wilcox WF (2007) Biological control of grape powdery mildew using mycophagous mites. Plant Dis 91:421–429

Ewing HE (1917) New Acarina. Part II. Descriptions of the new species and varieties from Iowa, Missouri, Illinois, Indiana and Ohio. Bull AMNH 37:401–417

Ewing HE, Webster RL (1912) Mites associated with oyster-shell scale (*Lepidosaphes ulmi* Linne). Psyche 19:121–134

Fan Q-H, Zhang Z-Q (2005) Raphignathoidea (Acari: Prostigmata), vol 52, Fauna of New Zealand. Manaaki Whenua Press, Lincoln/Canterbury, p 396

Faroni LRA, Guedes RNC, Matioli AL (2000) Potential of *Acarophenax lacunatus* (Prostigmata: Acarophenacidae) as a biological control agent of *Rhyzopertha dominica* (Coleoptera: Bostrichidae). J Stored Prod Res 36:55–63

Faroni LRA, Guedes RNC, Matioli AL (2001) Effect of temperature on development and population growth of *Acarophenax lacunatus* (Cross and Krantz) (Prostigmata: Acarophenacidae) on *Rhyzopertha dominica* (F.) (Coleoptera: Bostrichidae). Biocontrol Sci Tech 11:5–12

Ferragut F, Gallardo A, Ocete R, López MA (2008) Natural predatory enemies of the erineum strain of *Colomerus vitis* (Pagenstecher) (Acari: Eriophyidae) found on wild grapevine populations from southern Spain (Andalusia). Vitis 47:51–54

Fisher JR, Skvarla MJ, Bauchan GR, Ochoa R, Dowling APG (2011) *Trachymolgus purpureus* sp. n., an armored snout mite (Acari: Bdellidae) from the Ozark highlands: morphology, development, and key to *Trachymolgus* Berlese. ZooKeys 125:1–34

Flaherty DL, Hoy MA (1971) Biological control of Pacific mites and Willamette mites in San Joaquin Valley vineyards: part III. Role of tydeid mites. Res Pop Ecol 13:80–96

Flechtmann CHW, Arruda GP (1967) Sobre um ácaro parasitando trips, em Recife, Pernambuco. Rev Agric Piracicaba 42:123–124

Fleschner CA, Arakawa KA (1953) The mite *Tydeus californicus* on citrus and avocado leaves. J Econ Entomol 45:1092

Forest J, Pilon JG, Paradis RO (1982) Acariens des vergers de pommiers du sud-ouest du Québec. Ann Soc Entomol Qué 27:7–67

Foster WA, Treherne JE, Evans PD, Ruscoe CNE (1979) Short-term changes in activity rhythms in an intertidal arthropod (Acarina: *Bdella interrupta* Evans). Oecologia 38:291–301

Gauchat CA (1972) A note on *Podapolipoides grassi* Berlese (Acarina: Podapolipidae), a parasite of *Chortoicetes terminifera* Walker, the Australian plague locust. J Austr Entomol Soc 11:259

Gerson U, Smiley RL, Ochoa R (2003) Mites (Acari) for pest control. Blackwell Publishing, Oxford, p 539

Goka K, Okabe K, Yoneda M, Niwa S (2001) Bumblebee commercialization will cause worldwide migration of parasitic mites. Mol Ecol 10:2095–2099

Goldarazena A, Jordana R, Zhang ZQ (1997) *Adactylidium moundi* and *A. costarricensis*, two new species of Acarophenacidae (Acari: Tarsonemida) parasitic on Thysanoptera. Int J Acarol 23:261–268

Goldarazena A, Ochoa R, Jordana R, OConnor BM (2001) Revision of the genus *Adactylidium* Cross (Acari: Heterostigmata: Acarophenacidae), mites associated with thrips (Thysanoptera). Proc Entomol Soc Wash 103:473–516

Gonçalves JR, Faroni LRD'A, Guedes RNC (2002) Pyrethroid-*Acarophenax lacunatus* interaction in suppressing the beetle *Rhyzopertha dominica* on stored wheat. Exp Appl Entomol 26:231–242

Gonçalves JR, Faroni LRD'A, Guedes RNC, de Oliveira CRF (2004) Insecticide selectivity to the parasitic mite *Acarophenax lacunatus* (Prostigmata: Acarophenacidae) on *Rhyzopertha dominica* (Coleoptera: Bostrichidae). Neotrop Entomol 33:243–248

Gonçalves JR, Faroni LRD'A, Guedes RNC, de Oliveira CRF, Garcia FM (2006a) Associação de deltametrina com *Acarophenax lacunatus* e seu impacto sobre o desenvolvimento de *Rhyzopertha dominica*. Pesqui Agropec Bras 41:1235–1240

Gonçalves JR, Faroni LRD'A, Guedes RNC, de Oliveira CRF, Garcia FM (2006b) Interação entre *Acarophenax lacunatus* (Cross & Krantz) (Prostigmata: Acarophenacidae) e *Anisopteromalus calandrae* (Howard) (Hymenoptera: Pteromalidae) sobre *Rhyzopertha dominica* (Fabricius) (Coleoptera: Bostrichidae). Neotrop Entomol 35:823–827

Gonçalves JR, Faroni LRD'A, Guedes RNC, de Oliveira CRF, Silva RM (2007) Suscetibilidade de *Rhyzopertha dominica* (Fabricius) (Coleoptera: Bostrichidae) ao enxofre. Ciênc Rural 37:1145–1148

Gordon AJ (2011) Biological Control Endeavours Against Australian Myrtle, *Leptospermum laevigatum* (Gaertn.) F. Muell. (Myrtaceae), in South Africa. Afr Entomol 19:349–355

Grout TG, Ueckermann EA (1999) Predatory mites (Acari) found under citrus trees in the Southern African Lowveld. Int J Acarol 25(3):235–238

Gupta SK, Paul K (1985) Some mites associated with bird's nests in West Bengal, with descriptions of eleven new species. Bull Zoolog Surv India 7:1–23

Haitlinger R (1989) New species *Chrysomelobia donati* and *Coccipolipus arturi* (Acari: Prostigmata: Podapolipidae) connected with insects from Cameroon and Sumatra. Wiad Pamzytol 35:161–164

Halliday RB (1991) Taxonomic background of the redlegged earth mite *Halotydeus destructor* (Tucker) (Acarina: Penthaleidae). Plant Prot Q 6:162–165

Helle W, Bolland HR, Jeurissen SHM, van Seventer GA (1984) Chromosome data on the Actinedida, Tarsonemida and Oribatida. In: Griffiths DA, Bowman CE (eds) Acarology VI, vol 1. Ellis Horwood Ltd., Chichester, pp 449–454

Hernandes FA, Feres RJF (2006a) Diversidade e Sazonalidade de Ácaros (Acari) em Seringal (*Hevea brasiliensis*, Muell. Arg.) no Noroeste do Estado de São Paulo. Neotrop Entomol 35:523–535

Hernandes FA, Feres RJF (2006b) *Tetrabdella neotropica* (Acari: Bdellidae), a new genus and species from Brazil. Zootaxa 1135:57–68

Hernandes FA, Feres RJF, Nomura F (2006) Biological cycle of *Lorryia formosa* (Acari: Tydeidae) on rubber tree leaves: a case of thelytoky. Exp Appl Acarol 38:237–242

Hernandes FA, Bernardi LFO, Ferreira RL (2011) Snout mites from caves in Brazil, with description of a new species. J Nat Hist 45:799–812

Hessein NA, Perring TM (1986) Feeding habits of the Tydeidae with evidence of *Homeopronematus anconai* (Acari: Tydeidae) predation of *Aculops lycopersici* (Acari: Eriophyidae). Int J Acarol 12:215–221

Horn TB, Johann L, Ferla NJ (2011) Ecological interactions between phytophagous and predaceous mites in citrus agroecosystems in Taquari Valley, Rio Grande do Sul, Brazil. Syst Appl Acarol 16:133–144

Hoschele W, Tanigoshi LK (1993) *Pyemotes tritici* (Acari: Pyemotidae), a potential biological control agent of *Anagasta kuehniella* (Lepidoptera: Pyralidae). Exp Appl Acarol 17:781–792

Hostounsky Z (1971) Ácaros del género *Pyemotes*, parásitos de las crias de insectos en gran escala. Poeyana 85:1–4

Hummel NA, Castro BA, MacDonald EM, Pellerano MA, Ochoa R (2009) The panicle rice mite, *Steneotarsonemus spinki* Smiley, a rediscovered pest of rice in the United States. Crop Prot 28:547–560

Husband RW (1981) The African species of *Coccipolipus* with a description of all stages of *Coccipolipus solanophilae* (Acarina: Podapolipidae). Rev Zool Afr 95:283–299

Husband RW (1991) A preliminary investigation of the phylogeny of *Tarsopolipus*, a comparison with other early derivative podapolipid genera. In: Dusbábek F, Bukva V (eds) Modern acarology, vol 2. Academia, Prague and SPB Academic Publishing bv, The Hague, pp 193–205

Husband RW (1998) Two new species of *Eutarsopolipus* (Acari: Podapolipidae) from *Agonum extensionale* and *Pterostichus lucublandus* (Coleoptera: Carabidae) from Canada, including taxonomic keys of the 13 American species of Podapolipidae from carabid beetles. Ann Entomol Soc Am 91:279–287

Husband RW, Kurosa K (2000) Two new genera and a new species of mites (Acari: Podapolipidae) associated with weevils (Coleoptera: Curculionidae) in Argentina. Int J Acarol 26:247–255

Husband RW, Sinha RN (1970) A revision of the genus *Locustacarus* with a key to the genera of the family Podapolipidae (Acarina). Ann Entomol Soc Am 63:1152–1162

Ikegami T, Ebihara T (1980) On the control of *Pyemotes ventricosus* Newport attacking the silkworm, *Bombyx mori*. Bull Ibaraki Sericult Exp Sta 34:23–28

Ireson JE (1984) The effectiveness of *Bdellodes lapidaria* (Kramer) (Acari: Bdellidae) as a predator of *Sminthurus viridis* (L.) (Collembola: Sminthuridae) in North West Tasmania. J Austr Entomol Soc 23:185–191

Ireson JE, Webb WR (1995) Effectiveness of *Neomolgus capillatus* (Kramer) (Acarina: Bdellidae) as a predator of *Sminthurus viridis* (L.) (Collembola: Sminthuridae) in Northwestern Tasmania. J Austr Entomol Soc 34:237–240

Jeppson LR, Keifer HH, Baker EW (1975) Mites injurious to economic plants. University of California Press, Berkeley

Ji J, Zhang Y, Chen X, Lin J (2007a) Studies on the functional responses of *Bdellodes japonicus* (Ehara) (Acari: Bdellidae) to *Hypogastrura communis* Folsom (Collembola: Hypogastruridae). Syst Appl Acarol 12(1):13–17

Ji J, Zhang Y, Chen X, Lin J (2007b) Studies on the functional responses of *Bdella tropica* Atyeo (Acari: Bdellidae) to *Xenylla longauda* Folsom (Collembola: Hypogastruridae) at different temperatures. Syst Appl Acarol 12(3–4):189–193

Ji J, Tao L, Chen X, Zhang Y, Lin J (2012) The effect of different temperatures on the development of the predatory mite *Bdella tropica* Atyeo (Acari: Bdellidae) with the prey *Xenylla longauda* Folsom (Collembola: Hypogastruridae). Syst Appl Acarol 17:10–14

Kaźmierski A (1998) Tydeinae of the world: generic relationships, new and redescribed taxa and keys to all species. A revision of the subfamilies Pretydeinae and Tydeinae (Acari: Actinedida: Tydeidae) – part IV. Acta Zool Cracov 41:283–455

Khanjani M, Ueckermann EA (2003) Four new tydeid species from Iran (Acari: Prostigmata). Zootaxa 182:1–11

Khaustov AA (1999) Redescription of <*Pediculoides*> *Ipidarius* Redikortzev, 1947, and a description of a new species from the genus *Paracarophenax* (Acari: Heterostigmata: Acarophenacidae). Acarina 7:57–59

Khaustov AA (2014) A new genus and species in the mite family Eupodidae (Acari, Eupodoidea) from Crimea. ZooKeys 422:11–22

Knell RJ, Webberley KM (2004) Sexually transmitted diseases of insects: distribution, evolution, ecology and host behaviour. Biol Rev 79:557–581

Knop NF, Hoy MA (1983) Biology of a tydeid mite, *Homeopronematus anconai* (n. comb.) (Acari: Tydeidae), important in San Joaquin vineyards. Hilgardia 51:1–30

Krantz GW, Walter DE (2009) A manual of acarology, 3rd edn. Texas Tech Universiy Press, Texas, p 807

Krczal H (1959) Systematik und Ökologie der Pyemotiden. In: Stammer HJ (ed) Beiträge zur Systematik und Ökologie Mitteleuropäischer Acarina, vol 1. Akademische Verlagsgesselschaft, Leipzig, pp 1–625

Kurosa K, Husband RW (2013) A new genus and species, *Simalurapolipus hiraii* (Acari: Heterostigmatina: Podapolipidae) parasitic on *Simalura coerulea* (Coleoptera: Tenebrionidae) in Japan. Syst Appl Acarol 18:252–262

Lahiri S, Podder S, Saha GK, Gupta SK (2004) Diversity of phytophagous and predatory mites occurring on medical plants in Kolkata metropolis. Proc Zool Soc 57:47–52

Laing JE, Knop NF (1983) Potential use of predaceous mites other than Phytoseiidae for biological control of orchard pests. In: Hoy MA, Cunningham GL, Knutson L (eds) Biological control of pests by mites, Special Publications, No. 3304. University of California, Berkeley, pp 12–20

Le Fichoux Y, Rack G, Motte P, Dellamonica P, Marty P (1980) Dermatite prurigineuse due à *Pyemotes zwoelferi* Krczal, 1963. A propos de plusieurs cas dans les Alpes-Maritimes. Acta Trop 37:83–89

Liguori M, Simoni S, Castagnoli M (2002) Aspects of life history of *Tydeus californicus* (Banks) (Acari: Tydeidae). Redia 85:143–153

Lindquist EE (1983) Some thoughts on the potential for use of mites in biological control including a modified concept of 'parasitoids'. In: Hoy MA, Cunningham GL, Knutson L (eds) Biological control of pests by mites, Special Publications, No. 3304. University of California, Berkeley, pp 12–20

Lindquist EE (1986) The world genera of Tarsonemidae (Acari: Heterostigmata): a morphological, phylogenetic, and systematic revision, with a reclassification of family-group taxa in the Heterostigmata. Memoirs Entomol Soc Can 136:517

Linnaeus C (1758) Systema naturæ per regna tria naturæ, secundum classes, ordines, genera, species, cum characteribus, differentiis, synonymis, locis. Tomus I, 10th edn. pp 1–4, 1–824 *Holmiæ* (Salvius)

Magowski W (1994) Discovery of the first representative of the mite subcohort Heterostigmata in the Mesozoic Siberian Amber. Acarologia 35:229–241

Mallikarjunappa S, Nageshchandra BK (1990) Observations on predatory mites in association with pecan leaf scorch mite on guava. Curr Res Univ Agric Sci 19:31–33

McCoy CW, Selhime AG, Kanavel RF (1969) The feeding behavior and biology of *Pronematus accaciae* (Acarina: Tydeidae). Florida Entomol 52:13–19

McDaniel B, Morrill W (1969) A new species of *Tetrapolipus* from *Hippodamia convergens* from South Dakota (Acarina: Podapolipidae). Ann Entomol Soc Am 62:1465–1468

McDonald G, Hoffmann AA, Ridsdill-Smith TJ (1999) Managing earth mites as establishment pests of Australian canola: a gradual path towards IMP. From http://www.regional.org.au/au/gcirc/3/144.htm?print=1. Accessed 10 Oct 2014

Mejía-Recamier BE, Castaño-Meneses G (2007) Estructura de la comunidad de cunáxidos de una selva baja caducifolia en Chamela, México. Rev Biol Trop 55:911–930

Melidossian HS, Seem RC, English-Loeb G, Wilcox WF, Gadoury DM (2005) Suppression of grapevine powdery mildew by a mycophagous mite. Plant Dis 89:1331–1338

Mendel Z, Gerson U (1982) Is the mite *Lorryia formosa* Cooreman (Prostigmata: Tydeidae) a sanitizing agent in citrus groves? Acta Oecol 3:47–51

Meyer MKP, Ueckermann EA (1987) A taxonomic study of some Anystidae (Acari: Prostigmata), vol 68, Entomology Memoir. South Africa Department of Agriculture and Water Supply, Pretoria, p 37

Michocka S (1987) Polskie roztocze (Acari) z rodzin Bdellidae i Cunaxidae. Monogr Fauny Pol Krakow 14:1–130

Mineiro JLC, Raga A, Sato ME, Matioli AL, Berton LHC (2010) Ácaros de cafeeiro (*Coffea* spp.) no Estado de São Paulo, Brasil. Parte II. Prostigmata. Biota Neotrop 10:15–225

Moore JC, Walter DE, Hunt HW (1988) Arthropod regulation of micro- and mesobiota in below-ground detrital food webs. Annu Rev Entomol 33:419–439

Moser JC (1975) Biosystematics of the straw itch mite with special reference to nomenclature and dermatology. Trans R Entomol Soc Lond 127:185–191

Moser JC, Smiley RL, Otvos S (1987) A new *Pyemotes* (Acari: Pyemotidae) reared from the douglas-fir cone moth. Int J Acarol 13:141–147

Muma MH (1960) Predatory mites of the family Cunaxidae associated with citrus in Florida. Ann Entomol Soc Am 53:321–326

Nesbitt HHJ (1947) Studies on three predaceous mites and the role they play in the orchard economy. Annual report of the Dominion Entomological Laboratory, Annapolis Royal

Newstead R, Duvall HM (1918) Bionomic, morphological and economic report on the acarids of stored grain and flour – part I. Rep Grain Pests (War) Com R Soc 2:1–48, Abstract in the review of applied entomology, Ser A, 91–94

Niemczyk E, Kaźmierski A (2002) Check list of species, occurrence and feeding of the members of families Stigmaeidae and Tydeidae in the apple orchards. In: Ignatowicz S (ed) Postępy polskiej akarologii. SGGW, Warszawa, pp 265–274, In Polish

Norton RA, Kethley JB, Johnston DE, O'Connor BM (1993) Phylogenetic perspectives on genetic systems and reproductive modes of mites. In: Wrensch DL, Ebbert MA (eds) Evolution and diversity of sex ratio in insects and mites. Chapman & Hall Publishers, New York, pp 8–99

Norton AP, English-Loeb G, Gadoury D, Seem RC (2000) Mycophagous mites and foliar pathogens: leaf domatia mediate tritrophic interactions in grapes. Ecology 81:490–499

Nucifora A, Inserra R (1967) Il *Panonychus ulmi* (Koch) nei vegneti dell'Etna. Entomologica 3:177–237

Nucifora A, Vacante V (1986) Some considerations on the role of mites in the biological control of the citrus ecosystem. In: Cavalloro R, Di Martino E (eds) Integrated pest control in citrus-groves: proceedings of the experts' meeting. Balkema, Acireale/Rotterdam [etc], p 111. Publications for the Commission of the European Communities

Oliveira CRF, Faroni LRD'A, Guedes RNC (2002a) Host egg preference by the parasitic mite *Acarophenax lacunatus* (Prostigmata: Acarophenacidae). J Stored Prod Res 39:571–575

Oliveira CRF, Faroni LRD'A, Guedes RNC, Pallini A, Gonçalves JR (2002b) Parasitismo de *Acarophenax lacunatus* (Cross and Krantz) (Prostigmata: Acarophenacidae) sobre *Dinoderus minutus* (Fabr.) (Coleoptera: Bostrichidae). Neotrop Entomol 31:245–248

Oliveira CRF, Faroni LRD'A, Guedes RNC (2003a) Host egg preference by the parasitic mite *Acarophenax lacunatus* (Prostigmata: Acarophenacidae). J Stored Prod Res 39:571–575

Oliveira CRF, Faroni LRD'A, Guedes RNC, Pallini A (2003b) Parasitism by the mite *Acarophenax lacunatus* on beetle pests of stored products. BioControl 48:503–513

Oliveira CRF, Faroni LRD'A, Guedes RNC, Araújo PA (2006a) Sobrevivência do ácaro *Acarophenax lacunatus* (Cross & Krantz) (Prostigmata: Acarophenacidae) na ausência de alimento. Neotrop Entomol 35:506–510

Oliveira CRF, Faroni LRD'A, Guedes RNC, Pallini A, Gonçalves JR (2006b) Dispersão de *Acarophenax lacunatus* (Cross & Krantz) (Prostigmata: Acarophenacidae) em trigo armazenado, sob condições artificiais. Neotrop Entomol 35:536–541

Oliveira CRF, Faroni LRD'A, Guedes RNC, Pallini A, Gonçalves JR (2006c) Parasitism of the mite *Acarophenax lacunatus* on *Tribolium castaneum*. Pesqui Agropecuária Bras 41:1059–1061

Oliveira CRF, Faroni LRD'A, Guedes RNC, Gonçalves JR, Garcia FM (2007) Biologia de *Acarophenax lacunatus* (Cross and Krantz) (Prostigmata: Acarophenacidae) sobre *Tribolium castaneum* (Herbst) (Coleoptera: Tenebrionidae) e *Cryptolestes ferrugineus* (Stephens) (Coleoptera: Cucujidae). Neotrop Entomol 36:459–464

Olivier PAS, Theron PD (2000) Pentapalpidae, a new family of eupodoid mites (Prostigmata: Eupodoidea) from South Africa. Acarologia 40:385–392

Otto JC, Halliday RB (1991) Systematics and biology of a predatory mite (*Anystis* sp.) introduced into Australia for biological control of the redlegged earth mite. Plant Prot Q 6:181–185

Oudemans AC (1914) Acarologisches aus Maulwurfsnestern. Arch Naturg 79:68–136

Plischuk S, Pocco ME, Lange CE (2013) The tracheal mite *Locustacarus buchneri* in South American native bumblebees (Hymenoptera: Apidae). Int Parasitol 62:505–507

Pringle KL (1975) The seasonal occurrence of phytophagous mites (Acari) and their predators on apple trees in the Elgin area. In: Proceedings of the 1st congress of the Entomological Society of Southern Africa, Pretoria, pp 131–140
Pugh PJA, King PE (1985) Feeding in intertidal Acari. J Exp Mar Biol Ecol 94:269–280
Pugh PJA, King PE (1986) Activity rhythms in littoral acari under laboratory conditions. J Exp Mar Biol Ecol 100:37–45
Rack G (1959) *Acarophenax dermestidarum* sp. n. (Acarina, Pyemotidae), ein Eiparasit von *Dermestes*–Arten. Z Parasitenkd 19:411–431
Ramaraju K, Poorani J (2012) A new species of *Coccipolipus* (Acari: Podapolipidae) parasitic on the giant coccinellid beetle from India. Int J Acarol 38:290–296
Regenfuss H (1968) Unterschungen zür Morphologie, Systematik und Ökologie der Podapolipidae (Acarina: Tarsonemini). Z Wissenschaftliche Zool Leipzig 177:183–282
Rhule EL, Majerus MEN, Jiggins FM, Ware RL (2010) Potential role of the sexually transmitted mite *Coccipolipus hippodamiae* in controlling populations of the invasive ladybird *Harmonia axyridis*. Biol Control 53:243–247
Riddick EW (2010) Ectoparasitic mite and fungus on an invasive lady beetle: parasite coexistence and influence on host survival. Bull Insectol 63:13–20
Rizk GN, El Badry E, Hafez SM (1979a) The effectiveness of predacious and parasitic mites in controlling *Tribolium confusum* Duv. Mesopotamia J Agric 14:167–182
Rizk GA, Soliman ZR, Ali MA (1979b) Survey on mites associated with citrus and grapevine in Minia region, Egypt. Bull Soc Entomol d'Egypte 62:105–110
Roberts JMK, Weeks AR, Hoffmann AA, Umina PA (2011) Does *Bdellodes lapidaria* (Acari: Bdellidae) have a role in biological control of the springtail pest, *Sminthurus viridis* (Collembola: Sminthuridae) in south-eastern Australia? Biol Control 58:222–229
Rocha SL, Pozo–Velázquez E, Faroni LRDA, Guedes RNC (2009) Phoretic load of the parasitic mite *Acarophenax lacunatus* (Cross and Krantz) (Prostigmata: Acarophenacidae) affecting mobility and flight take–off of *Rhyzopertha dominica* (F.) (Coleoptera: Bostrichidae). J Stored Prod Res 45:267–271
Ryder JJ, Hathway J, Knell RJ (2007) Constraints on parasite fecundity and transmission in an insect–STD system. Oikos 116:578–584
Sadeghi H, Lanieka I, Kaźmierski A (2012) Tydeoid Mites (Acari: Triophtydeidae, Iolinidae, Tydeidae) of Razavi Khorasan Province, Iran, with description of three new species. Ann Zool 62:99–114
Samarasinghe S, LeRoux EJ (1966) The biology and dynamics of the oystershell scale, *Lepidosaphes ulmi* (L.) (Homoptera: Coccidae), on apple in Quebec. Ann Entomol Soc Que 11:206–259
Santos JC (2013) Ácaros (Arthropoda: Acari) edáficos do Estado de Alagoas, com ênfase nos Mesostigmata. Dissertação Mestrado – Universidade Estadual Paulista. Faculdade de Ciências Agrárias e Veterinárias, Jaboticabal, p 74
Santos PF, Phillips J, Whitford WG (1981) The role of mites and nematodes in early stages of buried litter decomposition in desert. Ecology 62:664–669
Sathiamma B (1995) Biological suppression of the white spider mite *Oligonychus iseilemae* (Hirst) on coconut foliage. Entomon 20:237–243
Schroder RFW (1979) Host specificity test of *Coccipolipus epilachnae*, a mite parasitic of the Mexican bean beetle. Environ Entomol 8:46–47
Schroder RFW (1981) Biological control of the Mexican bean beetle, *Epilachna varivestis* Mulsant, in the United States. In: Papavizas GC (ed) Biological control in crop production. Allanhead/Osmun, Totowa, pp 351–360
Schroder RFW (1982) Effect of infestation with *Coccipolipus epilachnae* Smiley (Acarina: Podapolipidae) on fecundity and longevity of the Mexican bean beetle. Int J Acarol 8:81–84
Schruft G (1971) *Haleupalus oliveri* nov. spec., eine Dornpalpenmilbe an Reben (*Vitis* spec.) (Acari; Cunaxidae). Dtsch Entomologishe Z 18:377–382
Schruft G (1972) Das Vorkommen von Milben aus der Familie Tydeidae (Acari) an Reben. VI. Beitrag uber Untersuchungen zur Fdunistik und Biologie der Milben (Acari) an Kulturreben (*Vitis* spec.). Z Angew Zool 71:124–133

Sepasgosarian H (1984) The world genera and species of the family Cunaxidae (Actinedida: Acarida). Z Angew Zool 71:135–150

Sidorchuk E, Bertrand M (2013) New fossil labitostomatids (Acari: Labidostomatidae) from Eocene amber and presence of an apustulate species in Europe. Acarologia 53:25–39

Silva GL, Cunha US, Ferla NJ (2014) Life cycle of *Tydeus californicus* (Acari: Tydeidae) on leaves of *Inga marginata* with and without pollen of *Typha angustifolia* under laboratory conditions. Int J Acarol. doi:10.1080/01647954.2014.953999

Skvarla MJ, Fisher JR, Dowling APG (2014) A review of Cunaxidae (Acariformes, Trombidiformes): histories and diagnosis of subfamilies and genera, keys to world species, and some new locality records. ZooKeys 418:1–103

Smiley RL (1974) A new species of *Coccipolipus* parasitic on the Mexican bean beetle. J Wash Acad Sci 64:298–302

Smirnoff WA (1957) La cochenille du palmier dattier (*Parlatoria blanchardi* Targ.) en Afrique du Nord, comportement, importance économique, prédateurs et lutte biologique. Entomophaga 2:1–98

Snetsinger R (1956) Biology of *Bdella depressa*, a predaceous mite. J Econ Entomol 49:745–746

Soliman ZR, Mohamed MI (1972a) Biological studies on the soil – inhabiting mite, *Spinibdella bifurcata* (Acarina: Bdellidae) in the U.A.R. Z Angew Entomol 70:15–23

Soliman ZR, Mohamed MI (1972b) On the development and biology of the predaceous mites, *Neomolgus aegyptiacus* Sol. (Acarina: Bdellidae). Z Angew Entomol 71:1–4

Soliman ZR, Zaher MA, El–Bishlawy SM (1975) Zur Biologie der Raubmilben *Cunaxa capreolus* Berl. (Acari: Prostigmata: Cunaxidae). Anz Schädlingskunde Pflanzenschutz Umweltschutz 48:124–126

Sorensen JT, Kinn DN, Doutt RL, Cate JR (1976) Biology of the mite *Anystis agilis* (Acari: Anystidae): a California vineyard predator. Ann Entomol Soc Am 69:305–910

Sorensen JT, Kinn DN, Doutt RL (1983) Biological observations on *Bdella longicornis*: a predatory mite in California vineyards (Acari: Bdellidae). Entomography 2:297–305

Southcott RV (1976) Arachnidism and allied syndromes in the Australian region. Rec Adelaide Child Hosp 1:97–186

Steinkraus DC, Cross EA (1993) Description and life history of *Acarophenax mahunkai*, n. sp. (Acari, Tarsonemina: Acarophenacidae), an egg parasite of the lesser mealworm (Coleoptera: Tenebrionidae). Ann Entomol Soc Am 86:239–249

Swan DC (1940) The lucerne flea: its life history and control in South Australia. J Dept Agric South Austr 43:462–471

Tagore A, Putatunda BN (2003) Mites associated with some ornamental plants at Hisar, Haryana. Pest Manag Econ Zool 11:37–44

Taha HA, El-Naggar MEE, Abou-El-Ngaga MM, Soliman SM (1988) Effect of different prey species on the development and fecundity of the predaceous mite, *Neocunaxoides andrei* Baker and Hoff. (Acari: Cunaxidae). Agric Res Rev 66:129–135

Thor S (1931) Bdellidae, Nicoletiellidae, Cryptognathidae. Das Tierreich 56:87

Tomalski M, Bruce WA, Travis J, Blum MS (1988) Preliminary characterization of toxins from the straw itch mite *Pyemotes tritici*, which induces paralysis in the larvae of a moth. Toxicon 26:127–132

Treat AE (1975) Mites of moths and butterflies. Cornell University Press, London

Ueckermann EA, Grout TG (2007) Tydeoid mites (Acari: Tydeidae, Edbakerellidae, Iolinidae) occurring on citrus in Southern Africa. J Nat Hist 41:2351–2378

Vaivanijkul P, Haramoto FH (1969) The biology of *Pyemotes boylei* Krczal (Acarina: Pyemotidae). Proc Hawaiian Entomol Soc 20:443–454

van der Schyff J, Theron PD, Ueckermann EA (2003) Polytrichinae, a new subfamily of Bdellidae (Acari: Prostigmata) from the Afrotropical region. Afr Plant Prot 9:19–22

Wadhi SR, Kishore P (1975) New host record of *Sciara hirtilineata* (Sciaridae: Diptera) and *Acarophenax tribolii* Newstead and Duvall (Scutacaridae: Acarina). Indian J Entomol 35:341–342

Wallace MMH, Mahon JA (1972) The taxonomy and biology of Australian Bdellidae (Acari). I. Subfamilies Bdellinae, Spinibdellinae and Cytinae. Acarologia 14:544–580

Wallace MMH, Mahon JA (1976) The taxonomy and biology of Australian Bdellidae (Acari). II. Subfamily Odontoscirinae. Acarologia 18:65–123

Wallace MMH, Walters MC (1974) The introduction of *Bdellodes lapidaria* (Acari: Bdellidae) from Australia into South Africa for the biological control of *Sminthurus viridis* (Collembola). Austr J Zool 22:505–517

Walter DE, Kaplan DT (1991) Observations on *Coleoscirus simplex* (Acarina: Prostigmata), a predatory mite that colonizes greenhouse cultures of root knot nematode (*Meloidogyne* spp.), and a review of feeding behaviour in the Cunaxidae. Exp Appl Acarol 12:47–59

Walter DE, Proctor HC (2013) Mites: ecology, evolution and behaviour. Life at a microscale, 2nd edn. Springer, Dordrecht, p 494

Walter DE, Hunt HW, Elliot ET (1988) Guilds or functional groups? An analysis of predatory arthropods from a shortgrass steppe soil. Pedobiologia 31:247–260

Walter DE, Kaplan DT, Davis EL (1993) Colonization of greenhouse nematode cultures by nematophagous mites and fungi. J Nematol 25:789–794

Walter DE, Lindquist EE, Smith IM, Cook DR, Kantz GW (2009) Order Trombidiformes. In: Krantz GW, Walter DE (eds) A manual of acarology, 3rd edn. Texas Tech University Press, Lubbock, pp 233–420

Weiser J, Hardy I (1962) *Pyemotes* – mites as parasites of termites. Z Angew Entomol 51:94–97

Willmann C (1951) Die hochalpine Milbenfauna der mittleren Hohen Tauern insbesondere der Groβlockner–Gebietes (Acari). Bonner Zool Beitr 2:141–176

Womersley H (1933) A possible biological control of the clover springtail or lucerne flea (*Sminthurus viridis* L.) in Western Australia. J Counc Sci Ind Res 6:83–91

Wrensch DL, Bruce WA (1991) Sex ratio, fitness and capacity for population increase in *Pyemotes tritici* (L.–F. and M.) (Pyemotidae). In: Schuster R, Murphy PW (eds) The Acari, reproduction, development and life history strategies. Chapman and Hall, London, pp 209–221

Wrensch D, Ebbert MA (1993) Evolution and diversity of sex ratio in insects and mites. Chapman and Hall, New York, p 630

Yu L, Zhang Z-Q, He L (2010) Two new species of *Pyemotes* closely related to *P. tritici* (Acari: Pyemotidae). Zootaxa 2723:1–40

Zacarias MS, de Moraes GJ (2002) Mite diversity (Arthropoda: Acari) on euphorbiaceous plants in three localities in the State of São Paulo. Biota Neotrop 2:1–12

Zacharda M (1980) Soil mites of the family Rhagidiidae (Actinedida: Eupodoidea). Morphology, systematics, ecology. Acta Universitatis Carolinae. Biologica 1978:489–785

Zaher MA, Shehata KK (1963) Biological studies on *Tydeus culifornicus* (Banks) in Egypt (U.A.R.). Bull Soc Entomol d'Égypte 47:297–300

Zaher MA, Soliman ZR, El-Bishlawy SM (1975) Feeding habits of the predacious mite, Cunaxa capreola (Acarina: Cunaxidae). Entomophaga 20:209–212

Chapter 7
Stigmaeidae

Qing-Hai Fan and Carlos H.W. Flechtmann

7.1 Taxonomic Position

The family Stigmaeidae belongs to the mite superorder Acariformes, order Trombidiformes, suborder Prostigmata, supercohort Eleutherengoides, cohort Raphignathina, Superfamily Raphignathoidea (Lindquist et al. 2009).

7.2 Origin

Stigmaeidae has a world wide range of distribution. Its species have been discovered from all biogeographical regions though most of them were discovered from the Palaearctic, Oriental, Nearctic, Afrotropical and Australian regions (Fan and Zhang 2005).

Members of the family have been collected in a large range of habitats (Table 7.1). More than 35 % of the species are free-living predators on plant leaves and branches. These species are mainly from two genera, *Agistemus* (Fig. 7.1) and *Zetzellia*. Approximately 27 % of the species such as those of *Eustigmaeus* (Fig. 7.2) and *Ledermulleriopsis* inhabit the soil-litter habitats and are most likely phytophages, feeding on moss or lichen (Gerson 1972). A few species are found in association with sand flies.

Q.-H. Fan (✉)
Plant Health & Environment Laboratory, Ministry for Primary Industries, Auckland, New Zealand
e-mail: qinghai.fan@mpi.govt.nz

C.H.W. Flechtmann
Departamento de Entomologia e Acarologia, Escola Superior de Agricultura "Luiz de Queiroz", Universidade de São Paulo, 13418-900 Piracicaba, SP, Brazil
e-mail: chwflech@usp.br

D. Carrillo et al. (eds.), *Prospects for Biological Control of Plant Feeding Mites and Other Harmful Organisms*, Progress in Biological Control 19,
DOI 10.1007/978-3-319-15042-0_7

Table 7.1 Number (approx.) of stigmaeid species in different types of habitat

Genus	Foliage, buds and flowers	Soil/litter, moss and lichen	Fallen leaves and soil/litter	Fresh water surface	Nest of birds and rats	Sand flies	Total
Agistemus	77	3	3		1		84
Caligohomus				2			2
Cheylostigmaeus	8	16		6	1		31
Eryngiopus	22	5	1				28
Erynglpusopis	1						1
Eustigmaeus	2	100		5	1	5	113
Farajimaeus	1						1
Gymnostigmaeus	1						1
Indostigmaeus	1						1
Ledermuelleriopsis	1	31		1			33
Macrostigmaeus			1				1
Makilingeria			2				2
Mediolata	20	10	4				34
Mediostigmaeus	2						2
Mendanaia			1				1
Mullederia	7			1			9[a]
Muellederiopsis	1	1					2
Neilstigmaeus	1						1
Parastigmaeus	1	1					2
Paravillersia		1					1
Pilonychiopus	3						3
Postumius		2					2
Primagistemus	1		1				2
Prostigmaeus		5					5
Pseudostigmaeus	4	2					6
Scutastigmaeus		3					3
Stigmaeus	8	102	2	1	2	2	117
Storchia		11					11
Summersiella	2						2
Villersia		2					2
Villersiella		1					1
Zetzellia	26	2	1				29
Zetzelliopsis		1					1
Total (33 genera)	190	299	26	16	5	7	534[a]

[a]The habitat of one species is unknown

Fig. 7.1 Line drawing of *Agistemus* sp.

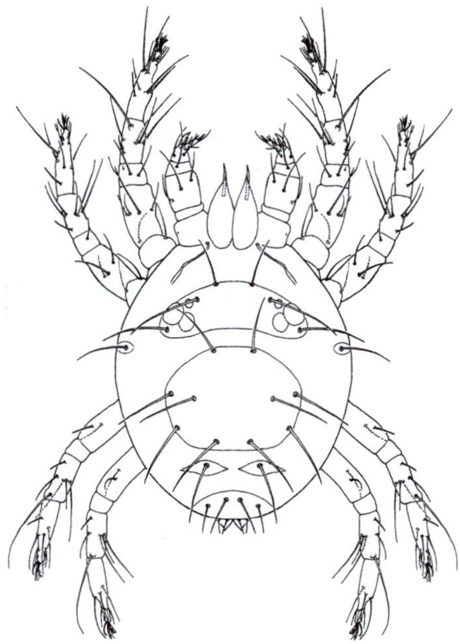

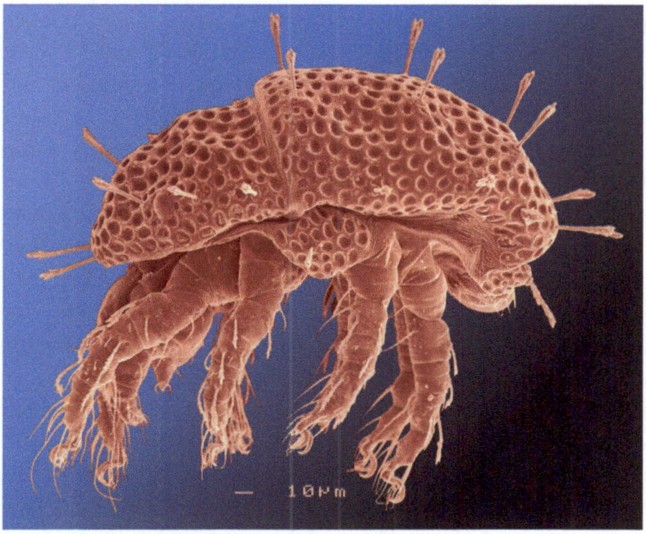

Fig. 7.2 Scanning electron micrograph image of *Eustigmaeus* sp.

7.3 Basic Biology

Stigmaeid mites have five life stages, egg, larva, protonymph, deutonymph and adult. There are also three quiescent periods after larval, protonymphal and deutonymphal stages.

7.3.1 *Development and Reproduction*

A stigmaeid mite generally completes a generation within no more than 4 weeks. *Agistmeus exsertus* completes its immature stages in 6.7 days feeding on *Tetranychus cinnabarinus* and 9–10.5 days on pollen (Abo Elghar et al. 1969). *Agistemus olivi* completes a generation in 14–15 days feeding on eriophyid mites *Aceria oleae* and *Tegolophus hassani* at 25 °C and 70–75 % rh. The eggs take the longest development time among immature stages (3.8 days) (Abou-Awad et al. 2010). The males usually reach their maturity slightly earlier than females and mate with immediately moulted adult females. After 2 days of pre-oviposition the females start to lay eggs and each female can lay 58–59 eggs in 19–21 days of oviposition period when fed on eriophyid mites *Aceria oleae* or *Tegolophus hassani* at 25 °C and 70–75 % rh (Abou-Awad et al. 2010).

Temperature, humidity and food sources are the three main factors that affect the development and reproduction (White and Laing 1977). *Agistemus exsertus* had shorter development time but lower fecundity at 30 °C and 35 °C than at 25 °C, 20 °C and 15 °C (Yue and Childers 1994). *Agistemus terminalis* laid more eggs at 25 °C than at 30 °C and 20 °C (Inoue and Tanaka 1983). Low humidity reduces the oviposition rate, egg hatchability and survival of all stages of *A. terminalis* (Hu et al. 2001).

Suitable prey shortens the development time and increases the oviposition rate. *Agistemus buntex* developed much faster when fed on *Tetranychus urticae* or *Eutetranychus orientalis* than on *Panonychus citri* (Khan and Afzal 2005). The total number of eggs laid by a female *Agistemus exsertus* varies from 60 to 67 when fed on *Bemisia tabaci*, *Brevipalpus pulcher* and *Eutetranychus orientalis*, while 91 when fed on *Tetranychus urticae* (Abou-Awad and Elsawi 1993).

Females start to produce eggs after half day to 4 days. Unmated females produce only male eggs (arrhenotokous) (Gerson et al. 2003). The sex ratio of offspring is female-biased (Table 7.3). Multiple-mated females produce more eggs than the single-mated females (Abou-Awad and Reda 1992; Rasmy and Hussein 1996). Females usually produce 1–4 eggs/day and the number of eggs produced is strongly affected by food type and quantity (Abou-Awad and Elsawi 1993; Momen 2001; Goldarazena et al. 2004; El Sawi and Momen 2006; Momen 2011, 2012) and temperature (Inoue and Tanaka 1983; Yue and Childers 1994).

7.3.2 Diapause

There have been very few records on the diapause in Stigmaeidae. White and Laing (1977) noticed that *Zetzellia mali* had diapause which could be terminated by the treatment of cold reactivation. *Agistemus fleschneri* overwinters in litter at the base of trees (Childers and Enns 1975b) or under bark scales on apple (Holdsworth 1972a, b). It became active and started feeding immediately at temperatures above 5 °C but didn't lay eggs until after 11 days at room temperature (White and Laing 1977).

Most stigmaeid species are active throughout the year. Although no oviposition occurred in winter months *Zetzellia graeciana* was active on the bark and fed on a variety of prey in Italy. *Agistemus longisetus* didn't stop breeding throughout the year and *A. novazelandicus* were active in winter in Auckland (Collyer 1964).

7.3.3 Predation

The predatory species of Stigmaeidae are generalist predators mainly feeding on gall and rust mites (Eriophyoidea), spider mites (Tetranychidae), false spider mites (Tenuipalpidae), broad mites (Tarsonemidae) and other mites in the field (Gerson et al. 2003) and green houses (Zhang 2003). They are also known to feed on pollen of various plants, eggs and crawlers of small insects such as Hemiptera and eggs of Lepidoptera (Table 7.2). At least 27 species have been recorded preying on various prey. These include 20 species of *Agistemus*, 1 species of *Caligohomus*, 1 species of *Mediostigmaeus* and 5 species of *Zetzellia* (Table 7.2).

Agistemus and *Zetzellia* usually prefer to feed on immatures than adults of spider mites and small insects. With the aid of the first pair of legs and palps they insert the chelicerae into the body of the prey and then suck the fluids. *Agistemus exsertus* punctures tetranychid eggs but does not necessarily suck their contents completely (Zaher and El-Badry 1961; ElBadry et al. 1969a).

An adult female *Agistemus exsertus* can consume 17, 12, 9 or 8 eggs/day of *Tetranychus urticae*, *Eutetranychus orientalis*, *Bemisia tabaci* or *Brevipalpus pulcher*, respectively (Abou-Awad and Elsawi 1993). It can also consume six larvae/day of *Tetranychus urticae* (Nawar 1992) or 5.6 immature mites/day of *Tetranychus arabicus* (Hanna et al. 1984; Hafez et al. 1983). When fed on eggs *A. exsertus* developed faster, lived longer and laid more eggs than those fed on postembryonic stages (Elbadry et al. 1969b). *Zetzellia mali* prefers to feed on active eriophyid mite *Aculus schlechtendali* than eggs of spider mites *Panonychus ulmi*, and eggs than quiescent nymphs of *P. ulmi* (Clements and Harmsen 1993).

Plant pollen is also one of the food sources of stigmaeid mites. The development time of the immature stages of *Agistemus exsertus* feeding on palm pollen were similar to those on *Tetranychus cinnabarinus* (Wafa et al. 1969). *Zetzellia malvinae* can complete its lifecycle when fed on *Typha angustifolia* but its oviposition rate was lower than those fed on spider mite eggs (De Vis et al. 2006).

Table 7.2 Stigmaeid species and their prey

Stigmaeid species	Prey	References
Agistemus africanus Gonzalez-Rodriguez	TETRANYCHIDAE: *Panonychus ulmi* on apple	Pringel (1975)
Agistemus aimogastaensis Leiva, Fernandez, Theron & Rollard	ERIOPHYIDAE: *Aceria oleae* and *Oxycenus maxwelli*	Leiva et al. (2013)
Agistemus arcypaurus Gonzalez-Rodriguez	TENUIPALPIDAE: *Brevipalpus phoenicis* on tea	Oomen (1982)
Agistemus brasiliensis Matioli, Ueckermann & Oliveira	TENUIPALPIDAE: *Brevipalpus phoenicis* TETRANYCHIDAE: *Panonychus citri* and *Tetranychus urticae* TYPHACEAE: *Typha* sp. pollen	Matioli et al. (2002)
Agistemus buntex Chaudhri	TETRANYCHIDAE: *Eutetranychus orientalis*	Khan and Afzal (2005)
	TETRANYCHIDAE: *Panonychus citri* and *Tetranychus urticae*	Chaudhri and Akbar (1985), Khan and Afzal (2005)
Agistemus collyerae Gonzalez-Rodriguez	ERIOPHYIDAE: *Aculus fockeui* on peach	Castagnoli and Nannelli (1987)
Agistemus cyprius Gonzalez-Rodriguez	ERIOPHYIDAE: *Aculus* spp. on nectarine and peach	Villaronga et al. (1993)
	TETRANYCHIDAE: *Panonychus citri*	Goldarazena et al. (2004)
	AIZOACEAE: ice plant (*Malephora crocea*) pollen	Goldarazena et al. (2004)
Agistemus denotatus Gonzalez-Rodriguez	TENUIPALPIDAE: *Brevipalpus phoenicis* on tea	Oomen (1982)
Agistemus exsertus Gonzalez-Rodriguez	ACARIDAE: *Tyrolichus casei* in laboratory	Rasmy et al. (1987)
	ERIOPHYIDAE: *Aceria dioscoridis* on sweet potato	El-Bagoury and Reda (1985), Rasmy et al. (1987)
	Aceria ficus on fig	El-Halawany et al. (1990)
	Aceria sheldoni on citrus	Sternlicht (1969)
	Aculops lycopersici on tomato	Osman and Zaki (1986)
	Aculops schlechtendali on apple	El-Halawany et al. (1990)
	Aculus ligustri on *Ligustrum lucidum*	Hu et al. (1997)
	Colomerus vitis on grapes	Osman et al. (1991)
	PHYTOSEIIDAE: *Typhlodromus negevi*	Abdel-Khalek and Momen (2013)
	TENUIPALPIDAE: *Brevipalpus obovatus*	Wang (1981)

(continued)

Table 7.2 (continued)

Stigmaeid species	Prey	References
	Brevipalpus pulcher	Abou-Awad and Elsawi (1993)
	Tenuipalpus granati	Yousef et al. (1982)
	TETRANYCHIDAE: *Eotetranychus kankitus*	Wang (1981)
	Eutetranychus orientalis	Elbadry et al. (1969a, b), Abou-Awad and Elsawi (1993)
	Oligonychus biharensis	Wang (1981)
	Oligonychus mangiferus	Hanna et al. (1984)
	Panonychus citri on citrus	Ehara (1964), González-Rodriguez (1965), Tanaka (1966), Wang (1981), Yue and Childers (1994), Ji et al. (2013)
	Panonychus ulmi	Wang (1981)
	Tetranychus arabicus	Soliman et al. (1976), Hanna et al. (1984)
	Tetranychus cinnabarinus	Abo Elghar et al. (1969), Afify et al. (1969), El Badry et al. (1969a, b)
	Tetranychus cucurbitacearum	Hafez et al. (1983)
	Tetranychus kanzawai	Ehara (1964)
	Tetranychus urticae	Hafez et al. (1983), Abou-Awad and Reda (1992), Nawar (1992), Abou-Awad and Elsawi (1993), Momen (2011)
	ALEYRODIDAE: *Bemisia tabaci*	Abou-Awad and Elsawi (1993)
	DIASPIDIDAE: *Chrysomphalus ficus*, *Lepidosaphes baledi* and *Lepidosaphes beckii*	Soliman et al. (1976)
	Chrysomphalus ficus, *Parlatoria blanchardi*	El Sawi and Momen (2006)
	Parlatoria zizyphus	Momen (2001)
	GELECHIIDAE: *Phthorimaea operculella*	Momen and El-Sawy (2006)
	NOCTUIDAE: *Agrotis ipsilon*, *Spodoptera littoralis*	Momen and El-Sawy (2006)
	PYRALIDAE: *Ephestia kuehniella*	Momen (2001)
	not specified	Romeih et al. (2004)
	Galleria mellonella	Momen (2011)
	ARECACEAE: *Phoenix dactylifera* pollen	Abo Elghar et al. (1969), Wafa et al. (1969)

(continued)

Table 7.2 (continued)

Stigmaeid species	Prey	References
	EUPHORBIACEAE: *Ricinus communis* pollen	Abo Elghar et al. (1969), Rasmy et al. (1987)
	POACEAE: *Zea mays* pollen	Abo Elghar et al. (1969), Rasmy (1975)
Agistemus fanari Dosse	TENUIPALPIDAE: *Cenopalpus* sp. TETRANYCHIDAE: *Eotetranychus cernuus*, *Tetranychus cinnabarinus* and *T. urticae* complex TYDEIDAE: *Brachytydeus* sp.	Dosse (1967)
Agistemus fleschneri Gonzalez-Rodriguez	ERIOPHYIDAE: *Aculus schlechtendali* on apple	Nelson et al. (1973)
	TENUIPALPIDAE: *Brevipalpus phoenicis* on tea	Oomen (1982)
	TETRANYCHIDAE: *Eotetranychus sexmaculatus* on citrus	Muma (1975)
	Eutetranychus banksi on sweet potato	Zaher and El-Badry (1961)
	Panonychus ulmi in orchard	Nelson et al. (1973)
	Tetranychidae on apple	Oatman (1973), Childers and Enns (1975a), Forest et al. (1982)
	Tetranychus cinnabarinus on cotton	Zaher and El-Badry (1961)
	Tetranychus kanzawai on tea	Osakabe (1963)
	Tetranychus urticae in orchard	Nelson et al. (1973)
Agistemus floridanus Gonzalez-Rodriguez	ERIOPHYIDAE: *Calacarus heveae*	Ferla and De Moraes (2003)
	Phyllocoptruta oleivora on citrus	Muma and Selhime (1971)
	TARSONEMIDAE: *Fungitarsonemus peregrinus* on citrus	Muma et al. (1961)
	P *olyphagotarsonemus latus*	Ferla and De Moraes (2003)
	TENUIPALPIDAE: *Brevipalpus phoenicis*	Muma and Selhime (1971)
	T *enuipalpus heveae*	Ferla and De Moraes (2003)
	TETRANYCHIDAE: *Eotetranychus sexmaculatus* on citrus	Muma et al. (1961)
	O *ligonychus gosspii*	Ferla and De Moraes (2003)
	Panonychus citri	Goldarazena et al. (2004)

(continued)

Table 7.2 (continued)

Stigmaeid species	Prey	References
	Tetranychus mexicanus	Ferla and De Moraes (2003)
	TYDEIDAE: *Tydeus* sp. on citrus	Muma et al. (1961)
	ALEYRODIDAE: *Aleurothrixus floccosus*	Muma (1960)
	Dialeurodes citrifolii on citrus	Muma and Selhime (1971)
	DIASPIDIDAE: *Chrysomphalus ficus* and *Lepidosaphis beckii* on citrus	Muma (1952)
	TYPHACEAE: *Typha angustifolia* pollen	Ferla and De Moraes (2003)
Agistemus herbarius Kuznetzov & Wainstein	TETRANYCHIDAE: *Tetranychus turkestani*	Sizova and Kuznetsov (1980)
Agistemus industani Gonzalez-Rodriguez	ERIOPHYIDAE: *Phyllocoptruta oleivora*	Matioli et al. (2002)
	TENUIPALPIDAE: *Brevipalpus phoenicis*	Matioli et al. (2002)
	TETRANYCHIDAE: *Panonychus citri*	Matioli et al. (2002)
	Panonychus citri	Goldarazena et al. (2004)
	AIZOACEAE: ice plant (*Malephora crocea*) pollen	Matioli et al. (2002), Goldarazena et al. (2004)
Agistemus longisetus Gonzalez-Rodriguez	ERIOPHYIDAE: unidentified species on fruit tree	Collyer (1964)
	TENUIPALPIDAE: *Brevipalpus* sp. on fruit tree	
	TETRANYCHIDAE: *Bryobia rubrioculus*, *Panonychus citri*, *Panonychus ulmi*, *Tetranychus lambi* and *Tetranychus urticae* on fruit tree	
	Panonychus ulmi on apple	Gonzalez (1971)
	TYDEIDAE: unidentified species on fruit tree	Collyer (1964)
Agistemus novazelandicus Gonzalez-Rodriguez	TENUIPALPIDAE: *Brevipalpus* sp.	González-Rodriguez (1963)
Agistemus olivi Romeih	ERIOPHYIDAE: *Aceria mangiferae*, *Aculus fockeui* and *Aculops lycopersici*	Momen (2012)
	Aceria oleae and *Tegolophus hassani*	Abou-Awad et al. (2010)
Agistemus pallinii Matioli, Tavares & Pallini	TETRANYCHIDAE: *Panonychus citri*	Matioli et al. (2007)
	ACARIDAE: *Tyrophagus* sp. in laboratory	

(continued)

Table 7.2 (continued)

Stigmaeid species	Prey	References
	TETRANYCHIDAE: *Oligonychus ilicis*	
	TYPHACEAE: *Typha* sp. pollen	
Agistemus terminalis	ERIOPHYIDAE: eriophyids	Ehara (1964)
	TENUIPALPIDAE: *Brevipalpus phoenicis* on tee	Oomen (1982)
	TETRANYCHIDAE: *Eotetranychus sexmaculatus*	Wang (1981)
	Panonychus citri on citrus	Ehara (1964), Razaq et al. (2000)
	DIASPIDIDIAE: *Unaspis yanonensis* on citrus	Ehara (1964)
Agistemus vulgaris Soliman & Gomaa	ERIOPHYIDAE: *Aculops lycopersici*	Mohamed (2014)
	TETRANYCHIDAE: *Oligonychus sayedi* and *Tetranychus urticae*	
Caligohomus durus Fan & Walter	Nematodes in laboratory	Fan and Walter (2004)
Mediostigmaeus citri Rakha & McCoy	TYDEIDAE: *Pseudolorryia mumai* eggs (as *Tydeus mumai* Baker) on citrus	Rakha and McCoy (1984)
Zetzellia graeciana Gonzalez-Rodriguez	TETRANYCHIDAE: *Panonychus ulmi* and *Tetranychus urticae*	Inserra (1973)[1970]
Zetzellia mali (Ewing)	ERIOPHYIDAE: *Aculus cornutus* on stonefruits	Rice et al. (1976)
	Aculus schlechtendali on apple	Nesbitt (1946), Hoyt (1969), Herbert and Butler (1973), Delattre (1974), Santos (1976a, 1991), White and Laing (1977), Laing and Knop (1983), Vogt et al. (1990), Clements and Harmsen (1993), Denizhan 2011
	Calepitrimerus vitis on grape	Schruft (1969), Liguori (1987), Vacante and Tropea Grazia (1989), Lászlóné and György (1995)
	Colomerus vitis on grape	Liguori (1987), Vacante and Tropea Grazia (1989)
	PHYTOSEIIDAE: *Amblyseius fallacis*	MacRae and Croft (1996)
	Metaseiulus occidentalis	Croft and MacRae (1992, 1993), MacRae and Croft (1996)

(continued)

Table 7.2 (continued)

Stigmaeid species	Prey	References
	Metaseiulus pomi eggs	Nesbitt (1946)
	Typhlodromus pyri	Croft and MacRae (1992, 1993)
	Typhlodromus caudiglans eggs	Clements and Harmsen (1990)
	TENUIPALPIDAE: *Brevipalpus lewisi* in vineyard	Göven et al. (2009)
	Cenopalpus irani	Darbemamieh et al. (2009)
	Tenuipalpus granati in vineyard	Göven et al. (2009)
	TETRANYCHIDAE: *Amphitetranychus viennensis*	Zahedi-Golpayegani et al. (2007)
	Bryobia praetiosa eggs on apple	Nesbitt (1946)
	Bryobia rubrioculus	Darbemamieh et al. (2009)
	Eotetranychus frosti	Khodayari et al. (2010)
	Panonychus ulmi on apple	Parent and Leroux (1956), Parent (1961), Holdsworth (1972a, b), Delattre (1971, 1974), Santos (1976a, 1991), Laing and Knop (1983), Khajuria (2009), Rahmani et al. (2010)
	Tetranychus canadensis	Hoyt (1969), Herne and Lund (1973), White and Laing (1977), Laing and Knop (1983)
	Tetranychus mcdanieli	Hoyt (1969)
	Tetranychus turkestani	Khodayari et al. (2009)
	Tetranychus urticae on apple	Parent and LeRoux (1956), Hoyt (1969), Santos (1976a, b, 1991), Laing and Knop (1983), Khodayari et al. (2008)
	TYDEIDAE: *Tydeus longisetosus*	Khodayari et al. (2009, 2010)
Zetzellia malvinae Matioli, Ueckermann & de Oliveira	TENUIPALPIDAE: *Brevipalpus heveae* on rubber tree	de Vis et al. (2008)
Zetzellia maori Gonzalez-Rodriguez	TENUIPALPIDAE: *Brevipalpus phoenicis* on tea	Oomen (1982)
Zetzellia talhouki Dosse	ERIOPHYIDAE: *Acalitus phloecoptes*, *Aculus fockeui*, *Diptacus gigantorhynchus* and *Phyllocoptes abaenus*	Bayan (1988)
	TETRANYCHIDAE: spider mites (not specified)	Dosse (1967)

Predation rates are mainly affected by temperature, humidity and prey type and prey density. The number of prey consumed is positively correlated with the temperature and density of prey within a certain range (Afify et al. 1969; Yue and Childers 1994) and prey type (Inoue and Tanaka 1983; Nawar 1992; Abou-Awad and Elsawi 1993; Yue and Tsai 1995).

Cannibalism and intraguild predation commonly occur in Stigmaeidae and between Stigmaeidae and Phytoseiidae, especially when prey density is low and living space is limited (Clements and Harmsen 1992; Croft and MacRae 1993; Slone and Croft 2001). *Agistemus exsertus* and *Zetzellia mali* reproduced successfully when fed on their own eggs (Clements and Harmsen 1992; Abdel-Khalek and Momen 2013) but the intrinsic rates of increase and finite rates were lower than those fed on *Tetanychus urticae* and *Typhlodromus negevi* (Abdel-Khalek and Momen 2013). Stigmaeids develop more effectively than phytoseiids when prey density is low but phytoseiids are more effective at high prey densities (Clements and Harmsen 1992).

7.4 Target Pest

As generalist predators stigmaeid mites prey on a large range of organisms mainly from three families, Eriophyidae, Tenuipalpidae and Tetranychidae in orchards (Table 7.2). Some of the economically important species such as the apple rust mite *Aculops schlechtendali*, the tomato russet mite *Aculops lycopersici*, the citrus red mite *Panonychus citri*, the European red mite *Panonychus ulmi*, the twospotted spider mite *Tetranychus urticae*, the Oriental red mite *Eutetranychus orientalis*, the false spider mite *Brevipalpus phoenicis* and some related species may be considered as the target pests of the genera *Agistemus* and *Zetzellia.*

7.5 Efficacy

The functional responses have been studied for *Agistemus exsertus* (Abou-Awad and Elsawi 1993; Hu et al. 1997; El Sawi and Momen 2006; Saber 2012) and *Zetzellia mali* (Santos 1982; Lawson and Walde 1993). *Agistemus exsertus* approximated Holling's Type II functional response to different densities of nymphs of *Tetranychus urticae* (Saber 2012). The maximum predation of *A. exsertus* to nymphs and adults of *Aculus ligustri* were 50.01 and 30.14, respectively (Hu et al. 1997). *Zetzellia mali* consumes more eggs (4.1/day) of *Panonychus ulmi* when the prey density was 30/leaf disc (Santos 1982). It seemed that *Zetzellia mali* could adjust its predation and oviposition rates at low prey densities (Santos 1982).

The life table parameters of stigmaeid species are summarized in Table 7.3. Temperature and prey sources are known to be the two main factors affect these parameters.

Table 7.3 Life table parameters of stigmaeid species (*A. Agistemus*; *Z. Zetzellia*)

	Prey, temperature (°C), humidity (% rh)	Generation time (day)	R_0	r_m	λ	Sex ratio (% females)	References
A. cyprius	*Panonychus citri*, 25, 90–97	32.50	6.36	0.0562	1.10	60	Goldarazena et al. (2004)
A. cyprius	Ice plant pollen, 25, 90–97	45.20	3.58	0.0311	1.00	70	Goldarazena et al. (2004)
A. exsertus	*Agrotis ipsilon*, 27, 70–75	18.91	30.378	0.18	1.20	60	Momen and El-Sawy (2006)
	Bemisia tabaci, 27–29, 60–80	27.69	33.23	0.13	1.14	62.6	Abou-Awad and Elsawi (1993)
	Brevipalpus pulcher, 27–29, 60–80	28.30	31.75	0.12	1.13	54.1	Abou-Awad and Elsawi (1993)
	Chrysomphalus ficus, 27, 70	22.64	45.5	0.17	1.18	60	El Sawi and Momen (2006)
	Ephestia kuehniella, 27, 70	20.99	61.25	0.196	1.22	72	Momen (2001)
	Eriophyes olive, 27, 60–80	16.61	40.59	0.222	1.249	–	El-Laithy (1998)
	Eutetranychus orientalis, 27–29, 60–80	24.90	35.32	0.14	1.15	53.9	Abou-Awad and Elsawi (1993)
	Galleria mellonella, 30, 70–75	18.98	92.30	0.2348	1.2692	70	Momen (2011)
	Panonychus citri, 35, 90	12.6	11.98	0.2255	–	–	Yue and Childers (1994)
	Panonychus citri, 30, 90	11.4	17.73	0.2114	–	–	Yue and Childers (1994)
	Panonychus citri, 25, 90	14.5	24.00	0.2288	–	–	Yue and Childers (1994)
	Panonychus citri, 20, 90	18.1	25.30	0.1650	–	–	Yue and Childers (1994)
	Panonychus citri, 15, 90	35.9	23.21	0.0690	–	–	Yue and Childers (1994)
	Parlatoria blanchardi, 27, 70	20.58	7.7	0.098	1.11	50	El Sawi and Momen (2006)
	Parlatoria zizyphus, 27, 70	21.70	43.71	0.174	1.19	66	Momen (2001)
	Phthorimaea operculella, 27, 70–75	20.78	2.29	0.03	1.04	50	Momen and El-Sawy (2006)
	Spodoptera littoralis, 27, 70–75	17.89	44.07	0.21	1.23	64	Momen and El-Sawy (2006)
	Tetranychus urticae, 27–29, 60–80	26.21	57.91	0.15	1.16	65.5	Abou-Awad and Elsawi (1993)

(continued)

Table 7.3 (continued)

	Prey, temperature (°C), humidity (% rh)	Generation time (day)	R_0	r_m	λ	Sex ratio (% females)	References
	Tetranychus urticae, 30, 70–75	19.41	57.291	0.2084	1.2318	65	Momen (2011)
A. floridanus	*Tenuipalpus heveae*, 25, 80	19.2	13.2	0.16	1.1	–	Ferla and De Moraes (2003)
A. floridanus	*Panonychus citri*, 25, 90–97	19.90	7.9	0.1001	1.1	70	Goldarazena et al. (2004)
A. industani	*Panonychus citri*, 25, 90–97	37.80	18.70	0.1031	1.10	50	Goldarazena et al. (2004)
A. industani	Ice plant pollen, 25, 90–97	35.10	10.07	0.1201	1.10	70	Goldarazena et al. (2004)
A. orbicularis	*Brevipalpus phoenicis*, 17.0–23.1, –	31.9	14.6	0.0840	–	–	Oomen (1982)
A. olive	*Aceria mangiferae*, 30, 75	17.188	92.779	0.263	1.301	75	Momen (2012)
	Aculops fockeui, 30, 75	17.720	83.376	0.249	1.283	80	
	Aculops lycopersici, 30, 75	20.001	35.36	0.178	1.195	60	
A. terminalis	*Panonychus citri*, 20, –	30.9	13.2	0.083	–	–	Inoue and Tanaka (1983)
	Panonychus citri, 25, –	17.4	13.5	0.149	–	–	
	Panonychus citri, 30, –	12.9	9.1	0.171	–	–	
Z. mali	*Aculus schlechtendali*, 19, 56	21.0	10.04	0.109	1.12	–	White and Laing (1977)
Z. malvinae	*Tenuipalpus heveae*, 25, 82	44.1	29.6	0.077	1.08	69.7	De Vis et al. (2008)

7.6 Seasonality

A few species have been studied for their seasonal fluctuations. *Zetzellia mali* was frequently collected on plums from late August to early October in California (Rice et al. 1976). It peaked in July on apple trees in Woodside, New York (Santos 1984) and in early September on apple trees in Southern Ontario (Amano and Chant 1990). It first appears in the early July and reached the peak in the mid-September in apple orchards of Zanjan, Iran (Rahmani et al. 2010). *Agistemus longisetus* started to present in late December or early January, became abundant (160 mites/leaf) in February in unsprayed apple tree in Auckland (Southern Hemisphere) (Collyer 1964).

The population of *Agistemus exsertus* started to increase in February and reached a peak in September (13.75 mites/leaf), then decreased from November to January on guava in Qalyoubia Egypt. A positive correlation was observed between the predatory mite and temperature. The stigmaeid mite was relatively numerous and widely distributed associated with eriophyid and tenuipalpid mites.

7.7 Alternative Prey and Mass Rearing

Agistemus and *Zetzellia* have relatively broad range of prey (Table 7.2). Many species can successfully develop and reproduce on phytophagous mites, insects and pollen. For example, *A. exsertus* feeds on mites from five families, eggs or crawlers of diaspidid scale insects and white flis, eggs of pyralid and noctuid moths, pollens from three plant families (Table 7.2). It can also develop and reproduce on artificial diets composed of yeast, milk, amino acids, and sugar though the number of eggs laid per female per day is one-third less than that for mites reared on a natural diet of pollen (Reda 1990). The lifespan of females on the artificial diet is equal to that on the standard diet. Some alternative prey which can be cost effective may be considered as food source for mass rearing of stigmaeid mites, i.e., *Tyrolichus casei*, *Ephestia kuehniella*, *Bemisia tabaci* and pollen of *Typha angustifolia* and *Zea mays*.

7.8 Compatibility with Other IPM Measure

Stigmaeid mites are susceptible to a large number of miticides, insecticides and fungicides which increase the motality of all stages and reduce the egg production. Benomyl and sulphur are highly toxic to *Agistemus fleschneri* (Abo Elghar et al. 1971; Childers and Enns 1975b); fungicides captan, flusilazole and mancozeb are toxic to *Zetzellia mali* (Hagley and Biggs 1989). The pyrethroid applications affect the population development and the spatial distribution of *Z. mali* (Li and Harmsen

1992; Li et al. 1992). All chemicals except pure mineral oil and lime mixture are highly toxic to overwintering population of *Zetzellia mali* (Fischer-Colbrie and El-Borolossy 1988).

Some pesticides appeared non-toxic to stigmaeid mites, e.g., diflubenzuron to *Zetzellia mali* (Anderson and Elliott 1982), fenbuconazole 2 F at 585 ml/ha, malathion 57 EC at 5.85 l/ha, FC 435–66 petroleum oil at 46.8 l/ha, carbaryl 80 S at 3.36 kg/ha, chlorpyrifos 4 EC at 4.68 l/ha, and formetanate 92 SP at 1.12 kg/ha to *Agistemus industani* (Childers et al. 2001).

Comparing with phytoseiids stigmaeid mites are more tolerant to chemicals (Sanford 1967; Hluchy et al. 1991). *Zetzellia mali* has developed resistance to some chemicals, i.e., azinphos-methyl (Motoyama et al. 1970), organophosphate pesticides (Croft 1994), pyrethroid and cypermethrin (Villanueva and Harmsen 1998).

7.9 Economic and Commercial Aspects

The species of *Agistemus* and *Zetzellia* have great commercial potential in control of eriophyid mites, spider mites, false spider mites and small hemipteran insects in orchards especially when the densities of phytoseiid mites are low. Understanding the toxic effects of agricultural chemicals against stigmaeid mites and the sellection of the least disruptive chemicals will greatly contribute to the success conservation and application of stigmaeids in IPM.

The discovery of cost effective alternative or artificial food, the techniques reducing the cannibalistic behavior and rearing of the pesticide resistance strains are probably the three key factors for the commercial application of Stigmaeidae.

References

Abdel-Khalek A, Momen FM (2013) Cannibalism and intraguild predation of the predatory mite, Agistemus exsertus Gonzalez (Acari: Stigmaeidae). Acta Phytopathologica et Entomologica Hungarica 48(2):259–268

Abo Elghar MR, Elbadry EA, Hassan SM, Kilany SM (1969) Studies on the feeding, reproduction and development of *Agistemus exsertus* on various pollen species (Acarina: Stigmaeidae). Zeitschrift für angewandte Entomologie 63:282–284

Abo Elghar MR, Elbadry EA, Hassan SM, Kilany SM (1971) Effect of some pesticides on the predatory mite *Agistemus exsertus*. J Econ Entomol 64:26–27

Abou-Awad BA, Elsawi SA (1993) Biology and life table of the predacious mite, *Agistemus exsertus* Gonz. (Acari: Stigmaeidae). Anzeiger fuer Schädlingskunde, Pflanzenschutz, Umweltschutz 66:101–103

Abou-Awad BA, Reda AS (1992) Studies on copulation, egg production and sex-ratio of the predacious mite *Agistemus exsertus* Gonzalez (Acari: Stigmaeidae). J Appl Entomol 113:472–475

Abou-Awad BA, Hassan MF, Romeih AHM (2010) Biology of *Agistemus olivi*, a new predator of eriophyid mites infesting olive trees in Egypt. Arch Phytopathol Plant Protect 43:817–824

Afify AM, Gomaa EA, Zaher MA (1969) Effectiveness of *Agistemus exsertus* Gonzalez (Acarina: Stigmaeidae), as an egg-predator of the spider-mite, *Tetranychus cinnabarinus* Boisd., under varying room conditions. Zeitschrift für Angewandte Entomologie 63:48–52

Amano H, Chant DA (1990) Species diversity and seasonal dynamics of Acari on abandoned apple trees in Southern Ontario, Canada. Exp Appl Acarol 8:71–96

Anderson DW, Elliott RH (1982) Efficacy of diflubenzuron against the codling moth, *Laspeyresia pomonella* (Lepidoptera: Olethreutidae), and impact on orchard mites. Can Entomol 114:733–737

Bayan A (1988) Mites of plum trees in Lebanon. I. A general survey and a diagnostic key. Arab J Plant Prot 6:1–6

Castagnoli M, Nannelli R (1987) Further observations on population trends of mites in an experimental peach meadow orchard in central Italy. Redia 70:121–133

Chaudhri WM, Akbar S (1985) Studies on biosystematics and control of mites of field crops, vegetables and fruit plants in Pakistan. University of Agriculture. Faisalabad Tech Bull No. 2, p 314

Childers CC, Enns WR (1975a) Predaceous arthropods associated with spider mites in Missouri apple orchards. J Kansas Entomol Soc 48:453–471

Childers CC, Enns WR (1975b) Field evaluation of early season fungicide substitutions on tetranychid mites and the predators *Neoseiulus fallacis* and *Agistemus fleschneri* in two Missouri apple orchards. J Econ Entomol 68:719–724

Childers CC, Villanueva R, Aguilar H, Chewning R, Michaud JP (2001) Comparative residual toxicities of pesticides to the predator *Agistemus industani* (Acari: Stigmaeidae) on citrus in Florida. Exp Appl Acarol 25:461–474

Clements DR, Harmsen R (1990) Predatory behaviour and prey-stage preferences of stigmaeid and phytoseiid mites and their potential compatibility in biological control. Can Entomol 122:321–328

Clements DR, Harmsen R (1992) Stigmaeid-phytoseiid interactions and the impact of natural enemy complexes on plant-inhabiting mites. Exp Appl Acarol 14:327–341

Clements DR, Harmsen R (1993) Prey preferences of adult and immature *Zetzellia mali* Ewing (Acari: Stigmaeidae) and Typhlodromus caudiglans Schuster (Acari: Phytoseiidae). Can Entomol 125:967–969

Collyer E (1964) Phytophagous mites and their predators in New Zealand orchards. New Zealand J Agric Res 7:551–568

Croft BA (1994) Biological control of apple mites by a phytoseiid mite complex and *Zetzellia mali* (Acari: Stigmaeidae): long-term effects and impact of azinphosmethyl on colonization by *Amblyseius andersoni* (Acari: Phytoseiidae). Environ Entomol 23:1317–1325

Croft BA, MacRae IV (1992) Persistence of *Typhlodromus pyri* and *Metaseiulus occidentalis* (Acari: Phytoseiidae) on apple after inoculative release and competition with *Zetzellia mali* (Acari: Stigmaeidae). Environ Entomol 21:1168–1177

Croft BA, MacRae IV (1993) Biological control of apple mites: impact of *Zetzellia mali* (Acari: Stigmaeidae) on *Typhlodromus pyri* and *Metaseiulus occidentalis* (Acari: Phytoseiidae). Environ Entomol 22:865–873

Darbemamieh M, Fathipour Y, Kamali K (2009) Bionomics of *Cenopalpus irani*, *Bryobia rubrioculus* and their egg predator *Zetzellia mali* (Acari: Tenuipalpidae, Tetranychidae, Stigmaeidae) in natural conditions. Munis Entomol Zool 4:341–354

De Vis RMJ, De Moraes GJ, Bellini MR (2006) Initial screening of little known predatory mites in Brazil as potential pest control agents. Exp Appl Acarol 39:115–125

De Vis RMJ, Silva ES, Bellini MR, De Moraes GJ (2008 [2007]) Life cycle of *Metaseiulus camelliae* and *Zetzellia malvinae*, predators of the rubber tree pest mite, *Tenuipalpus heveae* (Acari: Phytoseiidae, Stigmaeidae, Tenuipalpidae). Acarologia 47:109–114

Delattre P (1971) Contribution a l'étude biologique de *Zetzellia mali* Ewing (Acarina: Stigmaeidae). Ann Zoologie-Écologie Anim 3:297–303

Delattre P (1974) Étude de l'éfficacité prédatrice de *Zetzellia mali* (Acarina: Stigmaeidae) vis-avis du tétranyque du pommier, *Panonychus ulmi* (Acarina: Tetranychidae). Entomophaga 19:13–31

Denizhan E (2011) Population development of *Aculus schlechtendali* (Nalepa) (Acarina: Eriophyidae) feeding on different apple varieties in Van province and effect of *Zetzelia mali* (Ewing) (Turkey). Bitki Koruma Bulteni 51:239–253

Dosse GV (1967) Schadmilben des Libanons und ihre Prädatoren. Zeitschrift für Angewandte Entomologie 59:16–48

Ehara S (1964) Predaceous mites of the genus *Agistemus* in Japan (Acarina: Stigmaeidae). Annotationes Zoologicae Japonenses 37:226–232

El Sawi S, Momen F (2006) *Agistemus exsertus* Gonzalez (Acarina: Stigmaeidae) as a predator of two scale insects of the family Diaspididae (Homoptera: Diaspididae). Arch Phytopathol Plant Protect 39:421–427

Elbadry EA, Abo Elghar MR, Hassan SM, Kilany SM (1969a) *Agistemus exsertus* as a predator of two tetranychid mites. Ann Entomol Soc Am 62:660–661

Elbadry EA, Abo Elghar MR, Hassan SM, Kilany SM (1969b) Life history studies on the predatory mite *Agistemus exsertus*. Ann Entomol Soc Am 62:649–651

El-Bagoury ME, Reda AS (1985) *Agistemus exsertus* Gonzalez (Acarina: Stigmaeidae) as a predator of the ploughman's spikenard gall mite, *Eriophyes dioscoridis* (Eriophyidae). Bull Fac Agr Cairo Univ 36:571–576

El-Halawany ME, Abou-El-Ela RG, Esmail HM, El-Ela RGA (1990) Population dynamics of mites and their natural enemies on apple and apricot trees. Agric Res Rev 68:59–66

El-Laithy AYM (1998) Laboratory studies on growth parameters of three predatory mites associated with eriophyid mites in olive nurseries. Zeitschrift für Pflanzenkrankheiten und Pflanzenschutz 105:78–83

Fan Q-H, Walter DE (2004) *Genus Caligohomus* Habeeb (Acari: Prostigmata: Stigmaeidae). Syst Appl Acarol 9:77–88

Fan Q-H, Zhang Z-Q (2005) Raphignathoidea (Acari: Prostigmata). Fauna of New Zealand 52:400

Ferla NJ, De Moraes GJ (2003) Oviposição dos ácaros predadores *Agistemus floridanus* Gonzalez, *Euseius concordis* (Chant) e *Neoseiulus anonymus* (Chant & Baker) (Acari) em resposta a diferentes tipos de alimentos. Rev Brasileira Zool 20:153–155

Fischer-Colbrie P, El-Borolossy M (1988) Nebenwirkungen von winter- und Austriebbehandlungen auf Raubmilbenpopulationen an Apfelbäumen. Pflanzenschutzberichte 49:125–131

Forest J, Pilon G, Paradis RO (1982) Acariens des vergers de pommiers du sud-ouest du Québec. Ann Soc Entomol Québec 27:7–67

Gerson U (1972) Mites of the genus *Ledermuelleria* (Prostigmata: Stigmaeidae) associated with mosses in Canada. Acarologia 13:319–343

Gerson U, Smiley RL, Ochoa R (2003) Mites (Acari) for pest control. Blackwell Science, Oxford, p 539

Goldarazena A, Aguilar H, Kutuk H, Childers CC (2004) Biology of three species of *Agistemus* (Acari: Stigmaeidae): life table parameters using eggs of *Panonychus citri* or pollen of *Malephora crocea* as food. Exp Appl Acarol 32:281–291

González-R RH (1971) Biología, ecología y control natural de la aranita roja europea, *Panonychus ulmi* (Koch), en manzanos y perales de Chile Central. Rev Peruana Entomol 14:56–65

González-Rodriguez RH (1963) Four new mites of the genus *Agistemus* Summers, 1960 (Acarina: Stigmaeidae). Acarologia 5:342–350

González-Rodriguez RH (1965) A taxonomic study of the genera *Mediolata*, *Zetzellia* and *Agistemus* (Acarina: Stigmaeidae). Univ Calif Publ Entomol 41:1–64

Göven MA, Çobanoğlu S, Güven B (2009) Ege Bölgesi bağ alanlarındaki avcı akar faunası. Bitki Koruma Bülteni 49:1–10

Hafez SM, Rasmy AH, ElSawy SA (1983) Effect of prey species and stages on predatory efficiency and development of the stigmaeid mite *Agistemus exsertus*. Acarologia 24:281–283

Hagley EAC, Biggs AR (1989) Effects of three fungicides on populations of a phytophagous and several predacious mites (Acarina) on apple. Exp Appl Acarol 6:253–256

Hanna MA, Shereef GM, Megali MK (1984 [1980–1981]) Effect of food type on longevity and fecundity of the predator mite, *Agistemus exsertus* Gonzalez (Acari: Prostigmata), with first description of its prelarva. Bull Soc Entomol d'Égypte 63:57–62

Herbert HJ, Butler KP (1973) Distribution of phytophagous and predacious mites on apple trees in Nova Scotia. Can Entomol 105:271–276

Herne DC, Lund CT (1973) Influence of acaricides on seasonal populations of European red mite and predaceous mites on apple. Pesticide Research Report, Canada Comm. on Pesticide Use in Agriculture, Agriculture Canada, pp 14–17

Hluchy M, Pospisil Z, Zacharda M (1991) Phytophagous and predatory mites (Acari: Tetranychidae, Eriophyidae, Phytoseiidae, Stigmaeidae) in South Moravian vineyards, Czechoslovakia, treated with various types of chemicals. Exp Appl Acarol 13:41–52

Holdsworth RP (1972a) *Zetzellia mali* and *Agistemus fleschneri*: difference in spatial distribution. Environ Entomol 1:532–533

Holdsworth RP (1972b) European red mite and its major predators; effects of sulfur. J Econ Entomol 65:1098–1099

Hoyt SC (1969) Population studies of five mite species on apple in Washington. Proceedings of the 2nd international congress of acarology, Sutton Bonington, 1967, pp 117–133

Hu S, Hu S, Gan M, Li H, Ding Q, Chen X (1997) The capture effect of several predacious mites to the eriophyid mites *Aculus ligustri* [sic] (Keifer). Chinese Sci Abstr 3:1386–1388

Hu S, Wu M-H, Chien X-W (2001) The effect of relative humidity on the development of the predator *Agistemus terminalis* (Quayle) (Acari: Stigmaeidae). Syst Appl Acarol 6:51–53

Inoue K, Tanaka M (1983) Biological characteristics of *Agistemus terminalis* (Quayle) (Acarina: Stigmaeidae) as a predator of the citrus red mite, *Panonychus citri* (McGregor). Japanese J Appl Entomol Zool 27:280–288

Inserra R (1973[1970]) Morphological observations and biological notes on *Zetzellia graeciana* Gonzalez (Acarina: Stigmaeidae). Bollettino di Zoologia Agraria e di Bachicoltura 10:85–119

Ji J, Lin T, Zhang Y-X, Li S, Saito Y, Lin J, Chen X (2013) Effects of starvation and humidity on the development and survival of *Amblyseius swirskii*, *Agistemus exsertus* and *Amblyseius eharai*. Syst Appl Acarol 18:321–328

Khajuria DR (2009) Predatory complex of phytophagous mites and their role in integrated pest management in apple orchard. J Biopesticides 2:141–144

Khan BS, Afzal M (2005) Comparison of life cycle of stigmaeid mite, *Agistemus buntex* on three tetranychid species. Pak Entomol 27:9–11

Khodayari S, Kamali K, Fathipour Y (2008) Biology, life table, and predation of *Zetzellia mali* (Acari: Stigmaeidae) on *Tetranychus urticae* (Acari: Tetranychidae). Acarina 16:191–196

Khodayari S, Kamali K, Fathipour Y (2009) Comparison of population density and spatial distribution of *Tydeus longisetosus* El-Bagoury & Momen, *Tetranychus turkestani* (Ugarov & Nikolskii) and their predator *Zetzllia mali* (Ewing) on two aplle varieties in Maragh region. J Entomol Res 1:197–208

Khodayari S, Fathipour Y, Kamali K, Naseri B (2010) Seasonal activity of *Zetzellia mali* (Stigmaeidae) and its preys *Eotetranychus frosti* (Tetranychidae) and *Tydeus longisetosus* (Tydeidae) in unsprayed apple orchards of Maragheh, Northwestern of Iran. J Agr Sci Tech 12(Suppl Issue):549–558

Laing JE, Knop NF (1983) Potential use of predaceous mites other than Phytoseiidae for biological control of orchard pests. In: Hoy MA, Cunningham GL, Knutson L (eds) Biological control of pests by mites, Special Publication No. 3304. University of California, Berkeley, pp 28–35

Lászlóné S, György B (1995) Az egri és a mátraaljai borvidék szoloültetvenyeiben végzett atkatelelés-vizsgálat tapasztalatai. Novenyvedelem 31(12):595–599

Lawson AB, Walde SJ (1993) Comparison of the responses of two predaceous mites, *Typhlodromus pyri* and *Zetzellia mali*, to variation in prey density. Exp Appl Acarol 17:811–821

Leiva S, Fernandez N, Theron P, Rollard C (2013) *Agistemus aimogastaensis* sp. n. (Acari: Actinedida: Stigmaeidae), a recently discovered predator of eriophyid mites *Aceria oleae* and *Oxycenus maxwelli*, in olive orchards in Argentina. ZooKeys 312:65–78

Li SY, Harmsen R (1992) Effects of low application rates of the pyrethroid PP_{321} on the apple orchard mite complex (Acari) in Ontario. Can Entomol 124:381–390

Li SY, Harmsen R, Thistlewood HMA (1992) The effect of pyrethroid lambdacyhalothrin applications on the spatial distribution of phytophagous and predatory mites in apple orchards. Exp Appl Acarol 15:259–269

Liguori M (1987) Population trends of phytophagous and predatory mites in two vineyards of the Chianti area, Italy. Redia 70:141–150

Lindquist EE, Krantz GW, Walter DE (2009) Chapter 8: Classification. In: Krantz GW, Walter DE (eds) A manual of acarology, 3rd edn. Texas Tech Univ Press, Lubbock, pp 97–103

MacRae IV, Croft BA (1996) Differential impact of egg predation by *Zetzellia mali* (Acari: Stigmaeidae) on *Metaseiulus occidentalis* and *Typhlodromus pyri* (Acari: Phytoseiidae). Exp Appl Acarol 20:143–154

Matioli AL, Ueckermann EA, de Oliveira CAL (2002) Some stigmaeid and eupalopsellid mites from citrus orchards in Brazil (Acari: Stigmaeidae and Eupalopsellidae). Int J Acarol 28:99–120

Matioli AL, Tavares MG, Pallini A (2007) *Agistemus pallinii* n. sp. (Acari: Stigmaeidae) from citrus orchards in Brazil. Int J Acarol 33:245–251

Mohamed OMO (2014) Biological aspects of the predaceous mite, *Agistemus vulgaris* Soliman and Gomaa and life table parameters on three host phytophagous mite species, (Acari: Stigmaeidae). Egyptian Acad J Biol Sci 7:165–171

Momen FM (2001) Effects of diet on the biology and life tables of the predacious mites *Agistemus exsertus* (Acari: Stigmaeidae). Acta Phytopathol Entomol Hungarica 36:173–178

Momen FM (2011) Natural and factitious prey for rearing the predacious mite *Agistemus exsertus* Gonzales (Acari: Stigmaeidae). Acta Phytopathol Entomol Hungarica 46:267–275

Momen FM (2012) Influence of life diet on the biology and demographic parameters of *Agistemus olivi* Romeih, a specific predator of eriophyid pest mites (Acari: Stigmaeidae and Eriophyidae). Trop Life Sci Res 23:25–34

Momen FM, El-Sawy SA (2006 [2005]) *Agistemus exsertus* (Acari: Stigmaeidae) predation on insects: life history and feeding habits on three different insect eggs (Lepidoptera: Noctuidae and Gelechiidae). Acarologia 46:203–209

Motoyama N, Rock GC, Dauterman WC (1970) Organophosphorus resistance in an apple orchard population of *Typhlodromus* (*Amblyseius*) *fallacis*. J Econ Entomol 63:1439–1442

Muma MH (1952) Insect parasitism and related biological factors as concerned with citrus insect and mite control. Fla Agr Exp Stn Annu Report, p 174

Muma MH (1960) Taxonomy and biology of mites associated with citrus. Fla Agr Exp Stn Annu Report, p 220

Muma MH (1975) Mites associated with citrus in Florida. Fla Agr Exp Stn Bull 640A:1–92

Muma MH, Selhime AG (1971) *Agistemus floridanus* (Acarina: Stigmaeidae), a predatory mite, on Florida citrus. Fla Entomol 54:249–258

Muma MH, Selhime AG, Denmark HA (1961) An annotated list of predators and parasites associated with insects and mites on Florida citrus. Fla Agr Exp Stn Bull 634:1–39

Nawar MS (1992) Effect of prey density on predaceous efficiency and oviposition of *Agistemus exsertus* (Acari: Stigmaeidae). Exp Appl Acarol 15:141–144

Nelson EE, Croft BA, Howitt AJ, Jones AJ (1973) Toxicity of apple orchard pesticides to *Agistemus fleschneri*. Environ Entomol 2:219–222

Nesbitt HHJ (1946) Three new mites from Nova Scotian apple trees. Can Entomol 78:15–22

Oatman ER (1973) An ecological study of arthropod populations on apple in Northeastern Wisconsin: population dynamics of mite species on the foliage. Ann Entomol Soc Am 66:123–131

Oomen PA (1982) Studies on population dynamics of the scarlet mite, *Brevipalpus phoenicis*, a pest of tea in Indonesia, Mededelingen Landbouwhogeschool 82–01, Wageningen, p 89

Osakabe M (1963) On the natural enemy of the tea red spider mite, *Tetranychus kanzawai* Kishida. Chagyo Kenkyu Hokoku 20:1–11

Osman AA, Zaki AM (1986) Studies on the predation efficiency of *Agistemus exsertus* Gonzalez (Acarina: Stigmaeidae) on the eriophyid mites *Aculops lycopersici* (Massee). Anzeiger fuer Schaedligskunde, Pflanzenschutz und Umweltschutz 59:135–136

Osman AA, Abou-Taka SA, Zaki AM (1991) *Agistemus exsertus* Gonzalez (Acari: Stigmaeidae) as a predator of the grapevine mite *Colomerus vitis* (Pgst) (Acarina: Actinedida). In: Dusbábek F, Bukva V (eds) Modern acarology, vol 2. Academia, Prague and SPB Academic Publishing bv, The Hague, pp 689–690

Parent B (1961) Effets de certains produits antiparasitaires sur *Typhlodromus rhenanus* (Oudms) et *Mediolata mali* (Ewing), deux acariens prédateurs du Tétranyque rouge du pommier. Annales de la Société Entomologique du Québec 6:55–58

Parent B, LeRoux EJ (1956) Note on *Mediolata mali* (Ewing) (Acarina: Raphignatidae) as a predator of the European red mite. Can Entomol 88:487

Pringel KL (1975) The seasonal occurrence of phytophagous mites and their predators on apple trees in the Elgm Area. Fruit & Fruit Tech Res Inst, Stellenbosh, pp 131–140

Rahmani H, Fathipour Y, Kamali K (2010) Spatial distribution and seasonal activity of *Panonychus ulmi* (Acari: Tetranychidae) and its predator *Zetzellia mali* (Acari: Stigmaeidae) in apple orchards of Zanjan, Iran. J Agric Sci Technol 12:155–165

Rakha MA, McCoy CW (1984) *Eryngiopus citri*, a new mite on Florida citrus with descriptions of the developmental stages (Stigmaeidae: Actinedida). Fla Entomol 67:504–507

Rasmy AH (1975) Eine Methode zur Massenzucht der Raubmilbe *Agistemus exsertus* Gonz. (Acarina: Stigmaeidae). Anzeiger fuer Schaedlingskunde, Pflanzenschutz, Umweltschutz 48:55–56

Rasmy AH, Hussein HE (1996) Effect of mating on egg production in two species of predatory mites, *Agistemus exsertus* Gonzalez and *Phytoseiulus persimilis* Athias-Henrio. Anzeiger fuer Schädlingskunde, Pflanzenschutz, Umweltschutz 69:88–89

Rasmy AH, ElBagoury ME, Reda AS (1987) A new diet for reproduction of two predaceous mites *Amblyseius gossipi* and *Agistemus exsertus* (Acari: Phytoseiidae: Stigmaeidae). Entomophaga 32:277–280

Razaq A, Ohbayashi N, Mohammad P, Shiraishi M (2000) SEM observations on the predation behavior of *Agistemus terminalis* (Quayle) (Acari: Stigmaeidae) and *Amblyseius sojaensis* Ehara (Acari: Phytoseiidae) on *Panonychus citri* (McGregor) (Acari: Tetranychidae). Pak J Biol Sci 3:576–579

Reda AS (1990) The use of artificial diets and natural diets in rearing *Agistemus exsertus* (Acari: Stigmaeidae). Ann Agric Sci 28:2633–2642

Rice RE, Jones RA, Hoffman ML (1976) Seasonal fluctuation in phytophagous and predaceous mite populations on stonefruits in California. Environ Entomol 5:557–564

Romeih AHM, El-Saidy EMA, El Arnaouty SA (2004) Suitability of *Ephestia kuehneilla* and *Corycera cephalonica* eggs as alternative preys for rearing predatory mites. Egypt J Biol Pest Control 14:101–105

Saber SA (2012) Predation, oviposition and longevity of the predacious mite, *Agistemus exsertus* Gonzalez at different prey densities of *Tetranychus urticae* Koch (Acari: Stigmaeidae: Tetranychidae). Arch Phytopathol Plant Protect 45:1341–1346

Sanford KH (1967) The influence of spray programs on the fauna of apple orchards in Nova Scotia XVII. Effects on some predacious mites. Can Entomol 99:197–201

Santos MA (1976a) Evaluation of *Zetzellia mali* as a predator of *Panonychus ulmi* and *Aculus schlechtendali*. Environ Entomol 5:187–191

Santos MA (1976b) Prey selectivity and switching response of *Zetzellia mali*. Ecology 57:390–394

Santos MA (1982) Effects of low prey densities on the predation and oviposition of *Zetzellia mali* (Acarina: Stigmaeidae). Environ Entomol 11:972–974

Santos MA (1984) Effects of host plant on the predator–prey cycle of *Zetzellia mali* (Acari: Stigmaeidae) and its prey. Environ Entomol 13:65–69

Santos MA (1991) Searching behavior and associational response of *Zetzellia mali* (Acarina: Stigmaeidae). Exp Appl Acarol 11:81–87

Schruft VG (1969) Das Vorkommen rauberischer Milben aus den Familien Cunaxidae und Stigmaeidae (Acari) an Reben. IV. Beitrag über Untersuchungen zur Faunistik und Biologie der Milben (Acari) an Kultur-Reben (Vitis spec.). Die Wein-Wissenschaft 24:320–326

Sizova IY, Kuznetsov NN (1980) Some biological characteristics of the predacious mite *Agistemus herbarius* Kuznetzov et Wainstein, 1977 (Acarina: Stigmaeidae). Uzbekskii Biologicheskii Zhurnal 3:64–65

Slone DH, Croft BA (2001) Species association among predaceous and phytophagous apple mites (Acari: Eriophyidae, Phytoseiidae, Stigmaeidae, Tetranychidae). Exp Appl Acarol 25:109–126

Soliman ZR, Shehata KK, Gomaa EA (1976) Über Nahrungskreis und wirtschaftliche Bedeutung der Raubmilbe Agistemus exsertus Gonz. (Acari: Prostigmata). Anzeiger für Schädlingskunde, Pflanzenschutz, Umweltschutz 49:87–90

Sternlicht M (1969) A study of fluctuations in the citrus bud mite population. Ann Zool-Écol Anim 1:127–147

Tanaka M (1966) Fundamental studies on the utilization of natural enemies in the citrus grove in Japan. I. The bionomics of natural enemies of the most serious pests. II. *Stethorus japonicus* H. Kamiya, a predator of the citrus red mites, *Panonychus citri* McGregor. Bull Hortic Res Stn 4:22–42

Vacante V, Tropea Grazia G (1989) Grape mites in Sicily. Contribution I. Cavalloro R, Pelerents, C (eds) Integrated pest management in protected vegetable crops: proceedings of the CEC/IOBC expert's group meeting. Balkema AA Publishers, Old Post Road, Brookfield, USA, pp 207–215

Villanueva RT, Harmsen R (1998) Studies on the role of the stigmaeid predator *Zetzellia mali* in the acarine system of apple foliage. Proc Entomol Soc Ontario 129:149–155

Villaronga P, Cosialls JR, Bonet J (1993) Mite fauna associated to peach orchards in Lleida (Spain). Bull OILB/SROP 16:14–21

Vogt H, Dickler E, Grauhan H (1990) Effects of an unique application of acaricides in apple orchards on the population dynamic of *Panonychus ulmi* (Acari: Tetranychidae) and *Aculus schlechtendali* (Acari: Eriophyoidea) with special regard to beneficials. J Appl Entomol 110:35–54

Wafa AK, Zaher MA, Afify AM, Gomaa EA (1969) Effect of diet on the development of the predaceous mite, *Agistemus exsertus* Gonzalez (Acarina: Stigmaeidae). Zeitschrift für Angewandte Entomologie 63:382–388

Wang H-F (1981) Some predatory species of Stigmaeidae from Chinese orchards. Insect Knowledge 18:81–82

White ND, Laing JE (1977) Some aspects of the biology and a laboratory life table of the acarine predator *Zetzellia mali*. Can Entomol 109:1275–1281

Yousef AEA, Zaher MA, El-Hafiez AMA (1982) Effect of prey on the biology of *Amblyseius gossipi* Elbadry and *Agistemus exsertus* Gonzalez (Acari: Phytoseiidae, Stigmeidae). Zeitschrift für Angewandte Entomologie 93:453–456

Yue B, Childers CC (1994) Effects of temperature on life table parameters of *Agistemus exsertus* Gonzalez (Acari: Stigmaeidae) and its attack rate on *Panonychus citri* eggs. Int J Acarol 20:109–113

Yue B, Tsai JH (1995) *Agistemus exsertus* Gonzalez (Acari: Stigmaeidae) as a predator of citrus red mite (*Panonychus citri* [McGregor]). J New York Entomol Soc 103:107–113

Zahedi-Golpayegani A, Saboori A, Sabelis MW (2007) Olfactory response of the predator *Zetzellia mali* to a prey patch occupied by a conspecific predator. Exp Appl Acarol 43:199–204

Zaher MA, El-Badry EA (1961) Life-history of the predator mite *Agistemus fleschneri* Summers and effect of nutrition on its biology. Bull Soc Entomol d'Égypte 45:375–385

Zhang Z-Q (2003) Mites of greenhouses: identification, biology and control. CABI Publishing, Wallingford, p 244

Chapter 8
The Erythraeoidea (Trombidiformes: Prostigmata) as Biological Control Agents, with Special Reference to the Genus *Balaustium*

Karen Muñoz-Cárdenas, Luz Stella Fuentes-Quintero, Diana Rueda-Ramirez, C. Daniel Rodríguez, and R. Fernando Cantor

8.1 Introduction

Mites of the superfamily Erythraeoidea are distributed worldwide. Larvae of most species are ectoparasites of different insects (as mosquitoes, house flies, plant hoppers, thrips, aphids, termites) and arachnids (as spiders, scorpions, etc.), whereas deutonymphs and adults prey upon phytophagous mites and other small arthropods. Different species of five genera have been reported as natural enemies of agricultural pests, namely *Balaustium*, *Charletonia*, *Erythraeus*, *Grandjeanella* and *Leptus*. However, their potential as biological control agents is poorly known. Species of the genus *Balaustium* von Heyden are the most widely studied in regard to their potential use in biological control of agricultural pests.

K. Muñoz-Cárdenas
IBED, Population Biology, University of Amsterdam,
Science Park 904, Amsterdam 1098 XH, The Netherlands
e-mail: K.A.MunozCardenas@uva.nl

L.S. Fuentes-Quintero (✉)
Departamento de Ciencias Biológicas y Ambientales, Facultad de Ciencias Naturales e Ingeniería, Universidad de Bogotá Jorge Tadeo Lozano,
Carrera 4 # 22-61, Bogotá, D.C., Colombia
e-mail: luz.fuentes@utadeo.edu.co; luzfue21@gmail.com

D. Rueda-Ramirez
Departamento de Entomologia e Acarologia, Escola Superior de Agricultura "Luiz de Queiroz" – Universidade de São Paulo, Av Pádua Dias, 11, Piracicaba 13418-900, SP, Brazil

C.D. Rodríguez • R.F. Cantor
Programa de Biología Aplicada, Facultad de Ciencias Básicas y aplicadas, Universidad Militar Nueva Granada, Km 3 via Cajicá-Zipaquirá, Cajicá, Colombia

D. Carrillo et al. (eds.), *Prospects for Biological Control of Plant Feeding Mites and Other Harmful Organisms*, Progress in Biological Control 19,
DOI 10.1007/978-3-319-15042-0_8

8.2 Morphological and Taxonomic Characterization of Erythraeoidea

This superfamily includes moderately large to very large mites, with flattened, globular or ellipsoidal bodies (Southcott 1961a; Walter et al. 2009), and with relatively long and slender legs (Wohltmann et al. 2007). The morphology of the post-larval stages is moderately to clearly distinct from the larva; many species are known only from the larval or the post-larval stages (Southcott 1961a; Welbourn 1983; Walter et al. 2009).

Eryothraeioid larvae have well-developed prodorsal sclerite (scutum), bearing two pairs of trichobothria. Urstigma (= Claparede's organ) always absent; anus usually absent; coxal plates usually separated and with one or two setae each (Meyer and Ryke 1959; Walter et al. 2009). Movable cheliceral digit blade-like.

Deutonymphs and adults have smooth cuticle; they are often reddish, sometimes iridescent with whitish spots on dark background, whitish, brownish or greenish, with scarce to numerous setae that vary from setiform to highly-modified in form (Wohltmann et al. 2007; Walter et al. 2009). The high density of hairs of many species give them a velvety appearance. The prodorsal area bears two pairs of trichobothria and one or two pairs of eyes. Chelicerae or the entire gnathosoma is retractable into the idiosoma; movable cheliceral digit stylet-like (Meyer and Ryke 1959; Southcott 1961a; Walter et al. 2009).

8.2.1 Taxonomy of Erythraeioidea

The taxonomy of the erythraeoids is complex because of the morphological and feeding behavioral differences between the larva and the post-larval stages. These differences make the correlation between stages difficult without experimental rearing. They often lead to independent designation of taxons that need to be correlated. But this has rarely been done, and thus the state taxonomy of the erythraeioids is presently rather confusing. It is unknown the number of misidentification or "double classification"-when a larva and a post-larval form of the same species are classified separately like they were different species.

The authorship of the superfamily name has been attributed to Robineau-Desvoidy (1828), although Grandjean (1947) was the first to list morphological attributes of this superfamily, also proposing diagnostic characteristics to separate the larval stage of the two families. Southcott (1961a, b, 1963) reviewed the systematic of Erythraeioidea and performed further studies and experimental rearing of larval Erythraeidae and Smarididae allowing correlation between various taxons that had been described based only on the larval or on the post-larval stages. Works conducted by Southcott (1961a, b) constituted an important base for subsequent taxonomic works about this superfamily. Between 1960 and 2000, several generic revisions (Southcott 1966, 1972, 1991a, 1992) and description of new species (e.g. Ahsan and Anwarullah 1970, Southcott 1961c, 1983, 1984, 1988a, b, c, d, 1989a, b, c, 1991b, 1993a, b, 1994a, b, 1995a, b, 1996, 1997, 1999; Gabryś 2000a, b, c; Mąkol 2010a; Mayoral and Barranco 2011) were published. Only some of the more recent experimental literature have been based on laboratory rearings to correlate the different developmental stages (e.g. Mąkol 2002, 2010b; Laydanowicza and Mąkol 2010; Mąkol et al. 2012; Fuentes et al. 2014). Catalogs of groups of the eryothraeioid species (Beron 2008; Mąkol and Wohltmann 2012) are now published to assist in the study and identification of species. Recently, a number of authors (eg. P. Beron, G. Gabryś, A. Goldarazena, R. Haitlinger, R.B. Halliday, M.L.I Judson, M. M. Kamran, J.S. Karimi Iravanlou, A. Saboori, M. Khanjani, J. Łaydanowicz, E. Ueckermann, A. Wohltmann, Z.-Q. Zhang) have published taxonomic papers about this group. More taxonomic and morphological studies are needed for a better understanding of this group so that other species can be evaluated for practical use in biological pest control programs.

8.2.2 Families of Erythraeoidea

This large group includes the families Erythraeidae and Smarididae, both terrestrial (Mąkol and Wohltmann 2012). Erythraeidae is a diverse and large family that includes seven subfamilies: Abrolophinae (three genera), Balaustiinae (14 genera), Callidosomatinae (12 genera), Erythraeinae (26 genera), Leptinae (one genus), Myrmicotrombiinae (one genus) and Phanolophinae (three genera) (Zhang et al. 2011; Mąkol and Wohltmann 2012). Larvae of this family have two (occasionally three) solenidia on tibia of leg I but no trichobothria (Meyer and Ryke 1959; Southcott 1961a). Deutonymphs and adults have a globular or ellipsoidal, often reddish body, and generally bear setiform dorsal setae. The prodorsum bears one or two pairs of eyes, two pairs of trichobothria and a median longitudinal crista metopica (prodorsal sclerite bearing the two pairs of trichobothria). The chelicerae of these stages is retractable into the idiosoma.

Smarididae includes two widely distributed subfamilies: Hirstiosomatinae (five genera) and Smaridinae (four genera) (Walter et al. 2009; Mąkol and Wohltmann 2012). The larvae of this family presents at least one trichobothria on genu, tibia, and/or tarsus of leg I and at least three solenidia on tibia of leg I (Meyer and Ryke 1959; Southcott 1961a; Walter et al. 2009). Adults and deutonymphs of this family have vaguely fish-like, ovoid, flattened body with moderate shoulders and bear generally highly modified (leaf-like, flattened, broad, serrated) dorsal setae (Meyer and Ryke 1959; Southcott 1961a, 1963). The prodorsum bears one (Hirstiosomatinae) or two (Smaridinae) pairs of eyes, with two pairs of trichobothria and occasionally with crista metopica. The gnathosoma of deutonymphs and adults is entirely retractable into the idiosoma.

8.3 Erythraeidae (Diversity, Distribution and Habitats)

According to Mąkol and Wohltmann (2012), the Erythraeidae comprises 823 species distributed in 60 genera of seven subfamilies (Table 8.1). The most specious genera are *Abrolophus*, *Balaustium*, *Caeculisoma*, *Callidosoma*, *Charletonia*,

Table 8.1 Number of species and distribution of subfamilies and genera of Erythaeidae based on Mąkol and Wohltmann (2012, 2013)

Subfamily	Genus	Number of species	Distribution
Abrolophinae	*Abrolophus* Berlese	114	Cosmopolitan
	Grandjeanella Southcott	6	Palearctic (Turkey, Australia and Indonesia)
	Marantelophus Haitlinger	7	Palearctic, Australasia (Croatia, Greece, Hungary, Romania, San Marino, Spain, Turkey, Ukraine, Iran, Indonesia)
Balaustiinae	*Balaustium* von Heyden	37	Cosmopolitan
	Bursaustium Haitlinger	2	Palearctic (Turkey and Iran)
	Fozustium Haitlinger	1	Brazil
	Italustium Haitlinger	1	Italia
	Lomeustium Haitlinger	1	Afrotropical (Benin, Ghana and Togo)
	Microsmarialla Khot	1	India
	Micromaris Hirst	6	Indo-Malay, Australasia (India, Australia, New Zeland)
	Moldoustium Haitlinger	1	Palearctic (Moldova, Ukraine)
	Mypongia Southcott	1	Australia
	Neobalaustium Willmann	2	Palearctic (Romania and Bulgaria)
	Pollux Southcott	3	Indo-Malay, Australasia (India, Indonesia, Australia)
	Wartookia Southcott	1	Australia
Callidosomatinae	*Andrevella* Southcott	2	Palearctic (France and Iran)
	Caeculisoma Berlese	20	Cosmopolitan
	Callidosoma Womersley	23	Cosmopolitan
	Carastrum Southcott	2	Australasia (Indonesia, Australia)
	Cecidopus Karsch	7	Indo-Malay, Afrotropical
	Charletonia Oudemans	115	Cosmopolitan
	Dambullaeus Haitlinger	1	Sri Lanka
	Harpagella Southcott	1	Australia
	Iguatonia Haitlinger	1	Brazil
	Momorangia Southcott	4	Southern Hemisphere (Kenya, New Zeland, Brazil)
	Neoabrolophus Khot	4	Afrotropical-Indo-Malay (Tanzania, Malaysia, India)
	Pussardia Southcott	1	Australia

(continued)

Table 8.1 (continued)

Subfamily	Genus	Number of species	Distribution
Erythraeinae Robineau-Desvoidy	*Abalakeus* Southcott	4	Palearctic-Afrotropical (Spain, China, Iran, Nigeria)
	Augustsonella Southcott	2	Nearctic (Mexico, USA)
	Claverythraeus Trägårdh	1	China
	Curteria Southcott	7	Palearctic
	Eatoniana Cambridge	5	Argentina, Algeria, Armenia, Egypt, France, Greece, Jersey, Mauritania, South Africa, Spain, Switzerland, Turkmenistan, Tunisia
	Erythraeus Latreille	115	Cosmopolitan
	Erythraxus Southcott	1	Mexico
	Erythrellus Southcott	1	Australia
	Erythrites Southcott	8	Australasia (Australia, New Zeland)
	Erythroides Southcott	5	Australia
	Forania Southcott,	2	France and Spain
	Helladerythraeus Beron	1	Greece
	Kakamasia Lawrence	2	South Africa
	Kamertonia Gabryś	1	Finland and Poland
	Lasioerythraeus Welbourn and Young	4 (+2)[a]	Nearctic and Netropical (USA, Dominican Republic, Venezuela) (+ Iran, Pakistan)[a]
	Makolia Saboori, Khaustov and Hakimitabar	1	Ukraine
	Neophanolophus Shiba	6	Afrotropical, Indo-Malay
	Neosmaris Hirst	1	New Zeland
	Paraphanolophus Smiley	2	Mexico and Belize
	Pararainbowia Dunlop	1	Brazil
	Podosmaridia Trägårdh	1	China
	Proterythraeus Vercammen-Grandjean	1	Canada
	Rainbowia Southcott	2	Australia
	Ramsayella Zhang	1	New Zeland
	Taranakia Southcot	1	New Zeland
	Tepoztlana Hoffmann and Mendez	1	Mexico
Leptinae Billberg	*Leptus* Latreille	276	Cosmopolitan

(continued)

Table 8.1 (continued)

Subfamily	Genus	Number of species	Distribution
Myrmicotrombiinae Southcott	*Myrmicotrombium* Womersley	2	Greece and Australia
Phanolophinae Southcott	*Iraniella* Karimi Iravanlou, Kamali and Talebi	1	Iran
	Nagoricanella Haitlinger,	2	Macedonia and Cape Verde
	Phanolophus André	1	Palearctic

[a]*Lasioerythraeus saboorii* Khanjani et al. (2011) and *L. setarius* Kamran and Bashir (2013) are not included on the catalogue

Erythraeus, *Leptus*, all of which have cosmopolitan distribution. The other genera have less than 10 species each (28 genera have only one reported species generally restricted to a single country) and their distribution is restricted to certain regions.

Erythraeids, like all representatives of terrestrial Parasitengona, present a particular life cycle, which includes an egg, inactive stages (calyptostasic), and active developmental stages with usually different feeding habits. The inactive and active stages alternate as follows: inactive prelarva, active heteromorphic larva, inactive protonymph, active deutonymph, inactive tritonymph, and active adult (Belozerov 2008). Although the larvae are known as parasites of arthropods, and the adults and deutonymphs are free predators, there are exceptions to this behavior for the genus *Balaustium*, and possibly for other erythraeids (some *Abrolophus*), that are predacious or pollen feeders on larva and post-larval stages (Southcott 1961a; Newell 1963; Welbourn 1983; Halliday 2001; Mąkol et al. 2012; Wohltmann et al. 2007; Muñoz-Cárdenas et al. 2014).

Mites of this family are found worldwide in a great variety of habitats, ranging from dry to temporarily flooded or amphibious biotypes, and at approximately all latitudes (Wohltmann 2000; Wohltmann et al. 2007). Adults and deutonymphs are present in different edaphic and arboreal habitats, while larvae are typically found in association with arthropods and apparently some vertebrates (Walter et al. 2009). Larval hosts of this family include: the Enthognatha Collembola and Thysanura; the insects Blattodea, Coleoptera, Dermaptera, Diptera, Hemiptera, Hymenoptera, Lepidoptera, Mantodea, Mecoptera, Neuroptera, Odonata, Orthoptera, Phasmida, Psocoptera, Thysanoptera; Arachnida of the subclasses Amblypygi, Araneae, Actinotrichida, Opiliones, Pseudoscorpiones, Scorpiones; and apparently vertebrates of the classes Reptilia, Aves, Chiroptera, Rodentia (Mąkol and Wohltmann 2012). According to Wohltmann (2000), larval parasitism might represent an evolution from post-larval predation, increasing dispersal ability and allowing synchronization of life cycles with availability of host.

8.4 Genus *Balaustium* von Heyden (Balaustiinae)

The genus *Balaustium* has a worldwide distribution. Mąkol and Wohltmann (2012, 2013) published an annotated checklist of terrestrial Parasitegona including by 41 members (nominal species) of the genus (Mąkol et al. 2012). However, the taxonomic group is subject to constant change because most of the species descriptions are based solely on larval characters and the level of synonymy and the likelihood of misidentifications is unknown (Halliday 2001).

Mites of this genus are relatively large (about 1.0–2.5 mm long as adults), with conspicuous intense reddish coloration, sometimes with whitish shagreening. They have been found on the soil surface, on plants or climbing the walls of buildings. All active instars of *Balaustium* and most likely of other Balaustiinae, are free-living, predacious or pollenivorous (Newell 1963; Childers and Rock 1981; Hayes 1985; Halliday 2001; Mąkol et al. 2012; Muñoz-Cárdenas et al. 2014). This is different from most other erythraeids in which the larval stage is parasitic on other organisms. Although this is the genus that has been studied the most in the family Erythaeidae, there are very few studies regarding their potential as biological control agents.

8.4.1 Life Cycle

The life cycles of three *Balastium* species have been studied in detail: *B. putmani* Smiley (Cadogan and Laing 1977, 1981), *B. hernandezi* von Heyden (Mąkol et al. 2012), and *B. leanderi* (Heitlinger) (Muñoz-Cardenas et al. 2014).

The eggs are usually spherical and initially red, later dark red, about 0.16 mm in diameter (Fig. 8.1a, b) and laid in batches. Cadogan and Laing (1977) reported that eggs of *B. putmani* must be in contact with water for embryonic development to proceed to a prelarval stage.

The prelarval stage is marked by the rupture of the egg chorion, exposing an orange membrane called the deutovarial membrane (Fig. 8.1c). This membrane expands, providing the developing embryo more space for growth before hatching (Grandjean 1946; Cadogan and Laing 1977).

The larva is generally orange in color (Fig. 8.1d); and complete development by preying upon small insects and mites (Grandjean 1946; Mąkol et al. 2012). Before molting, larvae cease feeding and search for protected sites to spend the protonymphal stage (Cadogan and Laing 1977). Larvae then enter a quiescent period in which the cuticle splits and the appendages degenerate. According to Cadogan and Laing (1977), the coloration of *B. putmani* larvae changes with the type of food. When fed only on eggs of *Panonychus ulmi* (Koch), larvae showed a light red coloration, but when fed on active stages they became dark red.

The protonymphs are scarlet, oval or sac-like, legless and quiescent (Fig. 8.2a).

Deutonymphs are oval, and their coloration is scarlet to dark red depending on the diet. Deutonymphs of *B. leanderi* have three distinctive longitudinal-parallel

Fig. 8.1 Developmental stages of *B. leanderi*: (**a**) fresh eggs, (**b**) mature eggs, (**c**) prelarva, (**d**) larva (Photos credit Jan van Arkel, IBED, University of Amsterdam)

Fig. 8.2 Developmental stages of *B. leanderi*: (**a**) protonymph, (**b**) deutonymph, (**c**) trytonymph, (**d**) adult (Photos credit Jan van Arkel, IBED, University of Amsterdam)

white dorsal lines (Fig. 8.2b), and are similar to the adults but smaller. This stage exhibits fast movement and voracious predation before entering a quiescent period prior to the tritonymphal stage.

Tritonymphs are orange-red or scarlet and are morphologically similar to but larger than the protonymph (Fig 8.2c).

Adults are bright to dark-red, and there is no apparent dimorphism between sexes (Fig. 8.2d). They are also active predators. A pronounced semelparity, where all eggs were laid in a single batch by each female, has been reported for several *Balaustium* species, including *B. hernandezi* (Mąkol et al. 2012), *B. leanderi* (Muñoz-Cárdenas et al. 2014) and *B. murorum* (Hermann) (Grandjean 1946). Putman (1966) reported that females of *B. putmani* usually oviposit in bark crevices, and die in the proximity of the oviposition site soon after completing oviposition. This was also reported for *B. murorum* (Grandjean 1946).

The complete post-ovarian development of *B. putmani* reared on *P. ulmi* at 24 °C and 75 % RH was 18.7 days (Cadogan and Laing 1977). Development of *B. hernandezi* reared at a similar temperature and humidity regime on a mixed prey diet, with or without pollen provision, was completed in a similar period (20.4–20.8 days) (Mąkol et al. 2012). The egg-to-egg period of *B. leanderi* was about 2 months when reared on mixed prey diet under greenhouse conditions with daily maximum and minimum temperatures of approximately 31.7 and 11.2 °C (Muñoz-Cárdenas et al. 2014).

Deutonymphs and adults of *Balaustium* are characterized by a pair of structures (urnulae) on the dorsolateral area of the idiosoma, just behind the eyes. These structures are taxonomically important in distinguishing *Balaustium* from other erythraeid genera. Yoder et al. (2006) described one of the functions of the urnulae after observing a red fluid leaking from them. When threatened, *Balaustium* mites released the fluid, which spread over the adult's body surface. Yoder et al. (2006) applied the fluid to larvae of the mealworm beetle (*Tenebrio molitor* Linnaeus) (Tenebrionidae), observing that treated beetles were less attacked by ants. This led the authors to suggest that the fluid had a defensive role, as an allomone. Moreover, Yoder et al. (2010) discovered that this secretion served as an alarm pheromone, causing an excited dispersal response among conspecific mites. The secretions released through the urnulae also form a waterproof barrier to reduce water loss (Yoder et al. 2010). *Balaustium* sp. nr. *putmani* was observed not to be able to absorb water vapor during its active instars, consequently it is presumed that this mite imbibes liquid as free water or from food sources (Yoder and Heydinger 2011). Alberti and Mąkol (2013) found that larvae are more sensitive and vulnerable to water stress because they do not have urnulae, lacking the protective waterproofing secretion released by these structures.

8.4.2 Life History and Behavior

Putman (1966) and Cadogan and Laing (1977) reported that *B. putmani* overwinters in the egg stage. They concluded that in Ontario, Canada, there are two distinct generations. In the early spring, humidity is adequate for egg hatch and thus temperature is the determining factor for larval eclosion in the first generation. During the mid-summer the temperature is adequate for development, and the second generation relies on rain events (Cadogan and Laing 1977). Another aspect of the life history of *B. putmani* that has been extensively studied is its ability to resist high temperatures, up to 50 °C (Yoder et al. 2007; Hedges et al. 2012). Putman and Herne (1966) noted that *B. putmani* became abundant in peach orchards in Ontario, Canada that had not received pesticides for a number of years. According to Putman (1966, 1970), adults of *B. putmani* aggregated on woody limbs of peaches for mating. Cadogan and Laing (1977) also reported aggregations of 7–20 mites on ca 3.5 square cm, and described how the mites performed a "mating dance". Yoder and Heydinger (2011) reported population bursts of thousands of *B.* sp. nr. *putmani* in the spring in Central Ohio, USA, and concluded that population bursts might be a response to abundance of plant pests in the landscape during this season. In addition, Childers and Rock (1981) described the behavior of *B. putmani* feeding on different prey on apple orchards in North Carolina, noting that the first pair of legs were elevated and remained raised throughout the feeding process. This behavior is congruent with reports by Grandjean (1946) on *B. murorum* feeding upon pollen grains.

Balaustium leanderi has been studied by the authors of this chapter during the last years. This predator has been found frequently on the vegetation close to or inside commercial greenhouses dedicated to cut-flower and vegetable production in the Bogotá Plateau, Colombia (Getiva and Acosta 2004). The predators were found feeding on important pest species – provided that there were not chemical applications inside the greenhouses (Muñoz et al. 2009). Zuluaga (2012) studied the dispersion of *B. leanderi* on tomato plants infested with *Trialeurodes vaporariorum* Westwood (Hemiptera: Aleyrodidea), observing that the predator either did not move at all, or abandoned the tomato plant upon being placed on it. This could be related to the high density of trichomes present on the plant. Tomato trichomes exude substances that may affect mites (Chatzivasileidis and Sabelis 1997) and also constitute a physical barrier impeding their movement (Sato et al. 2011; van Houten et al. 2013). Dispersion was also evaluated on bean plants infested with *Tetranychus urticae* Koch, separated by 30, 40 or 50 cm, and with different predator: prey ratios (1:135, 1:285 and 1:435). Six hours after releasing the predator on a plant without prey, the predator moved to all plants infested with prey regardless of the plant separation and the predator: prey ratio. However, the predators moved predominantly to plants placed closer to the release point (Zuluaga 2012). Little is known about the seasonality of this predator under natural field conditions.

Balaustium hernandezi was found on a greenhouse sweet pepper crop (*Capsicum annuum*) in the Province of Almeria, Spain and was later observed on other protected crops including cucumber (*Cucurbita pepo* L.) and aubergine (*Solanum melongena* L.) (Mąkol et al. 2012). This predator was observed throughout the year, with the highest abundance in spring and summer, but a second peak occurring during the early winter. It's unclear if this species was introduced with plant material, or whether it had escaped the attention of local acarologists until the recent broad adoption of IPM in Almeria, with a dramatic shift from intensive insecticide use to biocontrol approaches, and most likely, permitted this species to become abundant

(Mąkol et al. 2012). Coombs and Bale (2014) determined the lower and upper thermal activity thresholds of adult and larval *B. hernandezi* compared with those of *T. urticae.* Adult female *B. hernandezi* remained active at significantly lower temperatures compared to *T. urticae*, and showed significantly faster walking speeds than *T. urticae* across a range of temperatures. However, *B. hernandezi* ceased movement below the upper locomotory limits of *T. urticae*. Overall, reports by Coombs and Bale (2014) suggest that *B. hernandezi* would make an effective biological control agent of *T. urticae* in temperate climates.

8.4.3 Potential Prey and Other Food Sources

Muñoz-Cárdenas et al. (2014) evaluated the biological parameters of *B. leanderi* under controlled conditions offering different prey (first instar larvae of the western flower thrips, *Frankliniella occidentalis*, eggs of greenhouse whitefly, *T. vaporariorum*, and *T. urticae*) *ad libitum*, singly or in combinations. The life cycle was completed on all prey, but when offered separately, spider mite eggs and western flower thrips larvae resulted in the lowest values of intrinsic rate of population increase ($r_m \approx 0.024$). Females reached reproductive stage faster and laid more eggs on mixed diets, and consequently had higher population growth rates. The combination of whitefly and spider mite eggs resulted in shorter egg-to-egg developmental period and higher juvenile survival, leading to an intrinsic rate of population increase of 0.066 females/female/day, which is still low compared to the numerous values reported in the literature for other plant-inhabiting predatory mites. The different stages of *B. leanderi* preferred different prey species but in general all stages preferred *T. vaporariorum.* Larvae of *B. leanderi* showed preference for eggs of *T. urticae* over eggs of *T. vaporariorum*, deutonymphs showed no preference for either prey, and adults preferred eggs of *T. vaporariorum* (Muñoz et al. 2009). Regardless of the preference, a conspicuous ability of adults to capture adults of *T. vaporariorum* was observed. In addition, a reduced preference for nymphs and adults of *T. urticae* was attributed to the ability of those prey stages to avoid predation. It was also clear from direct observations that the larvae and deutonymph of *B. leanderi* struggled to feed on thrips larvae, which showed an aggressive anti-predation behavior (Muñoz et al. 2009). Recently Nolasco and Cedola (2014) observed the preference of an unidentified species of *Balaustium* for different stages of important greenhouse pests in pepper crops in Argentina. Nolasco and Cedola (2014) preformed experiments in petri dishes on pepper leaves under laboratory conditions and found that the deutonymphs of *Balaustium* sp. preferred in first place thrips (*Thrips* sp.), second whiteflies (*Bemisia tabaci*) and the less preferred were spidermites (*Tetranychus urticae*). Larvae and nymphs of thrips and whiteflies were the most attacked, respectively. *Balaustium* sp. deutonymphs did not show any preference for a specific stage of spidermites. In addition, *Balaustium* sp. adults did not show preference for a particular prey or prey-stage (Nolasco and Cedola 2014).

Laboratory rearing revealed that *B. hernandezi* preys on a wide variety of small arthropods, including whiteflies, spider mites and thrips (Mąkol et al. 2012). In this study, *B. hernandezi* failed to complete immature development on a pollen diet alone. The addition of pollen to the prey diet had no positive effect on the developmental times of the different stages, which suggests that *B. hernandezi* has a marked predatory life style before reaching sexual maturity. This latter result was congruent with the data provided by Cadogan and Laing (1977) referring to the food spectrum of *B. putmani*.

Reports of prey of *B. putmani* and other *Balaustium* species are summarized in Table 8.2. In addition, unidentified *Balaustium* species have been reported preying on agricultural pests, including species feeding greedily on citrus rust mites (Acari: Eriophydae) infesting grapefruit in Texas (Dean 1959), and species associated with spider mite infestations on corn (Pickett and Gilstrap 1986). However, most of these reports lack any specific information regarding the efficacy of these predators regulating pest populations.

Three species of the genus *Balaustium* are known to be harmful: *B. medicagoense* Meyer and Ryke from Australia is the only species reported both as agricultural pest (Halliday 2001, 2005; Arthur et al. 2008, 2010, 2011) causing retarded and stunted growth of seedlings (Bailey 2007), and attacking humans (James 1995; James et al. 1995). Similarly, *B. murorum* and an unidentified *Balaustium* species from Australia (Rack 1973, 1983), Japan (Ido et al. 2004), United States and Canada (Newell 1963), were also reported attacking humans. Grandjean (1946) reported interesting observations on the feeding behavior of *B. murorum* associated with plane trees (*Platanus* sp.) in France. Interestingly, he observed that several females of *B. murorum* repeatedly deposited egg clusters in the same site (on the bark of the tree one meter above the soil) for long periods of time, which could be mediated by species-specific attractants or acute habitat preferences. Grandjean (1946) found *B. murorum* feeding upon pollen and described its feeding behavior. Larvae, deutonymphs, and adults grabbed the pollen grains with their palps and pierced them to then suck pollen content. The mites threw away shriveled grains before selecting a new grain to continue feeding while remaining in the same place for periods as long as 15 min if the pollen provision was sufficient. Despite feeding on pollen, the mites were never observed interacting with flower anthers, but mostly on the flower calyx or dispersing on the stems and leaves of different plants, including several weeds (*Ranunculus acris*, *Bellis perennis*, and *Acer campestris*), as well as on tree trunks and on rocks on the soil (Grandjean 1946). However, as mentioned before, *B. murorum* was not observed preying on arthropods, and is sometimes considered a nuisance because of its presence in buildings and for sporadic attacks on humans.

8.4.4 Rearing Techniques and Reproduction

Cadogan and Laing (1977) and Muñoz-Cárdenas et al. (2014) established colonies of *B. leanderi* and *B. putmani* for experimental purposes. *Balaustium putmani* was reared on apple-leaf disks placed on saturated cotton inside petri dish arenas.

Table 8.2 Reports of *Balaustium* species as natural enemies of agricultural pests

Habitat/substrate	Target herbivore	Place	Species	Reference
Citrus trees and ground under the trees	Crawlers of California red scale [*Aonidiella aurantii* (Maskell)] (Hemiptera: Diaspididae), other small insects and mites	California, USA	*B. aonidophagus*	Ebeling (1934)
Apple orchard	Spider mite *Panonychus ulmi* (Koch) (Acari: Tetranychidae)	Ontario, Canada; Ohio, USA	*B. putmani*	Cadogan and Laing (1977, 1981), Yoder and Heydinger (2011)
Apple orchard	*Bryobia praetiosa, P. ulmi* and *T. urticae*	Missouri, USA	*B dowelli*	Childers and Enns (1975)
Cotton	*Heliothis zea* eggs	Arkansas, USA	*B dowelli*	Smiley (1964)
Apple orchard	Nymphs of *Aphis pomi* DeGeer; *Dysaphis plantaginea* (Passerini); *Eriosoma lanigerum* (Hausmann); eggs of *Spodoptera frugiperda* (J.E. Smith); Motile forms of *T. urticae* and *P. ulmi* Cecidomyiid larvae,	North Carolina, USA	*B. putmani*	Childers and Rock (1981)
Peach Orchards	*Panonychus ulmi* (Acari: Tetranychidae) *Quadraspidiotus perniciosus* Comstock (Hemiptera:Diaspididae) *Aphis pomi* De Geer Newly hatched nymphs of *Paraphlepsius irroratus* Say (Hemiptera Cicadellidae)	Ontario, Canada,	*B. putmani*	Putman (1970)
	Spruce budworm [*Choristoneura fumiferana* (Clemens)] (Lepidoptera: Tortricidae)	Maine, USA	*B. kendalli*	Welbourn and Jennings (1991), Welbourn (1995)
Avocado (*Persea americana* Miller)	Thrips *Scirtothrips perseae* Nakahara (Thysanoptera: Thripidae)	Guatemala state, Guatemala	*Balaustium* sp.	Hoddle et al. (2002)

(continued)

Table 8.2 (continued)

Habitat/substrate	Target herbivore	Place	Species	Reference
Saw palmettos, *Serenoa repens* (Bartr.) Small, and Japanese spurge, *Pachysandra terminalis* Sieb. and Zucc.	Scale insects (Hemiptera)	Ohio, USA	*Balaustium* sp.	Yoder et al. (2006)
Sweet pepper, cucumber, aubergine	Thrips (Thysanoptera: Thripidae), whiteflies (Hemiptera: Aleyrodidae), spidermites	Almeria province, Spain	*B. hernandezi*	Mąkol et al. (2012)
Rose and tomato plants, *Trifolium repens*	*Frankliniella occidentalis*, *Trialeurodes vaporiarorum* and *Tetranychus urticae*	Cundina marca, Colombia	*B. leanderi*	Muñoz et al. (2009), Muñoz-Cárdenas et al. (2014)
Melon and pepper, *Portulaca oleraceae*, *Convolvulus arvensis*, *Sonchus oleraceous*	*Tetranychus urticae, Bemisia tabaci* and *Thrip sp.*	La plata, Buenos Aires, Argentina	*Balaustium* sp.	Nolasco et al. (2012), Nolasco and Cedola (2014)

Predators were fed on *P. ulmi* eggs and a small piece of apple bark was placed on the leaf to allow mites to rest or hide. *Balaustium leanderi* was reared in plastic containers units (18 cm diam, 20 cm height) with an opening (10 cm diam.) covered with a mite-proof steel mesh for ventilation. The bottom of the container was covered with a layer of moist peat (about 3 cm deep) and a disc of paper towel (17 cm diam) was placed on top of the layer to maintain moisture. A total of 15 adult individuals were placed in the container and they were provided with plant material infested with all stages of whiteflies, thrips and spider mites twice a week. Occasional cannibalistic behavior of *B. leanderi*, *B. hernadezi* and *B. putmani* were observed for both immature mites and adults. However, cannibalism was observed mostly in laboratory colonies when mites were not provided with protected areas to rest and serve as shelter during quiescent stages. There are no records of mass-rearing of *Balaustium* mites. Mąkol et al. (2012) indicated that the biocontrol commercial company Biobest® had established a rearing culture of *B. hernandezi*, but no information is provided regarding their rearing methods.

Balaustium species appear to have different types of reproduction. Parthenogenetic reproduction is assumed for *B. leanderi*, *B. murorum* and other species known only from females (Mąkol et al. 2012; Muñoz-Cárdenas et al. 2014). By contrast, the presence of both sexes was observed for *B. putmani* (Putman 1966), *B. southcotti* (Feider and Chioreanu 1977), and *B. xerothermicum* (Gabryś 2000a). Moreover, production of externally deposited spermatophores has been reported for *B. hernandezi* and *B. putmani* (Cadogan and Laing 1977; Mąkol et al. 2012). More research is needed to understand the type of reproduction exhibited by *Balaustium* species.

8.4.5 Efficacy Data

The information is restricted to prey consumption studies using *P. ulmi* as prey for *B. putmani* (Cadogan and Laing 1977), and functional response studies offering several preys for *B. leanderi* (L.S. Fuentes, 2010 unpublished data).

Cadogan and Laing (1977) determined the total number of *P. ulmi* eggs consumed by *B. putmani* during each motile stage of development under controlled conditions. At 25 °C the daily average of *P. ulmi* eggs consumed were 2.8, 14.7 and 15.9 per larva, deutonymph and adult, respectively.

Prey consumption by motile stages of *B. leanderi* when offered different densities of *T. vaporariorum* and *T. urticae* eggs, and *F. occidentalis* larvae, is shown in Fig. 8.3. The maximum consumption rates by all stages of *B. leanderi* were obtained when *T. vaporariorum* was offered as prey, followed by *T. urticae*. Maximum consumption rates of *F. occidentalis* were considerably lower. *Balaustium leanderi* exhibited a type III functional response feeding on the three prey items, which implies that the rate of detection is reduced at low prey densities, and increases as prey density increases. The available information suggests that *B. leanderi* exhibits higher prey consumption rates compared with *B. putmani* (L.S. Fuentes, 2010 unpublished data).

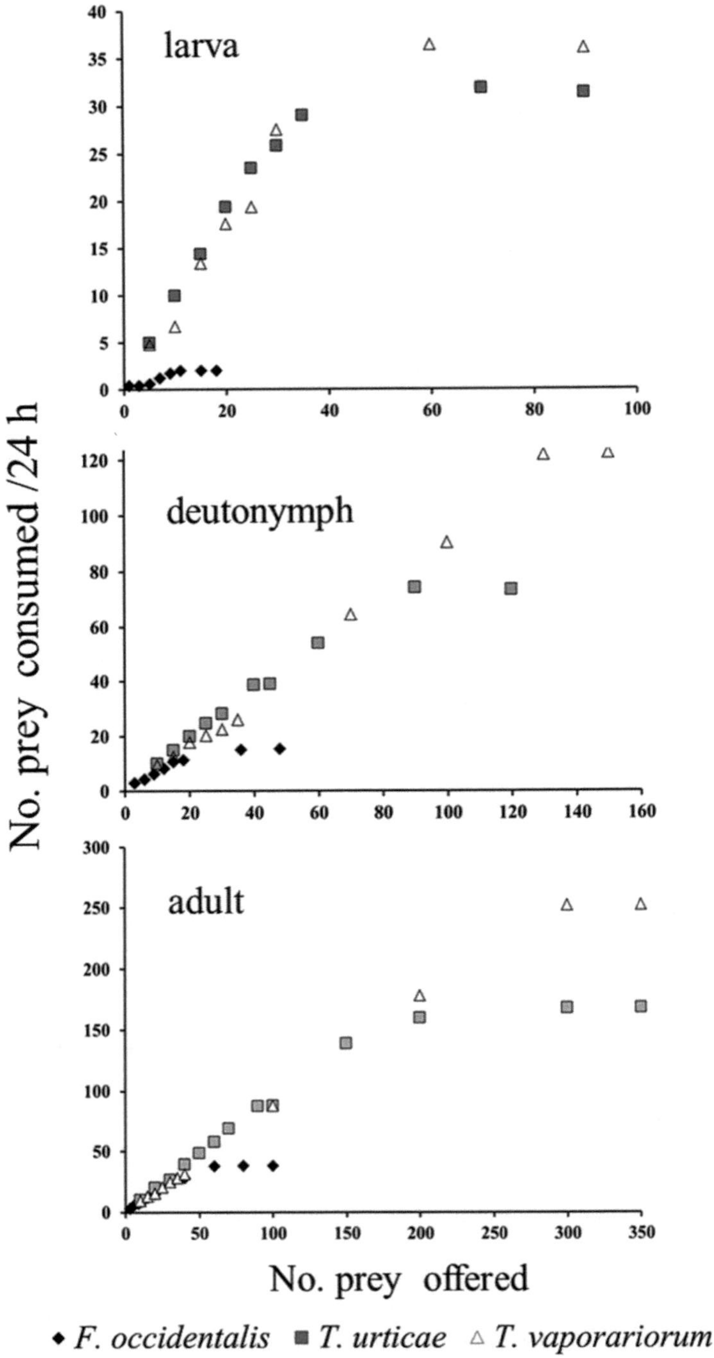

Fig. 8.3 Average numbers of prey eaten by motile stages of *B. leanderi* when offered different densities of *T. vaporariorum*, *T. urticae* and *F. occidentalis* (L.S. Fuentes, 2010 unpublished data)

8.4.6 Potential as Biocontrol Agents

As noted before, active post-larval forms of *B. putmani, B. hernandezi, B. leanderi* and several other *Balaustium* species have been regarded as beneficial predators in agricultural crops. However, very little is known about the potential of *Balaustium* species to regulate their prey. The polyphagous habits of some *Balaustium* species suggest that they could be used in biological control targeting several pest species. However, intraguild predation could be a limiting factor in their practical use, and there are no studies available to substantiate this hypothesis. *Balaustium* mites live in the ground and make excursions to the plant to feed upon herbivores (Muñoz-Cárdenas et al. 2014). Their above-ground habits have been investigated but there is not information regarding their impact in the below-ground arthropod web. Their role as biological controllers of prey that inhabits above and below-ground habitats must be studied. Yoder et al. (2009) presented evidence showing that an unidentified *Balaustium* species carries plant pathogens such as the fungi *Alternaria* sp., *Stigmella* sp. and *Verticillium* sp. They suggested that the *Balaustium* mites acquired the plant-pathogens from the soil and leaf litter where they live. However they did not show that plants with *Balaustium* individuals were more infected with the pathogens compared with plants without *Balaustium*. There are not reported observations correlating the presence of *Balaustium* mites on plants with the infection by plant-pathogens. More research is warranted to show if there are undesirable interactions regarding the above and below-ground movement of *Balaustium* mites.

Balaustium species have relatively long life cycles and slow rates of increase compared with their potential prey. Mass-rearing of *Balaustium* species appear to be difficult and costly. The available information suggests that *Balaustium* species could be used in conservative biological control projects but more research is needed to elucidate their potential use as effective biological control agents.

8.4.7 Compatibility with Chemical Control

Few studies have addressed the toxicity of some pesticides to *Balaustium* species. Herne and Putman (1966) studied the effects of pesticides used in Ontario peach orchards on *B. putmani*, concluding that sulphur was appreciably toxic but that Captan (fungicide) was not toxic. Hagley and Simpson (1983) determined that applications of permethrin and azinphosmethyl significantly reduced *B. putmani* numbers, which were preying on the pear psylla, *Cacopsylla pyricola* (Foerster) (Hemiptera: Psyllidae). Lastly, James et al. (1995) reported alphacypermethrin and bifenthrin causing high mortaility on *B. murorum* which was considered a predator of the pasture pest, *Halotydeus destructor* (Tucker) (Acari: Penthaleidae) in inland southern Australia. However, as previously reported, *B. murorum* is not considered predacious; thus, the true identity of the species reported by James et al. (1995) should be confirmed. Arthur (2010) confirmed that *B. medicagoense* is the only species of this genus present in grain crops, pastures and roadsides within southern Australia.

8.5 Other Erythraeoids

Several examples of the association between other erythraeoids and insects have been reported in numerous publications (see examples in Table 8.3, Welbourn 1983; Wohltmann 2000). Welbourn (1983) published a list of known erythraeoid parasites or predators of arthropods, and a list of genera with potential for use in biological control. Despite these multiple reports of associations with insects of economic importance, only a few studies have been conducted to evaluate the potential of these mites for biological control. This situation is especially true for the family Smarididae. Welbourn (1983) mentioned *Smaris prominens* (Banks) in association with two species of Pscoptera, *Troctes* sp. and *Pteroxanium kelloggi* (Ribaga). However, further research has not been conducted.

Reports show associations between thrips (including important pests as *F. occidentalis* Pergande and *Thrips tabaci* Lindeman) and *Leptus* sp. (Goldarazena and Zhang 1998), *Lasioerythraeus johnstoni* Welbourn and Young (Young and Welbourn 1988) some species of *Abrolophus* (Zhang and Goldarazena 1996; Welbourn and Young 1987; Goldarazena and Zhang 1998), *Erythraeus* (Zhang and Goldarazena 1996), and *Marantelophus* (Zhang and Goldarazena 1996; Goldarazena et al. 2000; Goldarazena and Zhang 1998). Eickwort (1983), Lewis (1973) and Ananthakrishnan (1984) cited by Sabelis and van Rijn (1997) reported two species of Erythraeids (*Erythrites* sp. and *Hauptmannia brevicollis*) predating and suppressing thrips populations. Species of *Leptus* are also found associated with Orthoptera (Southcott 1999; Haitlinger 2006), Diptera (Fain and Elsen 1972; Williams and Proctor 2002; Miranda and Bermudez 2008), Hemiptera (Tandon and Lal 1976; Southcott 1999; Saboori and Ostovan 2000; Chhillar et al. 2007; Kamran et al. 2009; Pereira et al. 2012) and Lepidoptera (Goldarazena and Zhang 1998). Other examples of associations can be found on Table 8.3.

Evaluations or observations of the potential for practical use of erythraeoids associated with pests have only been reported for some mites such as: *L. johnstoni* as predator and parasite of different insects, especially *Trigonotylus doddi* Distant and *Lygus lineolaris* (Palisot de Beauvois) (Hemiptera: Miridae) (Young and Welbourn 1988); *Leptus* spp. as parasite of *Choristoneura fumifera* (Clemens) (Lepidoptera: Tortricidae) affecting flight behavior and fecundity (Houseweart et al. 1980); and *Callidosoma metzi* as parasite and predator of *Platynota idaeusalis* (Walker) (Lepidoptera: Tortricidae) consuming a significative amount of eggs and moths (Sharma et al. 1983a; Adler and Browning 1986).

Although the changes in feeding preferences and habitat of these mites is a limiting factor for practical use to control certain pests, many species have great potential worth exploring. Species of *Abrolophus*, which similarly to *Balaustium* have apparently abandoned larval parasitism must be evaluated.

Table 8.3 Examples of erytraeoids (except *Balaustium* species) parasitizing and/or preying upon insects (orders Diptera, Coleoptera, Hemiptera, Lepidoptera and Orthoptera)

Erythraeidae species (known stage)[a]	Order host/prey	Host/prey species (family)	Host plant/substrate	Location	Reference
Abrolophus norvegicus (Thor) (larva, post-larval form)	Thysanoptera	*Chirothrips aculeatus* Bagnall (Thripidae)	Flower of *Carex hallerana*	Spain	Zhang and Goldarazena (1996), Goldarazena and Zhang (1998)
		Thrips trehernei Priesner (Thripidae)	Meadow, very abundant on flowers of *Sonchus* sp.	Basque Country (Spain)	Goldarazena and Zhang (1998)
		Aptinothrips rufus Haliday (Thripidae)	–	Basque Country (Spain)	Goldarazena and Zhang (1998)
		Thrips flavus Schrank (Thripidae)	–	Basque Country (Spain)	Goldarazena and Zhang (1998)
		Thrips angusticeps Heinrich-Uzel (Thripidae)	–	Basque Country (Spain)	Goldarazena and Zhang (1998)
		Frankliniella intosa Trybom (Thripidae)	–	Basque Country (Spain)	Goldarazena and Zhang (1998)
		Odontothrips sp. (Thripidae)	–	Basque Country (Spain)	Goldarazena and Zhang (1998)
		Thrips tabaci Lindeman (Thripidae)	–	Basque Country (Spain)	Goldarazena and Zhang (1998)
		Tenothrips frici (Uzel) (Thripidae)	–	Basque Country (Spain)	Goldarazena and Zhang (1998)
Abrolophus rudaensis (Haitlinger) (larva)	Thysanoptera	Unidentified Thysanoptera	–	Poland	Haitlinger (1986) cited by Zhang and Goldarazena (1996)
Abrolophus sp.	Thysanoptera	*Stenchaetothrips minutus* Van Deventer (Thripidae)	–	India	Jadhava and Varma (1996) cited by Zhang and Goldarazena (1996)

(continued)

Table 8.3 (continued)

Erythraeidae species (known stage)[a]	Order host/prey	Host/prey species (family)	Host plant/substrate	Location	Reference
Abrolophus sp.	Thysanoptera	other trips		Mississippi, USA	Welbourn and Young (1987)
Abrolophus sp.	Thysanoptera	*Thrips angusticeps* Heinrich-Uzel (Thripidae)	–	Basque Country (Spain)	Goldarazena and Zhang (1998)
		Thrips tabaci Lindeman (Thripidae)	–	Basque Country (Spain)	Goldarazena and Zhang (1998)
		Aptinothrips rufus Haliday (Thripidae)	–	Basque Country (Spain)	Goldarazena and Zhang (1998)
Abrolophus sp.	Thysanoptera	*Aelothrips intermedius* Bagnall (Aeolothripidae)	–	Basque Country (Spain)	Goldarazena and Zhang (1998)
Abrolophus sp.	Orthoptera	*Oedipoda caerulescens* Linnaeus (Acrididae)	–	Greece	Antonatos and Emmanouel (2014)
Caeculisoma mouldsi Southcott (larva)	Hemiptera	*Venustria superba* Goding and Froggatt (Cicadidae)	–	Australia	Southcott (1988b)
		Tamasa doddi (Goding and Froggatt) (Cicadidae)	–	Australia	Southcott (1988b)
		Chlorocysta suffusa (Distant) (Cicadidae)	–	Australia	Southcott (1988b)
		Psaltoda fumipennis Ashton (Cicadidae)	–	Australia	Southcott (1988b)
Callidosoma metzi Sharma, Drooz and Treat (larva, post-larval form)	Lepidoptera	*Platynota idaeusalis* (Walker) (Tortricidae)	–	Panama, USA	Sharma et al. (1983a), Adler and Browning (1986)
	Lepidoptera	*Anacamptodes vellivolata* (Hulst) (Geometridae)	Light traps	USA	Sharma et al. (1983b)
	Lepidoptera	*Lambdina pellucidaria* (Grote and Robinson) (Geometridae)	Light traps	USA	Sharma et al. (1983b)

Charletonia taiwanensis Tsai and Chow (larva, post-larval form)	Orthoptera	*Condracris rosea* DeGeer (Acridiidae)	–	Taiwan	Tsai and Chow (1988)
Charletonia bucephalia Beron (larva)	Orthoptera	Acrididae	–	Greece	Antonatos and Emmanouel (2014)
Charletonia dalegori Haitlinger (larva)	Orthoptera	Acrididae	–	Greece	Antonatos and Emmanouel (2014)
Charletonia venus Southcott (larva)	Orthoptera	Acrididae	–	Greece	Antonatos and Emmanouel (2014)
Charletonia sp.	Orthoptera	Acrididae	–	Greece	Antonatos and Emmanouel (2014)
Erythraeus sp.	Lepidoptera	*Heliothis zea* (Boddie) (Noctuidae)	Cotton		Whitcomb and Bell (1964)
Erythraeus spp. and *Charletonia* spp.	Hemiptera	*Nilaparvata* spp. (Delphacidae)	Rice		Barrion et al. (1981)
Erythraeus spp. and Charletonia spp.	Hemiptera	*Sogatella furcifera* (Horváth) (Delphacidae)	Rice		Barrion et al. (1981)
Erythraeus (*E.*) *kuyperi* (Oudemans) (larva)	Hemiptera	*Rhopalosiphum padi* (Linnaeus) (Aphidae)	–	Israel	André (1951)
		Schizaphis graminum (Rondani) (Aphidae)	–	Israel	André (1951)
		Sitobion fragariae (Walker) (Aphidae)	–	Israel	André (1951)
Erythrites (*G.*) *osmondensis* Southcott (larva, post-larval form)	Thysanoptera	Unidentified Thysanoptera	–	Australia	Greenslade and Southcott (1980) cited by Zhang and Goldarazena (1996)

(continued)

Table 8.3 (continued)

Erythraeidae species (known stage)[a]	Order host/prey	Host/prey species (family)	Host plant/substrate	Location	Reference
Erythrites (*U.*) *urrbrae* (Womersley) (post-larval form)	Thysanoptera	Unidentified Thysanoptera	–	Australia	Greenslade and Southcott (1980) cited by Zhang and Goldarazena (1996)
Lasioerythraeus johnstoni Welbourn and Young (larva, post-larval form)	Hemiptera	*Lygus lineolaris* (Palisot de Beauvois) (Miridae)	Margin of roads, cotton fields and woodlots	Mississippi, USA	Welbourn and Young (1987), Young and Welbourn (1988)
	Hemiptera	*Trigonotylus doddi* (Distant) (Miridae)	–	Mississippi, USA	Young and Welbourn (1988)
	Thysanoptera	*Frankliniella* sp.	–	Mississippi, USA	Young and Welbourn (1988)
Lasioerythraeus saboorii Khanjani, Raisi & Izadi (larva)	Hemiptera	*Aphis punicae* (Passerini) (Aphiidae)	*Punica protopunica* Balf. F	Iran	Khanjani et al. (2011)
Lasioerythraeus whitcombi (Smiley) (post-larval form)	Lepidoptera	*Heliothis zea* (Boddie) (Noctuidae)	Cotton	Arkansas USA	Smiley (1964, 1966)
Leptus (*L.*) *bankensis* Southcott (larva)	Orthoptera	Unidentified Orthoptera	–	Australia	Southcott (1999)
Leptus (*L.*) *bathypogonus* Womersley (larva)	Orthoptera	Unidentified Orthoptera	–	Australia	Southcott (1999), Beron (2008)
Leptus (*L.*) *benzaliensis* Fain et Elsen (larva)	Diptera	*Glossina fascipes* Newstead (Glossinidae)	–	Zaire	Fain and Elsen (1972)

Leptus (*L.*) *calcar* Southcott (larva, post-larval form)	Orthoptera	*Heteropternis obscurella* (Blanchard) (Acrididae)	–	Australia	Southcott (1999)
		Stenocantops vitripennis (Sjöstedt)(Acrididae)	–	Australia	Southcott (1999)
		Pardillana sp. (Acrididae)	–	Australia	Southcott (1999)
		Pycnosticnus seriatus Saussure (Acrididae)	–	Australia	Southcott (1999)
Leptus (*L.*) *carpenteri* Fain and Elsen (larva)	Diptera	*Glossina fascipes* Newstead (Glossinidae)	–	Zaire	Fain and Elsen (1972)
Leptus esmailii Saboori and Ostovan (2000) (larva)	Hemiptera	*Eurygaster integriceps* Puton (Scutelleridae)	Sunflower	Iran	Saboori and Ostovan (2000)
Leptus (*L.*) *elderi* Southcott (larva)	Hemiptera	*Heteropsylla cubana* Crawford (Psyllidae)	–	Australia	Southcott (1999)
Leptus (*L.*) *foliatus* Southcott (larva)	Orthoptera	Acrididae	Light traps	Australia	Southcott (1999)
Leptus (*L.*) *glossinarum* Fain and Elsen (larva)	Diptera	*Glossina fascipes* Newstead (Glossinidae)	–	Zaire	Fain and Elsen (1972)
Leptus (*L.*) *holgeri* Haitlinger (larva)	Orthoptera	Unidentified Orthoptera	–	Vietnam	Haitlinger (2013)
Leptus (*L.*) *lighti* Southcott (larva)	Orthoptera	*Chortoicetes terminifera* (Walker) (Acrididae)	–	Australia	Southcott (1999)
Leptus (*L.*) *maringensis* Fain and Elsen (larva)	Diptera	*Glossina fascipes* Newstead (Glossinidae)	–	Zaire	Fain and Elsen (1972)
Leptus (*L.*) *olafi* Haitlinger (larva)	Coleoptera	Unidentified Cicindellidae	–	Colombia	Haitlinger (1991) cited by Haitlinger (2006)
	Orthoptera	Unidentified Orthoptera	–	Venezuela	Haitlinger (2006)

(continued)

Table 8.3 (continued)

Erythraeidae species (known stage)[a]	Order host/prey	Host/prey species (family)	Host plant/substrate	Location	Reference
Leptus pakistanensis Kamran, Afzal, Bashir, Raza and Khan (larva)	Hemiptera	Aphidae	Sorghum halepence	Pakistan	Kamran et al. (2009)
Leptus sp.	Thysanoptera	*Frankliniella occidentalis* Pergande (Thripidae)	–	Basque Country (Spain)	Goldarazena and Zhang (1998)
Leptus sp.	Lepidoptera	*Leptocoris augur* (Fabricius) (Rhopalidae)	–		Dhidman and Singh (1988) cited by Goldarazena and Zhang (1998)
Leptus sp.	Orthoptera	Acrididae	–	Greece	Antonatos and Emmanouel (2014)
Leptus sp.	Hemiptera	*Edessa* sp. 1 (Pentatomidae)	Semideciduous seasonal forest	Brazil	Pereira et al. (2012)
		Edessa sp. 2 (Pentatomidae)	Semideciduous seasonal forest	Brazil	Pereira et al. (2012)
		Nezara ap. (Pentatomidae)	Semideciduous seasonal forest	Brazil	Pereira et al. (2012)
Leptus sp.	Diptera	*Chrysomya rufifacies* (Macquart) (Calliphoridae)	Traps	Panama	Miranda and Bermudez (2008)
		Chrysomya megacephala (Fabricius) (Calliphoridae)	Traps	Panama	Miranda and Bermudez (2008)
		Cochliomyia macellaria (Fabricius) (Calliphoridae)	Traps	Panama	Miranda and Bermudez (2008)
		Paralucilia adespota Dear (Calliphoridae)	Traps	Panama	Miranda and Bermudez (2008)

Leptus sp.	Diptera	*Hemilucilia semidiaphana* (Rondani) (Calliphoridae)	Traps	Panama	Miranda and Bermudez (2008)
Leptus sp.	Diptera	*Mesembrinella bicolor* (Fabricius) (Calliphoridae)	Traps	Panama	Miranda and Bermudez (2008)
Leptus sp.	Diptera	*Mesembrinella umbrosa* Aldrich (Calliphoridae)	Traps	Panama	Miranda and Bermudez (2008)
Leptus sp.	Diptera	*Culex annulirostris* Skuse (Culicidae)	Light traps	Australia	Williams and Proctor (2002)
Leptus spp.	Lepidoptera	*Choristoneura fumiferana* (Clemens)	Pheromone traps	USA	Houseweart et al. (1980)
Leptus sp. and *Erythraeus* sp.	Hemiptera	*Drosicha mangiferae* (Green) (Margarodidae)	Mango	India	Tandon and Lal (1976), Chhillar et al. (2007)
Leptus (*L.*) *tindalei* Southcott (larva)	Orthoptera	Unidentified Orthoptera	–	Australia	Southcott (1999)
Leptus (*L.*) *tiranicus* Haitlinger (larva)	Diptera	unidentified Asilidae	–	Venezuela	Haitlinger (2006)
Leptus (*L.*) *torresianus* Southcott 1988 (larva)	Hemiptera	*Venustria superba* Goding and Froggatt (Cicadidae)	–	Australia	Southcott (1988a, b, c, d)
		Tamasa doddi (Goding and Froggatt) (Cicadidae)	–	Australia	Southcott (1988a, b, c, d)
Marantelophus bella (Zhang) (larva)	Thysanoptera	*Sericothrips* (Thripidae)	*Glycyrhiza glabra*	Iran	Zhang and Goldarazena (1996)

(continued)

Table 8.3 (continued)

Erythraeidae species (known stage)[a]	Order host/prey	Host/prey species (family)	Host plant/substrate	Location	Reference
Marantelophus multisetosus (Zhang and Goldarazena) (larva)	Thysanoptera	Aeolothripidae larva	*Quercus rotundifolia*		Goldarazena et al. (2000), Goldarazena and Zhang (1998)
		Aeolothrips gloriosus Bagnall (Aeolothripidae)	*Quercus rotundifolia*	Croatia, Greece, Hungary, Romania, San Marino, Spain, Turkey, Ukraine	Goldarazena et al. (2000)
		Chirothrips aculeatus Bagnall (Thripidae)	*Deschampsia* sp.		Goldarazena et al. (2000)
		Frankliniella occidentalis Pergande (Thripidae)	*Eucaliptus* sp.		Goldarazena et al. (2000)
		Melanthrips fuscus (Sulzer) (Aeolothripidae)	*Brassica* sp.		Goldarazena et al. (2000)
		Rhipidothrips sp.	*Deschampsia* sp.		Goldarazena et al. (2000)
		Thripidae larva	*Cistus* sp.		Goldarazena et al. (2000)
		Thrips angusticeps Uzel	*Anacyclus* sp.		Goldarazena et al. (2000)
		Thrips major Uzel	*Bellis* sp.		Goldarazena et al. (2000)
		Thrips tabaci Lindeman (Thripidae)	Flower of *Galium aparine*, *Allium* sp.	Spain	Zhang and Goldarazena (1996), Goldarazena et al. (2000), Goldarazena and Zhang (1998)
		Odontothrips ignobilis Bagnall (Thripidae)	–	Basque Country (Spain)	Goldarazena and Zhang (1998)
		Tenothrips frici (Uzel) (Thripidae)	–	Basque Country (Spain)	Goldarazena and Zhang (1998)
Marantelophus haitlingeri (Goldarazena et Zhang) (larva)	Hemiptera	Unidentified Hemiptera	Flowers of *Gallium mollugo*	Spain	Goldarazena and Zhang (1997)

[a]Information taken from Mąkol and Wohltmann (2012)

8.6 Final Remarks

There are several limitations to the practical use of the members of this group as biological controllers in Integrated Pest Management programs. The principal limitation relates to the limited taxonomic, biological and ecological knowledge of most species.

The key aspect for the success of a biological control agent is proper identification, so clarifying the classification of most of the species of this group is necessary. This clarification can only be done by correlating the different stages through lab rearing, and, thus, doing more accurate descriptions. However, in many cases, the maintenance of laboratory colonies is difficult, requering the development of other techniques to assess taxonomic identification such as molecular tools. DNA barcoding technique is a tool already used for Trombididae specimens collected in Sweden (Stålstedt et al. 2014), that could be extended for Erythraids species for which rearing is not feasible.

The difficulty of maintaining colonies of these mites in the laboratory is not only a limitation for classification and identification, but also for mass production. Therefore, it is necessary to establish which species can be effectively used for inundative releases and which species should be treated with conservative biological control strategies. For the ectoparasites Ertythraeids it would be specially important to keep natural habitats in the vicinity of crop lands in order to provide a source of altemative hosts. It is well known that these areas are important as refugia for natural enemy populations (Letorneau 1998). Also important would be to stimulate the establishment of flower strips to provide alternative food for the predators near to agricultural areas as presented by Balzan et al. (2014). All these aspects require further studies.

Knowledge of ecological and biological aspects is also needed to understand the interaction of these potential controllers with the environment and with their prey pests, especially for *Balaustium* species that are already successfully maintained in laboratory and showed predatory capacity. The following are some factors that need to be studied: Firstly, determine if the pheromones released through the urnulae have the same role in different *Balaustium* species as well as to determine the effect of this pheromone in other arthropods. Second, elucidate the prey of *Balautium* species on/in the soil/litter. This can be done by sampling and identifying potential prey on/in the soil/litter and conducting gut content analysis (Harwood et al. 2004) on *Balaustium* species. Third, investigate the role of plants food sources like nectar or pollen on oviposition and population growth of *Balaustium* species different than *B. hernandezi* and *B. putmani*. It is known that the addition of pollen to the diet of these two species does not increase their fitness but for other species of *Balaustium*, there is not data available. Fourth, elucidate the compatibility of *Balaustium* mites with other natural enemies and techniques of control in order to know the best way of utilization and potential intraguild predation.

The knowledge of this group is limited to only few species and regions. Further studies in other geographical regions are necessary to identify the species abundance and evaluate their potential for practical use in Integrated Pest Management programs.

References

Adler CQ, Browning MW (1986) Attachment and dispersion of *Callidosoma metzi* (Acari: Erythraeidae) parasitizing *Playnota idaeusalis* (Lepidoptera: Tortricidae). Ann Entomol Soc Am 79:56–59

Ahsan M, Anwarullah M (1970) A new species of *Cecidopus* from Pakistan (Acarina: Erythraeidae). Pak J Zool 2:47–49

Alberti G, Mąkol J (2013) Fine structure of the urnulae of *Balaustium* mites (Actinotrichida: Erythraeidae) representing peculiar defense organs. Arthropod Struct Dev 42:483–494

Ananthakrishnan TN (1984) Bioecology of thrips. Indira Publishing House, Michigan, pp 81–101

André M (1951) Nouvelles observations sur le *Bochartia kuyperi* Oudemans (Acarien). Bull Mus Nat 2:253–255

Antonatos SA, Emmanouel NG (2014) A qualitative study of Eutrombidiidae and Erythraeidae, ectoparasites on Orthoptera, in two grassland areas of Attica – Greece. Int J Acarol 40:23–30

Arthur AL (2010) The biology, ecology and population genetics of *Balaustium medicagoense* and *Bryobia* spp.; emerging mite pests of grains in southern Australia. PhD thesis, University of Melbourne, Australia

Arthur AL, Hoffmann AA, Umina PA, Weeks AR (2008) Emerging pest mites of grains (*Balaustium medicagoense* and *Bryobia* sp.) show high levels of tolerance to currently registered pesticides. Anim Prod Sci 48:1126–1132

Arthur AL, Weeks AR, Umina PA, Hoffman AA (2010) Survival and reproduction of the pest mites *Balaustium medicagoense* and *Bryobia* spp. on winter grain crops. Exp Appl Acarol 52:141–153

Arthur AL, Weeks AR, Hill MP, Hoffman AA (2011) The distribution, abundance and life cycle of the pest mites *Balaustium medicagoense* (Prostigmata: Erythraeidae) and *Bryobia* spp. (Prostigmata:Tetranychidae) in Australia. Aust J Entomol 50:22–36

Bailey P (2007) Pests of field crops and pastures: identification and control. Csiro Publishing, Collingwood, p 414

Balzan MV, Bocci G, Moonen AC (2014) Augmenting flower trait diversity in wildflower strips to optimize the conservation of arthropod functional groups for multiple agroecosystem services. J Insect Conserv 18:713–728

Barrion AT, Pantura PC, Bandong JP, Delacrus EG, Raymundo F, Lumaban MD (1981) Food web of rice brown planthoppers in the Philippines. Int Rice Res Newsl 6:13–16

Belozerov VN (2008) Calyptostasy: its role in the development and life histories of the Parasitengone mites (Acari: Prostigmata: Parasitengona). Acarina 16:3–19

Beron P (2008) Acarorum Catalogus I. Acariformes: Calyptostomatoidea (Calyptostomatidae), Erythraeoidea (Smarididae, Erythraeidae). Edition of Pensoft Publishers and the National Museum of Natural History, Sofia

Cadogan BL, Laing JE (1977) A technique for rearing the predaceous mite *Balaustium putmani* (Acarina: Erythraeidae), with notes on its biology and life history. Can Entomol 109:1535–1544

Cadogan BL, Laing JE (1981) A study of *Balaustium putmani* (Acarina: Erythraeidae) in apple orchards in southern Ontario. Proc Entomol Soc Ont 112:13–22

Chatzivasileiadis EA, Sabelis MW (1997) Toxicity of methyl ketones from tomato trichomes to *Tetranychus urticae* Koch. Exp Appl Acarol 21:473–484

Chhillar BS, Gulati R, Bhatnagar P (2007) Agr acarol. Daya Publ House, Delhi, p 355

Childers CC, Enns WS (1975) Predaceous arthropods associated with spider mites in Missouri apple orchards. J Kansas Entomol Soc 48(4):453–471

Childers CC, Rock GC (1981) Observations on the occurrence and feeding habits of *Balaustium putmani* (Acari: Erythraeidae) in North Carolina apple orchards. Int J Acarol 7:63–68

Coombs MR, Bale JS (2014) Thermal thresholds of the predatory mite *Balaustium hernandezi*. Physiol Entomol 39:120–126

Dean HA (1959) Seasonal distribution of mites on Texas grapefruit. J Econ Entomol 52:228–232

Ebeling W (1934) A new predaceous mite from southern California (Acarine: Erythraeidae). Pan Pac Entomol 10:33–34

Eickwort GC (1983) Potential use of mites as biological control agents of leaf-feeding insects. In: Hoy MA, Cunningham GL, Knutson L (eds) Biological control of pests by Mites. Division of Agriculture and Natural Resources, University of California, Special Publication 3304, pp 41–52

Fain A, Elsen P (1972) Notes sur les acariens parasites ou commensaux des mouches tsé-tseś. 1. Famille Erythraeidae (Trombidiformes). Acta Zool Pathol Antverp 56:61–72

Feider Z, Chioreanu H (1977) Erythraeidae. In: La faune terrestre de l'ille de Sainte-Helene. Annales du Musée royal de l'Afrique centrale (Sciences Zoologiques), vol 220. Musée royal de l'Afrique centrale, Tervuren, pp 269–290

Fuentes LS, Muñoz-Cárdenas K, Combita O, Jimeno E, Getiva JC, Cantor F, Rodríguez D, Mąkol J (2014) A redescription of *Balaustium leanderi* comb. nov. (Actinotrichida: Erythraeidae) including a first report of all stages and some taxonomic notes on the genus. Fla Entomol 97:937–951

Gabryś G (2000a) *Balaustium xerothermicum* sp. nov. from Poland with remarks on all world species of the genus (Acari: Actinedida: Erythraeidae). Ann Zool 50:47–56

Gabryś G (2000b) *Kamertonia polonica* gen. and sp. nov. from Poland with a key to the world genera of "conalaeless" Erythraeinae (Acari: Actinedida: Erythraeidae). Ann Zool 50:57–63

Gabryś G (2000c) Four new homonyms in Erythraeidae (Acari: Actinedida). Ann Zool 50:65–66

Getiva J, Acosta A (2004) Taxonomia de acaros asociados a cultivos de flores. Asocolflores 66:59–68

Goldarazena A, Zhang Z-Q (1997) First record of larval *Grandjeanella* (Acari: Erythraeidae) from Heteroptera and description of a new species from Spain. Syst Appl Acarol 2:231–236

Goldarazena A, Zhang Z-Q (1998) Seasonal abundance of *Abrolophus neobrevicollis* larvae (Acari: Erythraeidae) ectoparasitic on *Thrips trehernei* (Thysanoptera: Thripidae) in the Basque Country (Spain). Syst Appl Acarol 3:93–96

Goldarazena A, Zhang Z-Q, Gordana R (2000) A new species and a new record of ectoparasite mites from thrips in Turkey (Acari: Trombidiidae and Erythraeidae). Syst Parasitol 45:75–80

Grandjean F (1946) Observations sur les Acariens (9e serie). Bulletin du Muséum national d'histoire naturelle Paris, ser 2 18(4):337–344

Grandjean F (1947) Au sujet des Erythroides. Bull Mus Nat Hist Nat Paris 19:327–334

Hagley EAC, Simpson CM (1983) Effect of insecticides on predators of the pear psylla, *Psylla pyricola* (Hemiptera: Psyllidae), in Ontario. Can Entomol 115:1409–1414

Haitlinger R (2006) *Dasitrombium margeritanum* sp. n., *Leptus tiranicus* sp. n. and the first record of *L. olafi* Haitlinger (Acari: Prostigmata: Neothrombiidae, Erythraeidae) ectoparasitic on Orthoptera and Diptera (Insecta) from Margerita, Venezuela. Pol J Entomol 75:347–357

Haitlinger R (2013) First record of *Leptus* (*Leptus*) *holgeri* (Acari: Prostigmata: Erythraeidae) from Vietnam, with redescription of the species. Pers J Acarol 2:341–351

Halliday RB (2001) Systematics and biology of the Australian species of *Balaustium* von Heyden (Acari: Erythraeidae). Aust J Entomol 40:326–330

Halliday RB (2005) Predatory mites from crops and pastures in South Africa: potential natural enemies of redlegged earth mite *Halotydeus destructor* (Acari: Penthaleidae). Zootaxa 1079:11–64

Harwood JD, Sunderland KD, Symondson WOC (2004) Prey selection by linyphiid spiders: molecular tracking of the effects of alternative prey on rates of aphid consumption in the field. Mol Ecol 13:3549–3560

Hayes JL (1985) The predator–prey interaction of the mite *Balaustium* sp. and the pierid butterfly *Colias alexandra*. Ecology 66:300–303

Hedges BZ, Rosselot AE, Tomko PM, Yoder JA, Benoit JB (2012) High temperature and dehydration tolerance of the red velvet mite, *Balaustium* sp (Erythraeidae), permit the exploitation of extremely hot, dry microhabitats. Int J Acarol 38:89–95

Herne DHC, Putman WL (1966) Toxicity of some pesticides to predacious arthropods in Ontario peach orchards. Can Entomol 9:936–942

Hoddle MS, Nakahara S, Phillips PA (2002) Foreign exploration for Scirtothrips perseae Nakahara (Thysanoptera: Thripidae) and associated natural enemies on avocado (Persea americana Miller). Biol Control 24:251–265

Houseweart MW, Jennings DT, Berkett LP, Brann TB (1980) Parasitic mires (Acari: Erythraeidae) on spruce budworm moths (Lepidoptera: Tortricidae). Can Entomol 112:193–197

Ido T, Kumakiri M, Lao L, Yano Y, Takada M (2004) Dermatitis caused by *Balaustium murorum*. Acta Derm Venereol 84:80–81

James DG (1995) Biological control of earth mites in pasture using endemic natural enemies. Plant Prot Q 10:58–59

James DG, O'Malley K, Rayner M (1995) Effect of alphacypermethrin and bifenthrin on the survival of five acarine predators of *Halotydeus destructor* (Acari: Penthaleidae). Exp Appl Acarol 19:647–654

Kamran M, Bashir MH (2013) On a new species of genus *Lasioerythraeus* and New Record of *Pollux workandae* Southcott (Prostigmata: Erythraeidae) from Pakistan. Pak J Zool 45:721–725

Kamran M, Afzal M, Bashir MH, Raza AM, Khan BS (2009) A new species of the genus *Leptus* Latreille (Acari: Erythraeidae) parasitizing aphids in Pakistan. Pak J Zool 41:17–20

Khanjani M, Raisi H, Izadi H (2011) A new record of the genus *Lasioerythraeus* Welbourn and Young (Acari: Erythraeidae) from Iran and description of a new species. Int J Acarol 37:544–549

Łaydanowicz J, Mąkol J (2010) Correlation of heteromorphic life instars in terrestrial Parasitengona mites and its impact on taxonomy – the case of *Leptus molochinus* (C. L. Koch, 1837) and *L. ignotus* (Oudemans, 1903) (Acari: Actinotrichida: Prostigmata: Erythraeidae). J Nat Hist 44:669–697

Letorneau DK (1998) Chapter 3: Agroecosystem and conservation biological control. In: Barbosa P (ed) Conservation biological control. Academic Press, San Diego, pp 9–38

Lewis T (1973) Thrips: their biology, ecology and economic importance. Academic, London/New York, p 349

Mąkol J (2002) A redescription of *Trombidium latum* (C. L. Koch, 1837) (Acari: Actinotrichida, Trombidioidea) with characteristics of all active instars. Ann Zool 52:433–442

Mąkol J (2010a) Redescription of *Erythraeus tonsus* (L. Koch, 1879) (Acari: Trombidiformes: Erythraeoidea) with record on an unusual case of polymely in adult Erythraeidae collected in Alaska (USA). Int J Acarol 36:221–232

Mąkol J (2010b) A Redescription of *Balaustium murorum* (Hermann, 1804) (Acari: Prostigmata: Erythraeidae) with Notes on Related Taxa. Ann Zool 60:439–454

Mąkol J, Wohltmann A (2012) An annotated checklist of terrestrial Parasitengona (Actinotrichida: Prostigmata) of the World, excluding Trombiculidae and Walchiidae. Ann Zool 62:359–562

Mąkol J, Wohltmann A (2013) Corrections and additions to the checklist of terrestrial Parasitengona (Actinotrichida: Prostigmata) of the world, excluding Trombiculidae and Walchiidae. Ann Zool (Warszawa) 61:15–27

Mąkol J, Arijs Y, Wäckers F (2012) A new species of *Balaustium* von Heyden, 1826 (Acari: Actinotrichida, Erythraeidae) from Spain. Zootaxa 3178:1–21

Mayoral JG, Barranco P (2011) A new species of Larval *Leptus* (Parasitengona: Erythraeidae) and new records of larval Erythraeidae parasitizing Orthoptera from French Guiana. Acarologia 51:411–417

Meyer MKP, Ryke PAJ (1959) Nine new species of the superfamily Erythraeoidea (Acarina: Trombidiformes) associated with plants in South Africa. Acarologia 1:304–323

Miranda R, Bermudez S (2008) Ácaros (Arachnida: Acari) asociados con moscas Calliphoridae (Diptera: Oestroidea) en tres localidades de Panamá. Rev Colomb Entomol 34:192–196

Muñoz K, Fuentes LS, Cantor F, Rodríguez D, Cure JR (2009) Preferencia alimenticia del ácaro depredador *Balaustium* sp. en condiciones controladas. Agronomía Colomb 27:95–103

Muñoz-Cárdenas K, Fuentes LS, Cantor F, Rodríguez D, Janssen A, Sabelis MW (2014) Generalist red velvet mite predator (*Balaustium* sp.) performs better on a mixed diet. Exp Appl Acarol 62:19–32

Newell IM (1963) Feeding habits in the genus *Balaustium* (Acarina: Erythraeidae), with special reference to attacks on man. J Parasitol 49:498–502

Nolasco V, Cedola C (2014) Selectividad alimentaria de Balaustium sp. (Acari: Erythraeidae) depredador de plagas hortícolas, en el cultivo de pimiento. In: Book of abstracts of the XXVI Reunión Argentina de Ecología, 2–5 Nov 2014, Comodoro Rivadavia, Chubut

Nolasco V, Cedola C, Polack A, Mezquiriz N (2012) Balaustium sp. (Acari: Erythraeidae) potencial depredador de plagas hortícolas, asociado al pimiento (Capsicum annuum L.) (Solanácea) y malezas aledañas. In: Book of abstracts of the XXXV Congreso Argentino de Horticultura, Corrientes, p 444

Pereira AIA, Fadini MAM, Pikart TG, Zanuncio JC, Serrão JE (2012) New hosts and parasitism notes for the mite *Leptus* (Acari: Erythraeidae) in fragments of the Atlantic Forest, Brazil. Braz J Biol 72:611–616

Pickett CH, Gilstrap FE (1986) *Balaustium* sp. occasional predator of spidermites on corn. J Kansas Entomol Soc 59:524–553

Putman WL (1966) Insemination im Balaustium sp. (Erythraeidae). Acarologia 8:424–426

Putman WL (1970) Life history and behavior of Balaustium putmani (Acarina: Erythraeidae). Ann Entomol Soc Am 63:76–81

Putman WL, Herne DHC (1966) The role of predators and other biotic agents in regulating the population density of phytophagous mites in Ontario peach orchards. Can Entomol 98:808–820

Rack G (1973) *Balaustium murorum* (Germann, 1804) (Acarina: Erythraeidae) in hausern. Eine wenig beachtete sommerliche Milbenplague. Anzeiger für Schädlingskunde, Pflanzenschutz, Umweltschutz 46:129–132

Rack G (1983) Systematics, morphology and biology of mites Acari in houses and stored products and mites of medical importance. Part 1: general introduction. *Balaustium murorum* Hermann, 1804. Praktische Schadlingsbekampfer 35:157–160

Sabelis MW, Van Rijn PCJ (1997) Predation by insects and mites. In: Lewis T (ed) Thrips as crop pests. CAB International, Wallingford, pp 259–354

Saboori A, Ostovan H (2000) A new species of genus *Leptus* Latreille, 1796 (Acari: Erythraeidae) ectoparasite of sun pest, *Eurygaster integriceps* Puton (Hemiptra: Scutelleridae) from Iran. Syst Appl Acarol 5:143–147

Sato M, Moraes G, Haddad M, Wekesa V (2011) Effect of trichomes on the predation of *Tetranychus urticae* (Acari: Tetranychidae) by *Phytoseiulus macropilis* (Acari: Phytoseiidae) on tomato, and the interference of webbing. Exp Appl Acarol 54:21–32

Sharma GD, Drooz AT, Treat AE (1983a) A new species of Callidosoma (Acari: Prostigmata: Erythraeidae) parasite on adults of *Anacamptodes vellivolata* (Lepidoptera: Geometridae) as a larva, and predator of eggs of Lepidoptera as deutonymph and adult. Ann Entomol Soc Am 76:120–129

Sharma GD, Farrier MH, Drooz AT (1983b) Food, life-history, and sexual differences of *Callidosoma metzi* Sharma, Drooz, and Treat (Acarina: Erythraeidae). Int J Acarol 9(3):149–155

Smiley RL (1964) Two new erythraeids predaceous upon cottonbollworm eggs (Acarina: Erythraeidae). Proc Entomol Soc Wash 66:110

Smiley RL (1966) Further description of two erythraeids predaceous upon cotton bollworm eggs. Proc Entomol Soc Wash 68:110

Southcott RV (1961a) Studies on the systematics and biology of the Erythraeoidea (Acarina), with a critical revision of the genera and subfamilies. Aust J Zool 9:367–610

Southcott RV (1961b) Notes on the genus *Caeculisoma* (Acarina: Erythraeidae) with comments on the biology of the Erythraeoidea. Trans R Soc S Aust 84:163–178

Southcott RV (1961c) Description of two new Australian Smarididae (Acarina), with remarks on chaetotaxy and geographical distribution. Trans R Soc S Aust 85:133–153

Southcott RV (1963) The Smarididae (Acarina) of North and Central America and some other countries. Trans R Soc S Aust 86:159–245

Southcott RV (1966) Revision of the genus *Charletonia* Oudemans (Acarina: Erythraeidae). Aust J Zool 14:687–819

Southcott RV (1972) Revision of the larvae of the tribe Callidosomatini (Acarina: Erythraeidae) with observations on post-larval instars. Aust J Zool 13:1–84

Southcott RV (1983) A new Australian species of *Charletonia*. Proc Linnean Soc NSW 107(2):123–128

Southcott RV (1984) Acari from operation Drake in New Guinea. 4. a new species of *Leptus* (Erythraeidae). Acarologia 25(4):351–358

Southcott RV (1988a) Two new larval Erythraeinae (Acarina: Erythraeidae) from New Zealand, and the larval Erythraeinae revised. N Z J Zool 15:223–233

Southcott RV (1988b) Two new larval mites (Acarina: Erythraeidae) ectoparasitic on North Queensland cicadas. Rec S Aust Mus 22(2):103–116

Southcott RV (1988c) A new australian larval Callidosomatinae mite (Acarina: Erythraeidae) parasitic on flies, with notes on subfamily and tribe classification. Proc Linn Soc NSW 110:193–204

Southcott RV (1988d) Two new larval ectoparasitic mites (Acarina: Erythraeidae) from Sri Lankan tetrigid grasshoppers. Entomol Scand 30:151–159

Southcott RV (1989a) A balaustiine larva (Acarina, Erythraeidae) of a new genus from Iceland. Entomol Scand 20:173–177

Southcott RV (1989b) A larval mite (Acarina: Erythraeidae) parasitizing the European honey bee in Guatemala. Acarologia 30:123–129

Southcott RV (1989c) Larval *Leptus* (Acarina: Erythraeidae) ectoparasitic on a North American ant. Acarologia 30:241–248

Southcott RV (1991a) A further revision of *Charletonia* (Acarina: Erythraeidae) based on larvae, protonymphs and deutonymphs. Invertebr Taxon 5:61–131

Southcott RV (1991b) Descriptions of larval *Leptus* (Acarina: Erythraeidae) ectoparasitic on Australian Diptera, and two earlier described Australian larvae. Invertebr Taxon 5:717–763

Southcott RV (1992) Revision of the larvae of *Leptus* Latreille (Acarina, Erythraeidae) of Europe and North America, with descriptions of post-larval instars. Zool J Linn Soc 105:1–153

Southcott RV (1993a) A new larval Charletonia (Acarina: Erythraeidae) from Spain. Acarologia 34:51–56

Southcott RV (1993b) Larvae of *Leptus* (Acarina: Erythraeidae) ectoparasitic on higher insects of Australia and New Guinea. Invertebr Taxon 7:1473–1550

Southcott RV (1994a) A new larval Erythraeine mite (Acarina: Erythraeidae) from West Africa. Int J Acarol 20:81–85

Southcott RV (1994b) Two new larval Erythraeidae (Acarina) from Thailand, with keys to the larvae of *Leptus* for Asia and New Guinea, and world larvae of *Hauptmannia*. Steenstrupia 20:165–176

Southcott RV (1995a) A new larval erythraeine mite (Acarina: Erythraeidae) from Spain. Acarologia 36:223–228

Southcott RV (1995b) A new larval smaridid mite (Acarina: Smarididae) from Costa Rica. Acarologia 36:57–64

Southcott RV (1996) On some Australian and other larval Callidosomatinae (Acari: Erythraeidae). Int J Acarol 21:253–278

Southcott RV (1997) Description of two new larval Smarididae (Acarina) from Australia. Rec S Aust Mus 30:1–12

Southcott RV (1999) Larvae of *Leptus* (Acarina: Erythraeidae), free-living or ectoparasitic on arachnids and lower insects of Australia and Papua New Guinea, with description of reared post-larval instars. Zool J Linn So 127:113–276

Stålstedt J, Mąkol J, Bergsten J (2014) Matching adults and larvae with the help of DNA. In: Abstract book of the XIV Intl congress of Acarol. Tokyo, Japan, 61p

Tandon PL, Lal B (1976) New record of predatory mites on mango mealy bug, *Drosicha mangiferae* Green (Margarodidae: Hemiptera). Curr Sci 45:566–567

Tsai RS, Chow YS (1988) *Charletonia taiwanensis* n.sp. (Acari: Prostigma, Erythraeidae) an ectoparasite of the grasshopper *Condracris rosea* deGeer (Orthoptera: Acridiidae) in Taiwan. Int J Acarol 14:171–181

Van Houten YM, Glas JJ, Hoogerbrugge H, Rothe J, Bolckmans KJF, Simoni S, van Arkel J, Alba JM, Kant MR, Sabelis MW (2013) Herbivory-associated degradation of tomato trichomes and its impact on biological control of Aculops lycopersici. Exp Appl Acarol 60:127–138

Walter DE, Lindquist EE, Smith IM, Cook DR, Krantz GW (2009) Order trombidiformes. In: Krantz GE, Walter DE (eds) A manual of acarology, 3rd edn. Texas Tech University Press, Texas, pp 233–420

Welbourn WC (1983) Potential Use of Trombidioid and Erythraeoid Mites as Biological Control Agents of Insect Pests. In: Hoy MA, Cunninham GL, Knuson L (eds) Biological control of pests by Mites. Proceedings of a conference help, 5–7 Apr 1982, University of California, Berkeley, pp 103–40

Welbourn WC (1995) *Balaustium* spp. in Florida (Acari: Erythraeidae). Gainesville (FL): Florida Department of Agriculture and Consumer Services. Entomol Circ 368:1–2

Welbourn WC, Jennings D (1991) Two new species of Erythraeidae (Acari: Prostigmata) associated with the spruce budworm, *Choristoneura fumiferana* (Clemens) (Lepidoptera: Tortricidae), in Maine. Can Entomol 123:567–580

Welbourn WC, Young OP (1987) New genus and species of Erythraeinae (Acari, Erythraeidae) from Mississippi with a key to the genera of North American Erythraeidae. Ann Entomol Soc Am 80:230–242

Whitcomb WH, Bell K (1964) Predacious insects, spiders and mites of Arkansas cotton fields. Univ Ark Bull 690:50–51

Williams CR, Proctor HC (2002) Parasitism of mosquitoes (Diptera: Culicidae) by larval mites (Acari: Parasitengona) in Adelaide, South Australia. Aust J Entomol 41:161–163

Wohltmann A (2000) The evolution of life histories in Parasitengona (Acari: Prostigmata). Acarologia 41:145–204

Wohltmann A, Gabrys G, Mąkol J (2007) Acari: terrestrial parasitengona inhabiting transient biotopes. In: Moltmann UG, Saglio B (eds) Chelicerata: Araneae, Acari I. Süßwasserfauna von Mitteleuropa 7/2-1. Elsevier GmbH, München, pp 158–240

Yoder JA, Heydinger DJ (2011) Proliferation of red velvet mite *Balaustium* sp. nr. *putmani* (Erythraeidae) in Ohio based on water relations of active instars. Int J Acarol 37:166–175

Yoder JA, Ark JT, Benoit JB, Bellinger EJ, Gribbins KM (2006) Water balance components in Adults of Terrestrial Red Mite *Balaustium* sp. (Acarina: Erythraeidae). Ann Entomol Soc Am 99:560–566

Yoder JA, Glenn BD, Schumaker LK, Tank JL (2007) High temperature resistance of the terrestrial red mite (Balaustium sp.) as a product of suppressed heat induced water permeability. Int J Acarol 33:275–281

Yoder JA, Schumaker LK, Tank JL (2009) Potential for spread of plant pathogens by presence of red velvet mite (*Balaustium* sp.) in an Ohio landscape site. Int J Acarol 35:19–24

Yoder JA, Condon MR, Hart CE, Collier MH, Patrick KR, Benoit JB (2010) Use of an alarm pheromone against ants for gaining access to aphid/scale prey by the red velvet mite *Balaustium* sp. (Erythraeidae) in a honeydew-rich environment. J Exp Biol 213:386–392

Young OP, Welbourn WC (1988) Parasitism of *Trigonotylus doddi* (Heteroptera: Miridae) by *Lasioerythraeus johnstoni* (Acari: Erythraeidae), with notes on additional hosts and distribution. J Entomol Sci 23:269–273

Zhang Z-Q, Goldarazena A (1996) *Abrolophus* and *Grandjeanella* larvae (Acari: Erythraeidae) ectoparasitic on thrips (Thysanoptera: Thripidae). Syst Appl Acarol 1:127–144

Zhang Z-Q, Fan QH, Pesic V, Smit H, Bochkov AV, Khaustov AA, Baker A, Wohltmann, A, Wen TH, Amrine JW, Beron P, Lin JZ, Gabrys G., Husband R (2011). Order Trombidiformes Reuter, 1909. In: Zhang Z-Q (ed) Animal biodiversity: an outline of higher-level classification and survey of taxonomic richness. Zootaxa 3148:129–138

Zuluaga A (2012) Capacidad de dispersión de *Balaustium* n. sp. al interactuar con poblaciones de *Trialeurodes vaporariorum* y *Tetranychus urticae*. BSc thesis, Environ Biol Major. Univ Jorge Tadeo Lozano, Colombia

Chapter 9
The Cheyletoidea (Prostigmata), with Special Reference to the Potential of *Cheyletus malaccensis* Oudemans as Biological Control Agent of Post-harvest Pests

Christos G. Athanassiou and Nickolas E. Palyvos

9.1 Introduction

Cheyletidae is a family of the suborder Prostigmata, order Trombidiformes and superorder Acariformes. It belongs to the superfamily Cheyletoidea. Part of the Cheyletidae and all members of the other six families constituting this superfamily (Cloacaridae, Demodicidae, Epimyodicidae, Harpirhynchidae and Psorergatidae) are vertebrate parasites (Walter et al. 2009); these will not be treated in this chapter. The cheyletoids are characterized by having movable cheliceral digits styliform and approximate; peritreme present in free-living groups. Idiosoma with weakly defined dorsal shields, which may be distinguished by their different texture in relation to that of the surrounding cuticle; usually with contiguous or fused genital and anal openings.

Cheyletidae was mentioned by Volgin (1987) to include about 150 species, but many species were described since then. It is now estimated that over 370 species of approximately 74 genera belong to this family (Fain 1979; Evans 1992; Gerson et al. 1999; Walter et al. 2009). A revision of the cheyletids, with a key to separate the genera and a list of all species was presented by Gerson et al. (1999).

The body of cheyletids is often colorless, yellowish and sometimes red. The gnathosoma is large and distinctly separated from the idiosoma, moving widely from side to side. The chelicerae are fused to each other and to the subcapitulum to

C.G. Athanassiou (✉)
Laboratory of Entomology and Agricultural Zoology, Department of Agriculture, Crop Production and Rural Environment, University of Thessaly, Phytokou str., N. Ionia, Magnesia 38446, Greece
e-mail: athanassiou@agr.uth.gr

N.E. Palyvos
Laboratory of Agricultural Zoology and Entomology, Department of Plant Sciences, Agricultural University of Athens, 75 Iera Odos, Athens 11855, Greece

D. Carrillo et al. (eds.), *Prospects for Biological Control of Plant Feeding Mites and Other Harmful Organisms*, Progress in Biological Control 19,
DOI 10.1007/978-3-319-15042-0_9

form a single structure. The palp often with fewer than five segments; femur elbowed, tibia enlarged, bearing a distal claw-like structure which may contain a variable number of teeth along the inner margin; tarsus usually with 1–2 comb-like setae. The palps can be used as a pair of pincers for catching prey. Peritremes located on the gnathosoma. The dorsal surface of the idiosoma in the adult may be covered by one or two shields. Some groups with a pair of ocelli on the anterolateral region of the idiosoma. Some genera, especially *Cheyletus* Latreille, produce one or more heteromorphic males in which the length of the palp and sculpturing on the gnathosoma are exaggerated (Hughes 1976).

This is the most common and abundant family of Prostigmata found in stored products. Most cheyletids are free-living predators of various micro-arthropods associated with stored-products, including insect eggs (Zdarkova 1979, 1998). Cheyletids are commonly found in granaries, warehouses and barns (Hughes 1976), but can also be found in nests of birds and rodents, in association with Acariformes mites of the cohort Astigmatina, order Sarcoptiformes. In general, Cheyletidae are classified among the most frequently found mites at the post-harvest stages of agricultural commodities.

Cheyletus spp. are often found to be associated with pestiferous mites and insects in stored-products and are considered as biological control agents of these pests. The natural occurrence of *Cheyletus* species and their association with multiple arthropods suggests that they have a broad prey range (Cebolla et al. 2009; Athanassiou et al. 2011). Earlier studies showed that they occurred naturally in various commodities and in different types of storage facilities, feeding on both mites and moth eggs (Athanassiou et al. 2003, 2011). Athanassiou et al. (2011) noted that the distribution of *Cheyletus malaccensis* Oudemans was mostly spatially associated with moths (Pyralidae) rather than with astigmatids. At the same time, the authors indicated in that work that the densities of astigmatids were not related with most of the other species in the tested facility, which may indicate that their spatio-temporal distribution was mostly affected by abiotic factors, such as temperature and moisture, and not by predation. However, this should not be considered as a clear indicator of *C. malaccensis* prey preferences. Regarding its biology, the type of prey seriously affects *C. malaccensis* life table characteristics (Palyvos and Emmanuel 2009, 2011) (Fig. 9.1).

One of the most common *Cheyletus* species in northern Europe is *Cheyletus eruditus* (Schrank), considered as the first mite to be used for biological control at the post-harvest stages of agricultural products, such as grains and related amylaceous materials (Zdarkova 1998). *Cheyletus eruditus* has been found in association with other predatory mites, such as *Blattisocius* species (Acari: Mesostigmata: Blattisociidae) (Athanassiou et al. 2003; Hubert et al. 2006). Even from the 1970s, the noticeable activity of *C. eruditus* as a predator of other major stored product insect and mite species had been recorded, and its tolerance to low temperatures made Zdarkova and Pulpan (1973) to underline its potential as a biological control agent in less warm areas. Some years later, the same research group developed a mass rearing technique for this species (Zdarkova 1986), and developed the

Fig. 9.1 Different commodities infested by stored product mites

formulation known as Cheyletin, which was, practically, the first formulation for macro-biological control at the post-harvest stages (Zdarkova 1998). Cheyletin was tested with success in various cases, while its application was mostly an "enhancement" of naturally occurring populations, rather than an introduction of a novel biocontrol agent (Zdarkova and Fejt 1999; Hubert et al. 2006) This is why, in many cases, *C. eruditus* was found to coexist with *C. malaccensis*, which may suggest that there may be some competition for the same prey items (Hubert et al. 2006; Athanassiou et al. 2011). It is generally considered that *C. eruditus* is more adapted in less temperate climates, while *C. malaccensis* is abundant in temperate areas (Zdarkova and Fejt 1999; Hubert et al. 2006; Athanassiou et al. 2011). Hubert et al. (2006) noted that the first species was mostly recorded in grain residues, while the latter in grain bulks, a spatial segregation which partially explains co-occurrence.

There are indications that *C. eruditus* can also prey on psocids (Psocoptera) (Roesli et al. 1999; Kucerova 2004; Pascual-Villalobos 2005). This is particularly important and merits additional investigation, as psocids are emerging pests of stored products and not much is known about the potential of predators or parasitoids for their biological control.

9.2 Potential of *Cheyletus malaccensis* for Biological Control

Cheyletus malaccensis is the dominant cheyletid in storage facilities in tropical, and probably also in temperate regions (Palyvos et al. 2008; Athanassiou et al. 2011). This species is associated with both raw grains and grain residues (namely grain with qualitative deteriorations, which usually remains after the main grain mass is removed from the facility) (Hubert et al. 2006). According to a faunistic survey conducted in various types of commodities in storage and processing facilities, from 2000 to 2005, *C. malaccensis* was the most common cheyletid in Greece, accounting for 75.4 % of the collected cheyletids; moreover, it corresponded to a considerable proportion of all mite predators found (Palyvos et al. 2008). This predator was recovered from all different types of storage facilities found in Greece, being dominant in silos and prevalent in stores of agricultural cooperatives. It was particularly dominant in amylaceous products, but it was also found in commodities with high sugar content, such as dried fruit (Palyvos et al. 2008). Also, *C. malaccensis* has been found in stored seed cotton and cotton products, but in low numbers (Athanassiou et al. 2002).

Cheyletus malaccensis preys on multiple arthropods, including other mites (Hughes 1976; Pekar and Hubert 2008; Cebolla et al. 2009), eggs of pyralid moths [i.e. *Ephestia cautella* Walker, *Corcyra cephalonica* (Stainton) (Nangia et al. 1994), *Ephestia kuehniella* Zeller (Athanassiou and Palyvos 2006)], and eggs of tenebrionid beetles [i.e. *Tribolium confusum* Jacquelin Du Val (Rizk et al. 1979) and *Tribolium castaneum* (Herbst) (Cebolla et al. 2009)]. This predator can also attack young nymphs or larvae of other insects (Yousef et al. 1982; Sengonca et al. 2004; Kucerova 2004). Similarly to the procedure used by other stored product predatory mites (Athanassiou and Palyvos 2006), *C. malaccensis* inserts the chelicerae in the prey and immediately injects saliva that paralyzes it (Yoshikawa 1985).

Despite the scant published information about this mite, there are several contrasting reports about its biology (fecundity, parthenogenetic reproduction and prey consumption), which could be related to the possible existence of various biotypes of *C. malaccensis* in different parts of the world. Other sources of variation would be the ambient temperature(s) or differences in prey types offered as food to the predator. Males of this species are haploid and females diploid (Norton et al. 1993). It has been proven to reproduce by arrhenotokous parthenogenesis, virgin females producing males and fertilized females producing progeny of both sexes (Summers and Witt 1973). Fecundity is affected by prey species and the developmental stage at which they are offered to the predator (Palyvos et al. 2008; Palyvos and Emmanouel 2011). According to Palyvos and Emmanouel (2011), female fertilization may affect fecundity. Fertilized females were observed to deposit significantly more eggs than virgin females during their life span. Also, this species has heteromorphic males and belligerent behavior of nesting females, eggs being often laid in guarded clusters (Fig. 9.2).

As observed for other arthropods, temperature is one of the main environmental factors regulating development, reproduction and survival of *C. malaccensis*. This predator can complete development and reproduction at temperatures ranging between 17.5 and 35 °C, and can be used for control of pestiferous astigmatid mites occurring within this temperature range (Palyvos and Emmanouel 2009,

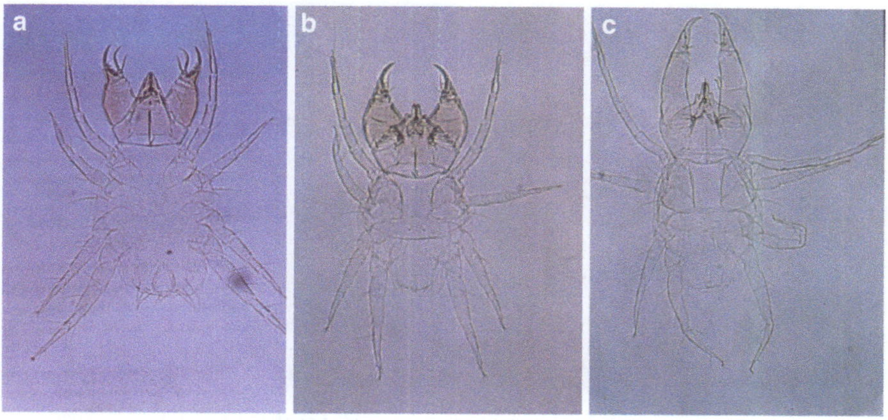

Fig. 9.2 (**a**) Female, (**b**) Homomorphic male, (**c**) Heteromorphic male

2011). The optimum temperature for development of *C. malaccensis* seems to be around 33.0–33.5 °C, which covers the upper zone for the development of several major stored product insects and mites (Palyvos and Emmanouel 2011). This experimental information could explain the higher frequency and abundance at which *C. malaccensis* is found in Greek stores compared with *C. eruditus* (Palyvos et al. 2008). Conditions in Greek stores are hotter and drier than stores in central Europe and other more temperate climates (Lukas et al. 2007). Thus, the use of *C. malaccensis* as a biological control agent seems better suited for drier and warmer storage ecosystems (Palyvos and Emmanouel 2009).

9.3 Combination of Cheyletids with Other Control Methods

It is generally assumed that the effect of chemicals on stored product mite pests is the same as the effect on their predators. However, available information suggests that some predatory mites may be more tolerant to pesticides than their prey.

Palyvos et al. (2006) tested a diatomaceous earth formulation against the cheese mite, *Tyrophagus putrescentiae* (Schrank), and its predator, *C. malaccensis*, under laboratory conditions. They found that while *T. putrescentiae* was very susceptible, *C. malaccensis* was not. In fact, *C. malaccensis* could survive for relatively long periods in grains that were treated with diatomaceous earth (which cause desiccation of insects and mites) that were lethal to *T. putrescentiae* for up to 48 days after the application. This high difference in susceptibility could be potentially attributed to three factors: (a) the predator has thicker cuticle than the prey, which means that desiccation was much slower, (b) although both prey and predator had body contact with the treated substrate, only the prey ingested it, so that internal desiccation could only occur in the latter; (c) as Prostigmata, *C. malaccensis* could moderate water loss through tracheal respiration (which does not occur in astigmatids as *T. putrescentiae*, which do not have tracheae). Moreover, *C. malaccensis* utilizes the

Fig. 9.3 Shrinked *T. putrescentiae* individuals and *E. kuehniella* eggs, after exposure to the predator *C. malaccensis*

grain kernels only for its movement and prey-seeking behavior, while *T. putrescentiae* infests the grain kernels, which means that this species has an increased contact with the treated substrate (Fig. 9.3). In fact, *C. malaccensis* tolerance was higher when prey was present, which suggests that access to food moderated water loss, as feeding mites can obtain water from the food. Cook and Armitage (2002) reported complete control of astigmatines, during their bioassays with diatomaceous earths, when a naturally occurring predatory mite population was recorded, which suggests that mite predators were less affected.

In a series of laboratory bioassays, Athanassiou and Palyvos (2006) also recorded that different life stages of *C. malaccensis* and *Blattisocius keegani* Fox (Acari: Mesostigmata) were more tolerant to diatomaceous earths than phytophagous mites. Similar trends were noted for a mixture of diatomaceous earth with natural pyrethrum and piperonil butoxide (Palyvos et al. 2006; Athanassiou and Palyvos 2006). For *B. keegani*, adult females were more susceptible than protonymphs, probably due to the removal of particles of the toxic agent during molting (Athanassiou and Palyvos 2006). Conversely, *C. malaccensis* females were less susceptible than protonymphs, which may be due to the longer duration of the protonymphal stage of this species in comparison with that of *B. keegani*. Also, *C. malaccensis* was observed to oviposit for some time after exposure to diatomaceous earth (Palyvos et al. 2006). The above studies suggest that although relatively vulnerable to specific toxic agents, those mite predators can be used in combination with low toxicity pesticides, as diatomaceous earth. Additional experimental work is needed to clarify this hypothesis.

9.4 Prospects for the Future

In augmentative biological control programs the initial objective is to increase the densities of the natural enemy being released, which in turn is expected to subsequently cause the death of the pest. For agricultural pests, the increased number of natural enemies in the environment and the presence of dead prey causes no other sort of problem. However in the case of pest control in storage, the release of macro-biological control agents should be carefully studied, as the individuals that are added, their excretions etc may not be eliminated through processing. In general, the presence of insects in raw grains is associated with increased fragments in processed commodities. For example, Perez-Mendoza et al. (2005) found that the increase of the presence of individuals of the lesser grain borer, *Rhyzopertha dominica* (F.) (Coleoptera: Bostrichidae) was directly associated with the increase of the insect fragments in flour. However, despite the fact that the same results are expected in the case of mites, it should be noted that mites are more difficult to be detected in the final product, which may "conceal" their actual presence.

The presence of arthropod fragments in food, is an issue or major importance and practically cannot be totally avoided. In the USA, the Food and Drug Administration has established a minimum of these contaminants, known as "Food Defect Action Levels", which also includes other substances. In contrast, in other parts of the world, including the European Union, require "zero tolerance", despite the fact that there is no "zero tolerance" in "real world" conditions.

Yet, two important aspects to consider are: (a) even the use of chemicals will not eliminate the presence of those organisms, especially if applied after the detection of the pest arthropods; (b) the introduction of new arthropods (efficient predatory mites) into the storage facility may reduce the growth of the pest organisms, which in turn will result in the reduction of the predator population, which is associated with the prey density. Still, information about the dynamics between predators and prey in storage facilities allowing the safe use of the latter to control the former (in terms of obtaining adequate control timely, so as to maintain to a minimum the levels attained by the pest and the predators) is insufficient. Thus, biological control of arthropods in storage facilities is considerably more complicated than in the field, given that it is not sufficient just to control the pest, but control with minimal levels of predator populations is of utmost importance. The development or the adoption of a mechanical mean to eliminate eventual fragments of pest and predatory mites would be highly desirable.

Despite this major limitation, a predator-based biological control should be planned and utilized, especially in the case of commodities on which astigmatines are more commonly found and for which the presence of predators may not be considered as essential. An example is animal feed, which usually has a very large variety of mite species that could be utilized as prey for *C. malaccensis*, as suggested by Zdarkova (1998) in relation to the use of *C. eruditus*.

Some mite species have been proven to be related to the occurrence of dermatitis, respiratory allergy etc. Yoshikawa (1985) and Hatsushika et al. (1989) noted several undesirable reactions after exposure to Cheyletidae. Nevertheless, even in this case, the number of predators is expected to be much lower than that of the prey in case no control action is taken. Still, these health problems should be seriously taken into account when a biological-based control strategy is planned at the post-harvest stages of agricultural products and food.

References

Athanassiou CG, Palyvos NE (2006) Laboratory evaluation of two diatomaceous earth formulations against *Blattisocius keegani* Fox (Mesostigmata, Ascidae) and *Cheyletus malaccensis* Oudemans (Prostigmata: Cheyletidae). Biol Control 38:350–355

Athanassiou CG, Palyvos NE, Eliopoulos PE, Papadoulis GP (2002) Mites associated with stored seed cotton and related products in Greece. Phytoparasitica 30:387–394

Athanassiou CG, Kavallieratos NG, Palyvos NE, Buchelos CT (2003) Three-dimensional distribution and sampling indices of insects and mites in horizontally-stored wheat. Appl Entomol Zool 38:413–426

Athanassiou CG, Kavallieratos NG, Sciarretta A, Palyvos N, Trematerra P (2011) Spatial associations of insects and mites in stored wheat. J Econ Entomol 104:1752–1764

Cebolla R, Pekar S, Hubert J (2009) Prey range of the predatory mite *Cheyletus malaccensis* (Acari: Cheyletidae) and its efficacy in the control of seven stored-product pests. Biol Control 50:1–6

Cook DA, Armitage DM (2002) Integrated pest management for stored grain in the U.K. incorporating diatomaceous earths to prevent surface infestations of insects and mites. IOBC Bull 25:221–229

Evans GO (1992) Principles of acarology. C.A.B. International, University Press, Cambridge, p 563

Fain A (1979) Idiosomal and leg chaetotaxy in the Cheyletidae. Int J Acarol 5:305–310

Gerson U, Fain A, Smiley RL (1999) Further observations on the Cheyletidae (Acari), with a key to the genera of the Cheyletinae and a list of all known species in the family. Bulletin de L'Institut Royal des Sciences Naturelles de Belgique Entomologie 69:35–86

Hatsushika R, Okino T, Miyoshi K (1989) A case study of itching dermatitis caused by *Cheyletus fortis* (Oudemans, 1904) (Acarina: Cheyletidae) found in Okayama Prefecture, Japan. Kawasaki Med J 15:151–157

Hubert J, Munzbergova Z, Kucerova Z, Stejskal V (2006) Comparison of communities of stored product mites in grain mass and grain residues in the Czech Republic. Exp Appl Acarol 39:149–158

Hughes AM (1976) The mites of stored food and houses. Tech Bull Minist Agric Fish Food 9:1–400

Kucerova Z (2004) Stored product psocids as one of the preys of the predatory mite *Cheyletus eruditus* (Schrank) (Acarina: Cheyletidae) Integrated Protection of Stored Products. IOBC Bull 27:191–195

Lukas J, Stejskal V, Jarosik V, Hubert J, Zdarkova E (2007) Differential natural performance of four *Cheyletus* predatory mite species in Czech grain stores. J Stored Prod Res 43:97–102

Nangia N, Channabasavanna G, Jagadish P (1994) A predator of primary acariforms in storage and its biology. Curr Res Univ Agric Sci Bangalore India 23:116–118

Norton RA, Kethley JB, Johnston DE, OConnor BM (1993) Phylogenetic perspectives on genetic systems and reproductive modes of mites. In: Wrensch DL, Ebbert MA (eds) Evolution and diversity of sex ratio in insects and mites. Chapman & Hall Publ, New York, pp 8–99

Palyvos NE, Emmanouel NG (2009) Temperature-dependent development of the predatory mite *Cheyletus malaccensis* (Acari: Cheyletidae). Exp Appl Acarol 47:147–158

Palyvos NE, Emmanouel NG (2011) Reproduction, survival, and life table parameters of the predatory mite *Cheyletus malaccensis* (Acari: Cheyletidae) at various constant temperatures. Exp Appl Acarol 54:139–150

Palyvos NE, Athanassiou CG, Kavallieratos NG (2006) Acaricidal effect of a diatomaceous earth formulation against *Tyrophagus putrescentiae* (Astigmata: Acaridae) and its predator *Cheyletus malaccensis* (Prostigmata: Cheyletidae) in four grain commodities. J Econ Entomol 99:229–236

Palyvos NE, Emmanouel NG, Saitanis CJ (2008) Mites associated with stored products in Greece. Exp Appl Acarol 44:213–226

Pascual-Villalobos MJ (2005) Occurrence of psocids and natural predators in organic rice in Callaspara (Murcia, Spain). J Stored Prod Res 41:231–235

Pekar S, Hubert J (2008) Assessing biological control of *Acarus siro* by *Cheyletus malaccensis* under laboratory conditions: effect of temperatures and prey density. J Stored Prod Res 44:335–340

Perez-Mendoza J, Throne JE, Maghirang EB, Dowell FE, Baker JE (2005) Insect fragments in flour: relationship to lesser grain borer (Coleoptera: Bostrichidae) infestation level in wheat and rapid detection using near-infrared spectroscopy. J Econ Entomol 98(6):2282–2291

Rizk GN, Badry E, Hafez SM (1979) The effectiveness of predacious and parasitic mites in controlling *Tribolium confusum* Duv. Mesopotamia J Agric 14:167–182

Roesli R, Jones R, Rees D (1999) Factors affecting outbreacks of Liposcelis (Psocoptera: Liposcelidae) populations in grain storage. In: Jin Z, Liang Q, Liang Y, Tan X, Guan L (eds) Proceedings of the 7th international working conference on stored-product protection, 14–19 Oct 1998, Beijing. Sichuan Publishing House of Science and Technology, Chengdu, pp 27–36

Sengonca C, Zegula T, Blaeser P (2004) The suitability of twelve different predatory mite species for the biological control of *Frankliniella occidentalis* (Pergande) (Thysanoptera: Thripidae). J Plant Dis Prot 111:388–399

Summers FM, Witt RL (1973) Oviposition and mating tendencies of *Cheyletus malaccensis* (Acarina: Cheyletidae). Fla Entomol 56:277–285

Volgin VI (1987) Acarina of the family Cheyletidae of the world. Akad Nauk SSSR Zool Inst Keys to the fauna of the USSR 101:1–532

Walter DE, Lindquist EE, Smkith IM, Cook DR, Krantz GW (2009) Order trombidiformes. In: Krantz GW, Walter DE (eds) A manual of acarology. Texas Tech University Press, Lubbock, pp 233–420

Yoshikawa M (1985) Ecology and control of mites in Japanese houses. J Environ Control Technol 3:4–11

Yousef AA, Zaher MA, Kandil MM (1982) Effect of prey and temperature on the development and biology of *Cheyletus malaccensis* Oudemans (Acari: Cheyletidae). J Appl Entomol 93:39–42

Zdarkova E (1979) Cheyletid fauna associated with stored products in Czechoslovakia. J Stored Prod Res 15:11–16

Zdarkova E (1986) Mass rearing of the predator *Cheyletus eriditus* (Schrank) (Acarina: Cheyletidae) for biological control of acarine mites infesting stored products. Crop Prot 5:122–124

Zdarkova E (1998) Biological control of storage mites by *Cheyletus eruditus*. Integr Pest Manag Rev 3:111–116

Zdarkova E, Fejt R (1999) Possibilities of biological control of stored product mites. In: Jin Z, Liang Q, Liang Y, Tan X, Guan L (eds) Proceedings of the 7th international working conference on stored-product protection, 14–19 Oct 1998, Beijing. Sichuan Publishing House of Science and Technology, Chengdu, pp 1243–1245

Zdarkova E, Pulpan J (1973) Low temperature storage of the predatory mite *Cheyletus eruditus* (Schrank) for future use in biological control. J Stored Prod Res 9:217–220

Chapter 10
Food Web Engineering to Enhance Biological Control of *Tetranychus urticae* by Phytoseiid Mites (Tetranychidae: Phytoseiidae) in Citrus

Josep A. Jaques, Ernestina Aguilar-Fenollosa, Mónica A. Hurtado-Ruiz, and Tatiana Pina

10.1 Introduction

Spain is the sixth largest world producer of citrus (7 million tons annually), and the largest for the fresh market of this commodity, producing mostly mandarins and oranges (around 2 and 3 million tons, respectively) (MAGRAMA 2013; FAOSTAT 2014). As a consequence of this market orientation, the fruits are subjected to very low cosmetic thresholds, and direct pests (those damaging fruits) are of crucial importance (Hare 1994; Jacas and Urbaneja 2010).

Several phytophagous mite species occur in Spanish citrus orchards (Table 10.1). Tetranychidae is the most important family, including *Tetranychus urticae* Koch, *Panonychus citri* (McGregor), *Eutetranychus banksi* (McGregor) and *Eutetranychus orientalis* (Klein). Tenuipalpidae is represented by *Brevipalpus californicus* (Banks) and *Brevipalpus phoenicis* (Geijskes). Tarsonemidae, by *Polyphagotarsonemus latus* (Banks), and Eriophyidae, by *Aceria sheldoni* (Ewing). From this list, only *T. urticae* and *A. sheldoni* are considered key pests (pest density always above economic injury level, EIL; Flint 2012) of clementine mandarins and lemons, respectively. The remaining tetranychids are considered secondary pests (pest density sometimes above EIL) and the remaining species of other families are considered occasional pests (pest density usually below EIL).

J.A. Jaques was formerly J.A. Jacas

J.A. Jaques (✉) • E. Aguilar-Fenollosa • M.A. Hurtado-Ruiz • T. Pina
Unitat Associada d'Entomologia Agrícola UJI-Institut Valencià d'Investigacions Agràries (IVIA), Universitat Jaume I (UJI), Campus del Riu Sec, E-12071 Castelló de la Plana, Spain
e-mail: josep.jaques@camn.uji.es; aguilare@uji.es; mhurtado@uji.es; pina@uji.es

D. Carrillo et al. (eds.), *Prospects for Biological Control of Plant Feeding Mites and Other Harmful Organisms*, Progress in Biological Control 19,
DOI 10.1007/978-3-319-15042-0_10

Table 10.1 Pest mite species occurring in citrus orchards in Spain

Family	Species	Common name	Hosts	Varieties attacked	Damage
Tetranychidae	*Eutetranychus banksi*	Texas citrus mite	Polyphagous	All varieties	Stippling and leaf drop
	Eutetranychus orientalis	Oriental mite	Polyphagous	All varieties	Stippling and leaf drop; spotted chlorosis on the rind
	Panonychus citri	Citrus red mite	Polyphagous	All varieties, especially oranges	Pale stippling on leaves and rind. Fruit may appear dull
	Tetranychus urticae	Two spotted spider mite	Highly polyphagous	All varieties, especially mandarins	Chlorotic spots on leaves; dark stains on fruit; defoliation
Tenuipalpidae	*Brevipalpus californicus*	Flat mite	Polyphagous	All varieties, especially oranges	Spotted grey scarring of the rind; light yellow spotting on mature leaves
	Brevipalpus phoenicis	Flat mite			
Tarsonemidae	*Polyphagotarsonemus latus*	Broad mite	Polyphagous	All varieties, especially lemons	Scarred tissue that cracks as fruit grows; grey scarring on leaves and twigs
Eriophyidae	*Aceria sheldoni*	Bud mite	Citrus only	All varieties, especially lemons	Severe distortion of all organs; reduced fruit set

Our research group is located in the Region of Valencia (eastern Spain), where most Spanish clementine mandarins are produced (MAGRAMA 2013). This group has focused on the sustainable management of clementine mandarin pests, particularly of *T. urticae*, also a key pest of citrus in other countries with Mediterranean climate (Jacas et al. 2010) as well as in many other citrus growing regions (Talhouk 1975; McMurtry 1985; Anon 1991; Smith et al. 1997).

The economic importance of *T. urticae* on clementine mandarins refers to the scarring it causes by feeding mostly in the stylar and apical areas of the fruit, severely reducing prices for the fresh market (Ansaloni et al. 2008; Pascual-Ruiz et al. 2014a). The Reduce Pest Population Strategy (Hammond 2002), using either preventive (cover crop management, pruning) or curative (acaricides) control methods can be used to reduce the densities of key and secondary pests below their EIL. With increasing demand for reduced pesticide use to avoid environmental impacts and effect on human health, the European Union (EU) endorsed in 2009 a Directive on the sustainable use of pesticides. This established the principles of Integrated Pest Management (IPM) as compulsory for EU member states starting in January 2014 (European Union 2009). Hence, preventive methods should become the cornerstone of modern IPM programs in the EU.

In this chapter, we summarize the work carried out by our group on the exploitation of cover crops as a means to enhance the impact of the main natural enemies of *T. urticae* in clementine mandarin orchards. In this agroecosystem, as in many other crops, these enemies are predatory mites of the family Phytoseiidae (Helle and Sabelis 1985; McMurtry 1992; Zhang 2003; Krantz and Walter 2009; Hoy 2011).

10.2 Conservation Biological Control

Conservation biological control (CBC) takes advantage of resident, either native or naturalized, natural enemies by modification of the environment or crop practices to protect and enhance their role to reduce pest damage (Eilenberg et al. 2001). The fact that natural enemies are resident distinguishes it from other biological control practices, where natural enemies are deliberately released. CBC is especially useful in perennial cropping systems where both pests and natural enemies are active and abundant all year round (Barbosa 1998; Landis et al. 2000). Citrus, as a permanent evergreen crop, is a system where CBC has many applications (Jacas and Urbaneja 2010).

One of the most popular CBC practices implemented in Spanish citrus and elsewhere (Ruberson et al. 1998; Jacas and Urbaneja 2010) has been the application of selective pesticides to reduce impact on beneficial arthropods. The use of banker plants and, more recently, the management of the ground cover and reservoir plants provide citrus growers new tools to enhance the performance of natural enemies (Jacas and Urbaneja 2010). These beneficial plants provide refuge, hibernation or aestivating sites, alternative hosts or prey, as well as plant resources, i.e., pollen or nectar, beneficial to (in some cases required by) some predatory and parasitic arthropods as food items (Landis et al. 2000; Boller et al. 2004; Jonsson et al. 2008).

10.2.1 Exploiting Trophic Cascades Through the Management of the Ground Cover

Herbivore populations are naturally regulated by "top-down" and "bottom-up" forces. On the one hand, biological control of agricultural pests takes advantage of the "top-down" control processes by exploiting natural enemies that may allow higher yield by consuming herbivores, an event known as a trophic cascade (Hairston et al. 1960; Oksanen et al. 1981). As mentioned earlier, phytoseiids are the key natural enemies of tetranychids in citrus agroecosystems (McMurtry 1985; Gerson 2003; Vacante 2010). On the other hand, "bottom-up" control strategies focus on managing the first trophic level to hamper the performance of herbivores, thus reducing pest impact on crop yield. According to the Resource Concentration Hypothesis (Root 1973), a highly diverse flora in the agroecosystem impairs the ability of herbivores to locate and therefore use host plants, among other things by diffusing the feeding resources in the environment. Such a diffusion effect may hinder colonization and reproduction of pests, therefore reducing the damage they cause (Gurr et al. 2000).

Thus, pest control could be upgraded with strategies combining the exploitation of natural enemies ("top-down" mechanisms) and plants ("bottom-up" mechanisms) (Speight et al. 2008), even in simple cropping systems. This constitutes the basis for modern food web engineering for sustainable crop protection.

10.2.2 Phytoseiids, the Key Natural Enemies of Tetranychid Mites

Phytoseiid mites are considered the most frequent and abundant natural enemies of tetranychid mites. Most of them also feed on other mite groups, small insects, as thrips, as well as on fungi, pollen and nectar (Helle and Sabelis 1985; Zhang 2003; Lundgren 2009). Interestingly, several species within this family may develop and reproduce feeding exclusively on pollen (Ferragut et al. 1987; Zhimo and McMurtry 1990; Grafton-Cardwell et al. 1999; Nomikou et al. 2002; Bouras and Papadoulis 2005; Ragusa et al. 2009). McMurtry and Croft (1997) and more recently McMurtry et al. (2013) categorized phytoseiids according to their food preferences, microhabitat and morphology, describing four main types: type I, specialized mite predators, including the genus *Phytoseiulus*; type II, selective predators of tetranychid mites, including some *Neoseiulus* species; type III, generalist predators, represented by most *Amblyseius* and *Typhlodromus* species; and type IV, pollen feeding generalist predators, exemplified mainly by *Euseius* species.

Given the generalist behavior of many phytoseiids, these may persist at relatively high densities even when the prey that they are expected to control (potential pest) is scarce or absent. Therefore, alternative food sources, including pollen and other prey, can prevent pest resurgence without the usual time lag normally associated with the numerical response of natural enemies at increasing pest populations (Hanski et al. 1991). Ground covers can provide alternative food items to generalist phytoseiids.

A growing number of studies deal with the effect of ground covers on phytoseiid mite populations and diversity (e.g. Alston 1994; Kim and Lee 2003; Mailloux et al. 2010). These covers can consist of either naturally growing plants (resident wild covers) or crops sown on purpose (cover crops). In the latter case, different plant species have been tested (Liang and Huang 1994; Grafton-Cardwell et al. 1999; Duso et al. 2004; González-Fernández et al. 2009; Maoz et al. 2008, 2011; Sahraoui et al. 2014; Zhao et al. 2014). However, enhancing the number of natural enemies in the system does not necessarily result in better yields. The importance of appraising the complex relationship between natural enemy diversity and effective biological control has been highlighted by many authors (Tscharntke et al. 2007; Griffiths et al. 2008; Jonsson et al. 2008; Straub et al. 2008; Moreno et al. 2010; Martin et al. 2013). One of the conclusions from these studies is that increased biodiversity *per se* is no guarantee of successful pest control (Landis et al. 2000) and this is attributed to the negative impact of enhanced intraguild predation on biological control (Jonsson et al. 2008). Generally, the basis for successful biological control combines the use of tactics that increase the relative abundance of the most effective natural enemies within the community (Straub and Snyder 2006).

10.3 Food Web Engineering for *T. urticae* in Clementine Mandarins

10.3.1 Starting Point

In the 1990s, the complex of natural enemies of *T. urticae* on citrus in eastern Spain was considered ineffective (García-Marí et al. 1991; Garrido and Ventura 1993; Ripollés et al. 1995). As no reliable sampling method and action threshold for *T. urticae* existed, growers' decisions to apply chemical control had no scientific basis. Chemical control was not always effective and up to seven applications could be done at each season (Martínez-Ferrer et al. 2005). Such an extensive use of pesticides was expected to decimate the citrus-inhabiting beneficial arthropods, inducing outbreaks of other pests, accelerating the development of pesticide-resistant spider mite populations and generating a plethora of problems associated with pesticide residues (Viñuela 1998; Urbaneja et al. 2008). As a consequence, this was an expensive strategy for farmers and the environment and there was an urgent need for developing more sustainable strategies against *T. urticae*.

10.3.2 Ground Cover Selection: Acarofauna of Plants Spontaneously Growing in Citrus Orchards

Because of the polyphagous and ubiquitous nature of *T. urticae* (Bolland et al. 1998), this species attacks both citrus and spontaneously growing plants of citrus orchards (García Marí et al. 1986; Aucejo et al. 2003). The persistence and pest status of *T. urticae* in many crops are due to its ability to exploit different suitable hosts that become progressively available along the seasons (Margolies and Kennedy 1985). When a cover crop contains plant species where *T. urticae* can complete its life cycle, mite populations can easily build up to explosive levels (Meagher and Meyer 1990; Flexner et al. 1991; Alston 1994; Kim and Lee 2003). Subsequently, different factors (Aucejo-Romero et al. 2004), including herbicide applications, that negatively affect ground cover can induce these explosive populations of *T. urticae* to move to the citrus trees (Johnson and Croft 1981; Hardman et al. 2005).

Application of herbicides constituted the most common weed management practice in citrus orchards in Spain until recently, when the use of ground covers was considered an important alternative, as the IP Guidelines and Endorsement of the IOBC/WPRS Commission do not encourage the use of bare soil (IOBC/WPRS 2004). Therefore, our first study was aimed at determining the presence of *T. urticae* and associated phytoseiids on plants spontaneously growing (usually referred to as weeds) in citrus orchards (Aucejo et al. 2003). This study paved the way for selecting suitable candidate plants as cover crops, our hypothesis being that the use of an unsuitable plant to *T. urticae* as cover crop would preclude canopy infestations (Unsuitable Host Hypothesis).

Our group (Aucejo et al. 2003) sampled 45 plant species naturally occurring in different commercial clementine orchards during the whole year. This study showed that grasses (Poaceae) harbored the lowest ratios between *T. urticae* and phytoseiid mite densities (Fig. 10.1). Furthermore, the grasses sheltered very low densities of *P. citri* and *Tetranychus evansi* Baker and Pritchard (an invasive tetranychid mite potentially harmful to different plants, albeit not to citrus). The unsuitability of grasses to *T. urticae* had already been mentioned by Flexner et al. (1991) and Alston (1994). Therefore, it was concluded that Poaceae should be considered appropriate candidates for ground covers in citrus. For additional agronomic reasons, Fibla-Queralt et al. (2000) had already selected the Poaceae *Festuca arundinacea* Schreber as a suitable cover crop for citrus orchards. Based on these results, the next goal was to compare the dynamics of tetranychid and phytoseiid mites in orchards having *F. arundinacea* as cover crop and orchards adopting the two most common ground management systems in Spain (the resident wild cover and bare soil). When *F. arundinacea* was selected, we were not aware of the unsuitability of its pollen to support development of generalist phytoseiid mites. Later on, this unsuitability was found to play a key role in the system (see Sect. 10.3.3.2).

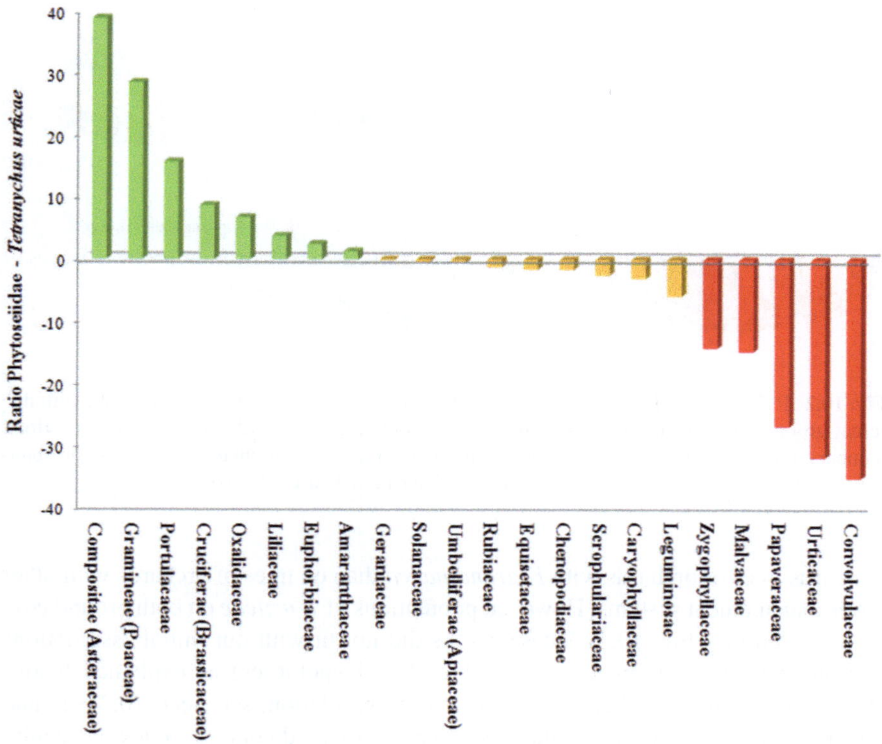

Fig. 10.1 Botanical families of weed species naturally occurring in citrus orchards classified according to the difference between Phytoseiidae and *T. urticae* mean densities. *Green bars* correspond to weeds sheltering higher densities of Phytoseiidae than *T. urticae*, *orange* ones to those where *T. urticae* were slightly higher than Phytoseiids and *red* ones those where *T. urticae* densities were clearly above *T. urticae* (Based on data from Aucejo et al. (2003)) (Color figure online)

10.3.3 Ground Cover Selection: Mite Dynamics

During three consecutive years, an experiment carried out in four different commercial orchards for each management system provided evidence that cover management affected *T. urticae* population densities on citrus trees (Aguilar-Fenollosa et al. 2011a, b). Tetranychid-specialist phytoseiids (*Phytoseiulus persimilis* Athias-Henriot and *Neoseiulus californicus* (McGregor)) were consistently found in higher number and frequency in the system with *F. arundinacea* than in the alternative ground management systems, considering both the citrus trees and the cover plants (Fig. 10.2). In addition, the densities of *T. urticae* and *P. citri* were much lower on

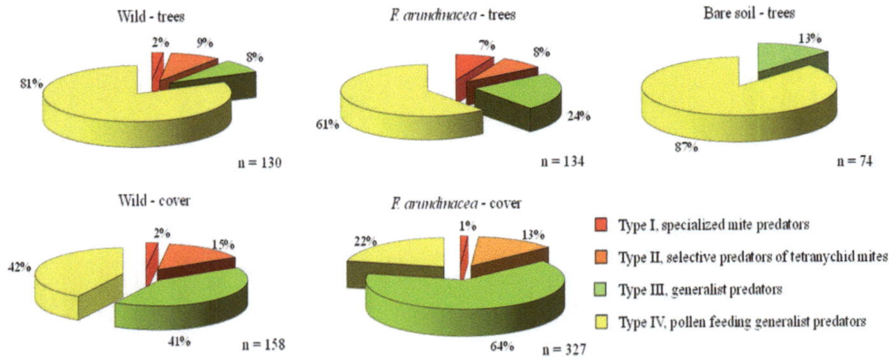

Fig. 10.2 Relative abundance of phytoseiid mites types, from 2006/2007 to 2008/2009, with different ground cover management systems, in four commercial orchards of clementines; n = total numbers of phytoseiids. Complete information about composition of phytoseiid species and abundance in each orchard and year is available in Aguilar-Fenollosa et al. (2011a)

the citrus trees of orchards with *F. arundinacea* than on trees of orchards with other cover management systems. However, populations of *T. urticae* on both ground covers were similar (Fig. 10.3). These results did not fit with our initial assumptions (Resource Concentration and Unsuitable Host Hypotheses) as explained below. Host specialization mechanisms ("bottom-up" regulation, see Sect. 10.3.3.1) and effective enhancement of populations of specialist predators ("top-down" regulation, see Sect. 10.3.3.2) are plausible explanations of the results observed that confirmed *F. arundinacea* as a key element in the sustainable management of phytophagous mites in citrus (Aguilar-Fenollosa et al. 2011c).

10.3.3.1 Ground Cover Selection: Bottom-up Regulation Mechanisms

According to the Unsuitable Host Hypothesis, extremely low populations of *T. urticae* should have been observed during the whole season in the *F. arundinacea* cover. These low densities should have precluded the infestation of the associated citrus trees. Nevertheless, this was not the case as both *F. arundinacea* and wild covers showed similar *T. urticae* densities during the whole season but only the grass cover reduced spider mite populations in the tree canopies. Should our system respond to the Resource Concentration Hypothesis (Root 1973), the most diverse system (the wild cover) should have better impaired the ability of *T. urticae* to attain high numbers during the whole season in the cover and therefore reduce the rate at which they successfully colonized citrus trees. Nevertheless, this did not occur either, because densities of *T. urticae* on the trees were higher in those associated with the wild cover relative to those associated with *F. arundinacea*. Therefore, the results did not support any of the initial hypotheses.

One of the key factors in determining the structure of herbivorous arthropod populations in crops is habitat fragmentation (Bailly et al. 2004). Temporal resource

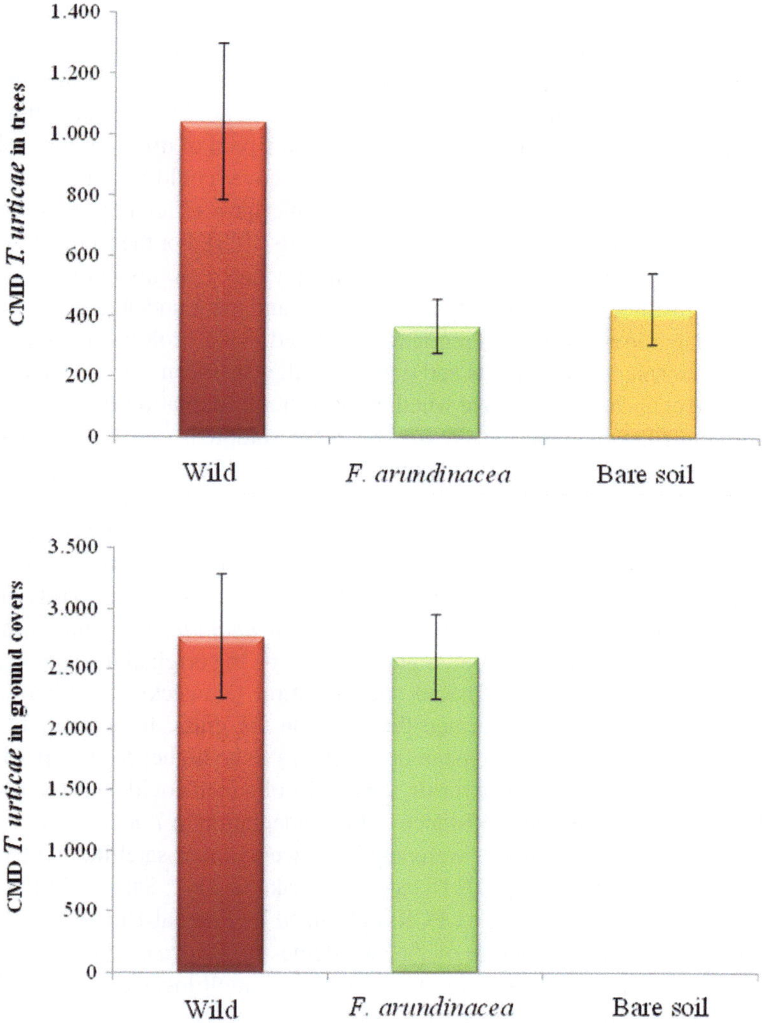

Fig. 10.3 Mean and standard error of cumulative mite days (CMD) for *T. urticae* in trees and in ground cover, from 2006/2007 to 2008/2009 with different ground cover management systems in four commercial orchards of clementines. Complete information about spider mite composition and abundance in each orchard and year is available in Aguilar-Fenollosa et al. (2011b)

availability as a consequence of natural or agricultural processes may result in local adaptation (which may relate to host adaptation when hosts have an irregular distribution) and result in the genetic differentiation of different subpopulations when gene flow between these subpopulations is reduced below a certain threshold (Williams et al. 2003). *Tetranychus urticae* can easily adapt to new plant hosts (Gould 1979; Fry 1990; Navajas et al. 1998; Agrawal 2000; Magalhães et al. 2007; Belliure et al. 2010; Grbić et al. 2011; Fellous et al. 2014). The reportedly high polyphagy of *T. urticae* deserves a consideration at this point. It could be the result of the ability of individuals of a population to use different plant species as hosts, or result from the coexistence of host-adapted strains, and this need further research (Navajas et al. 1998; Magalhães et al. 2007; Belliure et al. 2010). Should the populations of *T. urticae* occurring in citrus orchards be a pool of different host-adapted strains, the use of a mono-specific *F. arundinacea* cover could have resulted in the selection of a *T. urticae* specialist strain that could have efficiently exploited this grass but not the trees, as observed in our study (Fig. 10.3). For the same reason, the maintenance of the whole pool of *T. urticae* host-adapted strains in the wild cover would have resulted in high densities in the cover and the associated trees.

Local adaptation of mites is frequently explored using ecological studies using reciprocal transplant experiments and genetic studies by means of molecular markers (Belliure et al. 2010). To prove whether local (host) adaptation could explain the results obtained, two demes of *T. urticae* were sampled, one from a naturally infested clementine orchard and another one from *F. arundinacea* plants. We assessed the performance of these two demes on both their respective original host (sympatric combination) and the alternative one (allopatric combination) (Aguilar-Fenollosa et al. 2012). For most of the fitness parameters considered in this study, the *T. urticae* clementine population had higher mean fitness values on its original host than on the alternative one. However, the *F. arundinacea* population did not show differences among hosts. Better performance on the original host compared to alternative hosts denotes host-linked specialization (Kawecki and Ebert 2004). Nevertheless, both demes performed the same on the grass. In general statistical terms, local adaptation requires mean deme fitness to be higher for sympatric than for allopatric associateions. As a result, the results obtained could be taken as indicative but not conclusive of the existence of local adaptation in *T. urticae*. To confirm this result, molecular techniques were applied. Twenty microsatellite loci (Navajas et al. 1998; Nishimura et al. 2003; Uesugi and Osakabe 2007; Sabater-Muñoz et al. 2012) were used in four multiplex PCR, optimized by Pascual-Ruiz et al. (2014b), to ascertain the genetic structure of the two demes of *T. urticae*. Genotyping was performed and population structure was inferred for multilocus analysis including all scored loci. The optimal number of genetic clusters was two, which corresponded to the demes originally collected from clementine trees and *F. arundinacea* (Aguilar-Fenollosa et al. 2012).

In a nutshell, the patterns of local adaptation observed may contribute to the natural regulation of *T. urticae* populations noticed when *F. arundinacea* is used as cover crop. This adaptation may be the key bottom-up mechanism that prevents successful establishment of cover inhabiting *T. urticae* in the tree canopy.

10.3.3.2 Ground Cover Selection: Top-Down Regulation Mechanisms

In Spanish citrus orchards, the phytoseiid complex is represented by around 20 species (García Marí et al. 1986; Ferragut et al. 1988, 2010; Ferragut and Escudero 1997; Aucejo et al. 2003; Faraji et al. 2008; Abad-Moyano et al. 2009;

Table 10.2 Predacious phytoseiid mites from citrus orchards (on the citrus trees and on cover plants) in Spain

Species	Phytoseiid life-style category (McMurtry et al. 2013)	Reference
Amblyseiella setosa (Muma)	–	García Marí et al. (1986)
Amblyseius andersoni Chant	Type III, generalist predator	Aguilar-Fenollosa et al. (2011a)
Amblyseius bicaudus Wainstein	Type III	Aguilar-Fenollosa et al. (2011a)
Amblyseius obtusus (Koch)	Type III	Aguilar-Fenollosa et al. (2011a)
Anthoseius sp.	Type III	Aucejo et al. (2003)
Euseius scutalis (Athias-Henriot)	Type IV, pollen feeder generalist predator	Ferragut and Escudero (1997)
Euseius stipulatus (Athias-Henriot)	Type IV	Aguilar-Fenollosa et al. (2011a)
Kampimodromus aberrans Oudemans	Type III	Ferragut et al. (2010)
Neoseiulus barkeri Hughes	Type III	Aguilar-Fenollosa et al. (2011a)
Neoseiulus californicus (McGregor)	Type II, selective predator of tetranychids	Aguilar-Fenollosa et al. (2011a)
Neoseiulus cucumeris (Oudemans)	Type III	Aucejo et al. (2003)
Paraseiulus talbii (Athias-Henriot)	Type I-c, specialized predator of tydeoids	Boyero et al. (personal communication)
Phytoseiulus persimilis Athias-Henriot	Type I, specialized mite predator of *Tetranychus*	Aguilar-Fenollosa et al. (2011a)
Phytoseius plumifer (Canestrini and Fanzago)	Type III	García Marí et al. (1986)
Proprioseiopsis bordjelaini (Athias-Henriot)	Type III	García Marí et al. (1986)
Proprioseiopsis messor (Wainstein)	Type III	Aguilar-Fenollosa et al. (2011a)
Proprioseiopsis ovatus (Garman)	Type III	Faraji et al. (2008)
Typhlodromus phialatus Athias-Henriot	Type III	Aguilar-Fenollosa et al. (2011a)
Typhlodromus rhenanoides Athias-Henriot	Type III	García Marí et al. (1986)
Typhlodromus transvaalensis (Nesbitt)	Type III	Ferragut et al. (2010)

Aguilar-Fenollosa et al. 2011a) belonging to the four life-styles previously described (McMurtry et al. 2013) (Table 10.2). Most of these species were identified in our field experiment. However, phytoseiid diversity and relative abundance on citrus trees and the ground cover differed greatly, depending on the ground cover management strategy (Fig. 10.2). In agreement with previous reports (Ferragut et al. 1988;

Abad-Moyano et al. 2009), *Euseius stipulatus* (Athias-Henriot) was the most abundant species in the canopies irrespective of the ground cover management considered; in some orchards, it represented 100 % of total catches on trees grown either in bare soil or in the wild cover. Contrary to what would be expected, maximum diversity and abundance was not found on the wild cover but in *F. arundinacea*. Likewise, tetranychid-specialist predators (*P. persimilis* and *N. californicus*) were much more frequent and abundant in the mono-specific cover and this pattern was also observed in the associated trees. Bare soil resulted in the lowest diversity and abundance, and specialist predators were always missing from those trees (Fig. 10.2).

Pollen availability in the different ground covers could be behind the observed different phytoseiid composition (Aguilar-Fenollosa et al. 2011a). In fact, the extended pollination period of the wild multifloral cover (comprising most of the growing season) compared to *F. arundinacea* (blooming in spring only), could favor the higher abundance and frequency of the type-IV pollen feeding generalist predator *E. stipulatus* in the wild cover compared to *F. arundinacea. Euseius stipulatus* is considered a key natural enemy of *P. citri* (Ferragut et al. 1992; Abad-Moyano et al. 2009), and this species is able to successfully reproduce when feeding on pollen only (Ferragut et al. 1987; Bouras and Papadoulis 2005; Pina et al. 2012). Remarkably, Abad-Moyano et al. (2010a, b) described this species as a superior intraguild predator in the presence of *N. californicus* and *P. persimilis* in citrus. Hence, numerical increases of *E. stipulatus* related to pollen availability could trigger negative effects on the biological control of *T. urticae*, by the concurrent reduction of type I and type II specialist predatory mite populations. However, pollen quality could also explain the results of the reported work. Our group studied the contribution of two different pollen sources of presumably diverse quality on the efficacy of *E. stipulatus* and *N. californicus* against *T. urticae* (Pina et al. 2012). Pollen was obtained from *Carpobrotus edulis* (L.) L. Bolus (Aizoaceae), which is commonly used in the mass rearing of different phytoseiid species (Ferragut et al. 1987), and *F. arundinacea*, a pollen of presumed low quality (Ouyang et al. 1992). We proved that pollen supply did not increase the ability of those two phytoseiids in reducing *T. urticae* populations. However, pollen quality significantly affected the population numbers of the pollen feeder *E. stipulatus*. The constant provision of *C. edulis* pollen resulted in an explosive increase in populations of *E. stipulatus*, which did not occur with the provision of *F. arundinacea* pollen. That increase was never observed for *N. californicus*, regardless of the pollen source. This result would explain why pollen feeders never became overwhelmingly predominant when using *F. arundinacea*, but were possibly able to wipe specialists out of the system when in association with a wild cover, as observed in two of the locations included in the study (Aguilar-Fenollosa et al. 2011a)

While *F. arundinacea* pollen on its own is not likely to support long-term survival of phytoseiids (Pina et al. 2012), this cover crop may still be useful for these predators as it can provide non-pest prey as thrips (Aguilar-Fenollosa and Jacas 2013), frequently used by type II and III phytoseiid mites as an alternative food source (Rodríguez-Reina et al. 1992; van Baal et al. 2007). Our group has recently started a new research line on the role of the most abundant grass thrips species, *Anaphothrips obscurus* (Müller) (Thysanoptera: Thripidae), as a potential alternative prey for these predators. First results point out the suitability of this thrips in our system (Gómez-Martínez et al. 2013).

10.4 Final Remarks

The results of the studies conducted in eastern Spain, here summarized, demonstrate that the use of *F. arundinacea* as cover crop in citrus orchards can significantly reduce *T. urticae* populations on associated citrus trees and, consequently, the need for chemical applications. On the contrary, conventional alternatives as bare soil and wild cover might lead to increased *T. urticae* populations in citrus trees and hence to higher pesticide use, with the consequent enhanced production and environmental costs (Aguilar-Fenollosa et al. 2011c). As other CBC practices, the use *F. arundinacea* as a cover crop in clementine orchards should result in reduction of pesticide use and, importantly, in better market access, higher prices and better orchard agronomical features (Cullen et al. 2008; Hartwing and Ammon 2002).

Furthermore, in comparison with bare soil, ground cover can facilitate access of machinery to the orchard by accommodating the tractor wheeling, and by preventing the spread of important pathogens affecting citrus fruits, such as *Phythophthora* sp., by reducing splashing (Fibla-Queralt et al. 2000). Finally, implementation of *F. arundinacea* as cover crop may result in better biological control of other citrus key pests, such as the Mediterranean fruit fly, *Ceratitis capitata* (Wiedemann) (Diptera: Tephritidae), by increasing soil dwelling predators (Monzó et al. 2009, 2010, 2011), and aphids, by promoting the early arrival of natural enemies (Gómez-Marco et al. 2012). All these and other potential public benefits of reduced pesticide use (environmental and health benefits) are usually poorly computed. These benefits and the associated policies can be as, or even more relevant than, farmer benefits in determining ultimate acceptance (Griffiths et al. 2008). Hence, adoption of *F. arundinacea* as cover crop for the biological control of *T. urticae* in clementine mandarin orchards is strongly recommended.

Acknowledgments The work reported in this chapter was partially funded by the Spanish Ministerio de Ciencia e Innovación (projects AGL2000-1065-C02-01, AGL2004-07464-C03-01, AGL2005-07155-C03, AGL2008-05287-C04-03/AGR and AGL2011-30538-C03-01) and UJI through a pre- and post-doctoral grant to E.A.-F. and T.P., respectively.

References

Abad-Moyano R, Pina T, Dembilio O, Ferragut F, Urbaneja A (2009) Survey of natural enemies of spider mites (Acari: Tetranychidae) in citrus orchards in eastern Spain. Exp Appl Acarol 47:49–61

Abad-Moyano R, Urbaneja A, Schausberger P (2010a) Intraguild interactions between *Euseius stipulatus* and the candidate biocontrol agents of *Tetranychus urticae* in Spanish clementine orchards: *Phytoseiulus persimilis* and *Neoseiulus californicus*. Exp Appl Acarol 50:23–34

Abad-Moyano R, Urbaneja A, Hoffmann D, Schausberger P (2010b) Effect of *Euseius stipulatus* on establishment and efficacy in spider mite suppression of *Neoseiulus californicus* and *Phytoseiulus persimilis* in clementine. Exp Appl Acarol 50:329–341

Agrawal AA (2000) Host-range evolution: adaptation and trade-offs in fitness of mites on alternative hosts. Ecology 81:500–508

Aguilar-Fenollosa E, Jacas J (2013) Effect of ground cover management on Thysanoptera (thrips) in clementine mandarin orchards. J Pest Sci 86:469–481

Aguilar-Fenollosa E, Ibáñez-Gual MV, Pascual-Ruiz S, Hurtado M, Jacas JA (2011a) Effect of ground cover management on spider mites and their phytoseiid natural enemies in clementine mandarin orchards (II): top-down regulation mechanisms. Biol Control 59:171–179

Aguilar-Fenollosa E, Ibáñez-Gual MV, Pascual-Ruiz S, Hurtado M, Jacas JA (2011b) Effect of ground cover management on spider mites and their phytoseiid natural enemies in clementine mandarin orchards (I): bottom-up regulation mechanisms. Biol Control 59:158–170

Aguilar-Fenollosa E, Pascual-Ruiz S, Hurtado M, Jacas JA (2011c) Efficacy and economics of ground cover management as a conservation biological control strategy against *Tetranychus urticae* in clementine mandarin orchards. Crop Prot 30:1328–1333

Aguilar-Fenollosa E, Pina T, Gómez-Martínez MA, Hurtado M, Jacas JA (2012) Does host adaptation of *Tetranychus urticae* populations in clementine orchards with a *Festuca arundinacea* cover contribute to a better natural regulation of this pest mite? Entomol Exp Appl 144:181–190

Alston GD (1994) Effect of apple orchard floor vegetation on density and dispersal of phytophagous and predaceous mites in Utah. Agr Ecosyst Environ 50:73–84

Anon (1991) Integrated pest management for citrus, 2nd edn. University of California-Statewide IPM Project, Division of Agriculture and Natural Resources, Oakland

Ansaloni T, Pascual-Ruiz S, Hurtado MA, Jacas JA (2008) Can summer and fall vegetative growth regulate the incidence of *Tetranychus urticae* Koch on clementine fruit? Crop Prot 27:459–464

Aucejo S, Foó M, Gimeno E, Gómez-Cadenas A, Monfort R et al (2003) Management of *Tetranychus urticae* in citrus in Spain: acarofauna associated to weeds. OILB/WPRS Bull 26:213–220

Aucejo-Romero S, Gómez-Cadenas A, Jacas-Miret JA (2004) Effects of NaCl-stressed citrus plants on life-history parameters of *Tetranychus urticae* (Acari: Tetranychidae). Exp Appl Acarol 33:55–68

Bailly X, Migeon A, Navajas M (2004) Analysis of microsatellite variation in the spider mite pest *Tetranychus turkestani* (Acari: Tetranychidae) reveals population genetic structure and raises questions about ecological factors. Biol J Linn Soc 82:69–78

Barbosa P (ed) (1998) Conservation biological control. Academic, San Diego

Belliure B, Montserrat M, Magalhães S (2010) Mites as models for experimental evolution studies. Acarologia 50:513–529

Bolland HR, Gutierrez J, Flechtmann CHW (1998) World catalogue of spider mites family (Acari: Tetranychidae) with reference to taxonomy, synonymy, host pest, and distribution. Brill Academic Publishers, Leiden

Boller EF, Häni F, Poehling HM (2004) Ecological infrastructures: ideabook on functional biodiversity at the farm level temperate zones of Europe. IOBC/WPRS, LBL, Lindau

Bouras LS, Papadoulis GTH (2005) Influence of selected fruit pollen on life history of *Euseius stipulatus* (Acari: Phytoseiidae). Exp Appl Acarol 36:1–14

Cullen R, Warner KD, Jonsson M, Wratten SD (2008) Economics and adoption of conservation biological control. Biol Control 45:272–280

Duso C, Malagnini V, Paganelli A, Aldegheri L, Bottini M et al (2004) Pollen availability and abundance of predatory phytoseiid mites on natural and secondary hedgerows. BioControl 49:397–415

Eilenberg J, Hajek AE, Lomer C (2001) Suggestions for unifying the terminology in biological control. BioControl 46:387–400

European Union (2009) Directive 2009/128/EC of the European Parliament of 21 October 2009 establishing a framework for Community action to achieve sustainable use of pesticides. OJEU L 309:71–86

FAOSTAT (2014) Food and agriculture organization of the United Nations. Available via DIALOG. http://faostat.fao.org. Accessed 15 Apr 2014

Faraji F, Bakker F, Roig J (2008) A new species and two new records of Phytoseiidae (Acari: Mesostigmata) from Spain. Rev Iber Aracnol 16:105–111

Fellous S, Angot G, Orsucci M, Migeon A, Auger P et al (2014) Combining experimental evolution and field population assays to study the evolution of host range breadth. J Evol Biol 27:911–919

Ferragut F, Escudero A (1997) Taxonomía y distribución de los ácaros depredadores del género *Euseius* Wainstein 1962, en España (Acari: Phytoseiidae). Bol San Veg Plagas 23:227–235

Ferragut F, García-Marí F, Costa-Comelles J, Laborda R (1987) Influence of food and temperature on development and oviposition of *Euseius stipulatus* and *Typhlodromus phialatus* (Acari: Phytoseiidae). Exp Appl Acarol 3:317–329

Ferragut F, Costa-Comelles J, García-Mari F, Laborda R, Roca D et al (1988) Dinámica poblacional del fitoseido *Euseius stipulatus* (Athias-Henriot) y su presa *Panonychus citri* (McGregor) (Acari: Phytoseiidae, Tetranychidae), en los cítricos españoles. Bol San Veg Plagas 14:45–54

Ferragut F, Laborda R, Costa-Comelles J, García-Mari F (1992) Feeding behavior of *Euseius stipulatus* and *Typhlodromus phialatus* on the citrus red mite *Panonychus citri* (Acari: Phytoseiidae, Tetranychidae). Entomophaga 37:537–543

Ferragut F, Pérez-Moreno I, Iraola V, Escudero-Colomar A (2010) Ácaros depredadores en las plantas cultivadas. Familia Phytoseiidae. Ediciones Agrotécnicas, Madrid

Fibla-Queralt JM, Martínez-Ferrer MT, Pastor-Audí J, Pons-Mas J, Barceló-Salanguera F (2000) Establecimiento de cubiertas vegetales en parcelas de producción integrada de cítricos. Frutic Prof 112:67–72

Flexner JL, Westigard PH, Gonzalves P, Hilton R (1991) The effect of groundcover and herbicide treatment on two spotted spider mite density and dispersal in southern Oregon pear orchards. Exp Appl Acarol 60:111–123

Flint ML (2012) IPM in practice. Principles and methods of integrated pest management. 2nd edn. Publication 3418, University of California, Agriculture and Natural Resources, Oakland

Fry JD (1990) Trade-offs in fitness on different host – evidence from a selection experiment with a phytophagous mite. Am Nat 136:569–580

García Marí F, Ferragut F, Marzal C, Costa-Comelles J, Laborda R (1986) Ácaros que viven en las hojas de los cítricos españoles. Investig Agrar Prod Prot Veg 1:219–250

García-Marí F, Llorens JM, Costa-Comelles J, Ferragut F (1991) Ácaros de las plantas cultivadas y su control biológico. Ediciones Pisa, Alicante

Garrido A, Ventura JJ (1993) Plagas de los Cítricos. Bases para el Manejo Integrado. Ministerio de Agricultura, Pesca y Alimentación, Madrid

Gerson U (2003) Acarine pest of citrus: overview and non-chemical control. Syst Appl Acarol 8:3–12

Gómez-Marco F, Tena A, Jacas JA, Urbaneja A (2012) Ground cover management in citrus affects the biological control of aphids. Paper presented at the XII international citrus congress, Valencia, 18–23 Nov 2012

Gómez-Martínez MA, Fandos-Fernández D, Jacas JA, Pina T (2013) Principales parámetros biológicos de *Anaphothrips obscurus* (Müller) (Thysanoptera: Thripidae), presa potencial de *Neoseiulus californicus* (McGregor) (Acari: Phytoseiidae) en huertos de clementino. Paper presented at the VIII Congreso Nacional de Entomología Aplicada, Mataró, 21–25 Oct 2013

González-Fernández JJ, de la Peña F, Hormaza JI, Boyero JR, Vela JM et al (2009) Alternative food improves the combined effect of an omnivore and a predator on biological pest control. A case study in avocado orchards. Bull Entomol Res 99:433–444

Gould F (1979) Rapid host range evolution in a population of the phytophagous mite *Tetranychus urticae* Koch. Evolution 33:791–802

Grafton-Cardwell EE, Ouyang Y, Bugg RL (1999) Leguminous cover crops to enhance population development of *Euseius tularensis* (Acari: Phytoseiidae) in citrus. Biol Control 16:73–80
Grbić M, van Leeuwen T, Clark RM, Rombauts S, Rouze P et al (2011) The genome of *Tetranychus urticae* reveals herbivorous pest adaptations. Nature 479:487–492
Griffiths GJK, Holland JM, Bailey A, Thomas MB (2008) Efficacy and economics of shelter habitats for conservation biological control. Biol Control 45:200–209
Gurr GM, Wratten SD, Barbosa P (2000) Success in conservation biological control of arthropods. In: Gurr GM, Wratten SD (eds) Biological control: measures of success. Kluwer Academic Publishers, Dordrecht, pp 105–132
Hairston NG, Smith FE, Slobodkin LS (1960) Community structure, population control, and competition. Am Nat 94:421–425
Hammond RB (2002) Strategy. In: Pimentel D (ed) Encyclopedia of pest management. CRC Press, US, pp 802–804
Hanski I, Hansson L, Henttonen H (1991) Specialist predators, generalist predators, and the microtine rodent cycle. J Anim Ecol 60:353–367
Hardman JM, Jensen KIN, Franklin JL, Moreau DL (2005) Effect of dispersal, predators (Acari: Phytoseiidae), weather, and ground cover treatments on populations of *Tetranychus urticae* (Acari: Tetranychidae) in apple orchards. J Econ Entomol 98:862–874
Hare JD (1994) Sampling arthropod pests in citrus. In: Pedigo LP, Buntin GD (eds) Handbook of sampling methods for arthropods in agriculture. CRC Press, Boca Raton, pp 417–431
Hartwing NL, Ammon HU (2002) Cover crops and living mulches. Weed Sci 50:688–699
Helle W, Sabelis MW (1985) World crop pests. Spider mites: their biology, natural enemies and control, vol 1B. Elsevier, Amsterdam
Hoy M (2011) Agricultural acarology: introduction to integrated mite management. CRC Press (Taylor & Francis Group), Boca Raton
IOBC/WPRS (2004) Commission IP-guidelines and endorsement. Guidelines for integrated production of citrus. IOBC Technical Guideline III. IOBC wprs Bull. 28. Available via Dialog. http://iobc-wprs.org/ip_ipm/IOBC_Guideline_Citrus_2004_ENGLISH.pdf. Accessed 22 Sept 2014
Jacas JA, Urbaneja A (2010) Biological control in citrus in Spain: from classical to conservation biological control. In: Ciancio A, Mukerji KG (eds) Integrated management of arthropod pests and insect borne diseases, vol 5. Springer, The Netherlands, pp 57–68
Jacas JA, Karamaouna F, Vercher R, Zappalà L (2010) Citrus pest management in the Northern Mediterranean basin: Spain, Italy and Greece. In: Ciancio A, Mukerji KG (eds) Integrated management of arthropod pests and insect borne diseases, vol 5. Springer, The Netherlands, pp 3–26
Johnson DT, Croft BA (1981) Dispersal of *Amblyseius facillis* (Acarina: Phytoseiidae) in apple ecosystems. Environ Entomol 10:313–319
Jonsson M, Wratten SD, Landis DA, Gurr GM (2008) Recent advances in conservation biological control of arthropods by arthropods. Biol Control 45:172–175
Kawecki T, Ebert D (2004) Conceptual issues on local adaptation. Ecol Lett 7:1225–1241
Kim DS, Lee JH (2003) Oviposition model of overwintered adult *Tetranychus urticae* (Acari: Tetranychidae) and mite phenology on the ground cover in apple orchards. Entomol Exp Appl 31:191–208
Krantz GW, Walter DE (eds) (2009) A manual of acarology. Texas Tech University Press, Lubbock
Landis DA, Wratten SD, Gurr GM (2000) Habitat management to conserve natural enemies of arthropod pests in agriculture. Annu Rev Entomol 45:175–201
Liang W, Huang M (1994) Influence of citrus orchards ground cover plants on arthropod communities in China: a review. Agr Ecosyst Environ 50:29–37
Lundgren JG (2009) Relationships of natural enemies and non-prey foods, vol 7, Progress in biological control. Springer, The Netherlands
Magalhães S, Forbes MR, Skoracka A, Osakabe M, Chevillon C et al (2007) Host race formation in the Acari. Exp Appl Acarol 42:225–238

MAGRAMA (2013) Anuario de estadística. Ministerio de agricultura, alimentación y medio ambiente 2012. Available via DIALOG. http://www.magrama.gob.es/estadistica/pags/anuario/2012/AE_2012_Completo.pdf. Accessed 15 Apr 2014

Mailloux J, Le Bellec F, Kreiter S, Tixier MS, Dubois P (2010) Influence of ground cover management on diversity and density of phytoseiid mites (Acari: Phytoseiidae) in Guadeloupean citrus orchards. Exp Appl Acarol 52:275–290

Maoz Y, Gal S, Abrahams J, Gan-Mor S, Coll M, Palevsky E (2008) Pollen provisioning enhances *Euseius scutalis* (Phytoseiidae) populations and improves control of *Olygonychus persea* (Tetranychidae). In: Mason PG, Gillespie DR, Vincent D (eds) Proceedings of ISBCA, vol 3, Davos, Switzerland, pp 339–345

Maoz Y, Gal S, Argov Y, Coll M, Palevsky E (2011) Biocontrol of persea mite, *Olygonychus persea*, with an exotic spider mite predator and an indigenous pollen feeder. Biol Control 59:147–157

Margolies DC, Kennedy GG (1985) Movement of two spotted spider mite, *Tetranychus urticae*, among hosts in a corn-peanut agroecosystem. Entomol Exp Appl 37:55–61

Martin EA, Reineking B, Seo B, Steffan-Dewenter I (2013) Natural enemy interactions constrain pest control in complex agricultural landscapes. Proc Natl Acad Sci U S A 110:5534–5539

Martínez-Ferrer MT, Jacas-Miret JA, Ripollés-Moles JL, Aucejo-Romero S (2005) Sampling Plans for *Tetranychus urticae* (Acari: Tetranychidae) for IPM decisions on Clementines in Spain. IOBC/wprs Working Group "Integrated Control in Citrus Fruit Crops", Lisboa

McMurtry JA (1985) Citrus. In: Helle W, Sabelis MW (eds) World crop pests, vol 1B, Spider mites: their biology, natural enemies and control. Elsevier, Amsterdam, pp 339–347

McMurtry JA (1992) Dynamics and potential impact of "generalist" phytoseiids in agroecosystems and possibilities for establishment of exotic species. Exp App Acarol 14:371–382

McMurtry JA, Croft BA (1997) Life-styles of phytoseiid mites and their role in biological control. Annu Rev Entomol 42:291–321

McMurtry JA, De Moraes GJ, Sourassou NF (2013) Revision of the lifestyles of phytoseiid mites (Acari: Phytoseiidae) and implications for biological control strategies. Syst Appl Acarol 18:297–320

Meagher RL Jr, Meyer JR (1990) Influence of ground cover and herbicide treatments on *Tetranychus urticae* populations in peach orchards. Exp Appl Acarol 9:149–158

Monzó C, Molla O, Castañera P, Urbaneja A (2009) Activity-density of *Pardosa cribata* in Spanish citrus orchards and its predatory capacity on *Ceratitis capitata* and *Myzus persicae*. BioControl 54:393–402

Monzó C, Sabater-Muñoz B, Urbaneja A, Castañera P (2010) Tracking medfly predation by the wolf spider, *Pardosa cribata* Simon, in citrus orchards using PCR-based gut-content analysis. Bull Entomol Res 100:145–152

Monzó C, Sabater-Muñoz B, Urbaneja A, Castañera P (2011) The ground beetle *Pseudophonus rufipes* revealed as predator of *Ceratitis capitata* in citrus orchards. Biol Control 56:17–21

Moreno CR, Lewins SA, Barbosa P (2010) Influence of relative abundance and taxonomic identity on the effectiveness of generalist predators as biological control agents. Biol Control 52:96–103

Navajas MJ, Thistlewood HMA, Lagnel J, Hughes C (1998) Microsatellite sequences are underrepresented in two mite genomes. Insect Mol Biol 7:249–256

Nishimura S, Hinomoto N, Takafuji A (2003) Isolation, characterization, inheritance and linkage of microsatellite markers in *Tetranychus kanzawai* (Acari: Tetranychidae). Exp Appl Acarol 31:93–103

Nomikou M, Janssen A, Schraag R, Sabelis M (2002) Phytoseiid predators suppress populations of *Bemisia tabaci* on cucumber plants with alternative food. Exp Appl Acarol 27:57–68

Oksanen L, Fretwell SD, Arruda J, Niemela P (1981) Exploitation ecosystems in gradients of primary productivity. Am Nat 118:240–261

Ouyang Y, Grafton-Cardwell E, Bugg RL (1992) Effects of various pollens on development, survivorship, and reproduction of *Euseius tularensis* (Acari: Phytoseiidae). Environ Entomol 21:1371–1376

Pascual-Ruiz S, Aguilar-Fenollosa E, Ibáñez-Gual V, Hurtado-Ruiz MA, Martínez-Ferrer M et al (2014a) Economic threshold for *Tetranychus urticae* (Acari: Tetranychidae) in clementine mandarins *Citrus clementina*. Exp Appl Acarol 62:337–362

Pascual-Ruiz S, Gómez-Martinez MA, Ansaloni T, Segarra-Moragues JG, Sabater-Muñoz B et al (2014b) Genetic structure of a phytophagous mite species affected by crop practices: the case of *Tetranychus urticae* in clementine mandarins. Exp Appl Acarol 62:477–498

Pina T, Sá Argolo P, Urbaneja A, Jacas JA (2012) Effect of pollen quality on the efficacy of two different life-style predatory mites against *Tetranychus urticae* in citrus. Biol Control 61:176–183

Ragusa E, Tsolakis H, Jordà Palomero R (2009) Effect of pollens and preys on various biological parameters of the generalist mite *Cydnodromus californicus*. Bull Insectology 62:153–158

Ripollés JL, Marsà M, Martínez M (1995) Desarrollo de un programa de control integrado de las plagas de los cítricos en las comarcas del Baix Ebre-Montsià. Levante Agrícola 332:232–248

Rodríguez-Reina JM, García-Marí F, Ferragut F (1992) Actividad depredadora de varios ácaros fitoseidos sobre distintos estados de desarrollo del trips de las flores *Frankliniella occidentalis*. Bol San Veg Plagas 18:253–263

Root RB (1973) Organization of a plant-arthropod association in simple and diverse habitats: the fauna of collards (*Brassica oleracea*). Ecol Monogr 43:95–124

Ruberson JR, Nemoto H, Hirose Y (1998) Pesticides and conservation of natural enemies in pest management. In: Barbosa P (ed) Conservation biological control. Academic, San Diego, pp 207–220

Sabater-Muñoz B, Pascual-Ruiz S, Gómez-Martínez MA, Jacas JA, Hurtado MA (2012) Isolation and characterization of polymorphic microsatellite markers in *Tetranychus urticae* and cross amplification in other Tetranychidae and Phytoseiidae species of economic importance. Exp Appl Acarol 57:37–51

Sahraoui H, Tixier MS, Lebdi-Grissa K, Kreiter S (2014) Diversity and abundance of Phytoseiidae (Acari: Mesostigmata) in three crop management strategies of citrus orchards in Tunisia. Acarologia 54:155–169

Smith D, Beattie GAC, Broadley R (1997) Citrus pests and their natural enemies. IPM in Australia. Department of Primary Industries, Brisbane

Speight RM, Hunter DM, Watt DA (2008) Ecology of insects. Concepts and applications. Wiley-Blackwell, Oxford

Straub CS, Snyder WE (2006) Species identity dominates the relationship between predator biodiversity and herbivore suppression. Ecology 87:277–282

Straub CS, Finke DL, Snyder WE (2008) Are conservation of natural enemy biodiversity and biological control compatible goals? Biol Control 45:225–237

Talhouk AS (1975) Citrus pests throughout the world. In: Häfliger E (ed) Citrus. Ciba-Geigy, Basel, pp 23–27

Tscharntke T, Bommarcob R, Clougha Y, Cristc TO, Kleijnd D et al (2007) Conservation biological control and enemy diversity on a landscape scale. Biol Control 43:294–309

Uesugi R, Osakabe M (2007) Isolation and characterization of microsatellite loci in the two-spotted spider mite, *Tetranychus urticae* (Acari: Tetranychidae). Mol Ecol Notes 7:290–292

Urbaneja A, Pascual-Ruiz S, Pina T, Abad-Moyano R, Vanaclocha P et al (2008) Efficacy of five selected acaricides against *Tetranychus urticae* (Acari: Tetranychidae) and their effects on relevant natural enemies occurring in citrus orchards. Pest Manag Sci 64:834–842

Vacante V (2010) Citrus mites: identification. Bionomy and control. CABI, UK

van Baal E, van Houten YV, Hooger Brugge H, Bolckmans K (2007) Side effects on thrips of the spider mite predator Neoseiulus californicus. Proceedings of the Netherland Entomological Society meeting, vol 18, Amsterdam, The Netherlands, pp 37–42

Viñuela E (1998) Insecticide resistance in horticultural pests in Spain. In: Cuadrado IM, Viñuela E (eds) Pesticide resistance in horticultural crops. Fundación para la Investigación Agraria en la provincia de Almería, Almería, pp 19–29

Williams B, Brawn JD, Paige KN (2003) Landscape scale genetic effects of habitat fragmentation on a high gene flow species: *Speyeria idalia* (Nymphalidae). Mol Ecol 12:11–20

Zhang ZQ (2003) Mites of greenhouses: identification, biology and control. CABI, Wallingford

Zhao W, Zheng W, Zhang B, Yu G, Hu S, Xu X, Zhang H (2014) Effect of different ground cover management on spider mites (Acari: Tetranychidae) and their phytoseiid (Acari: Phytoseiidae) enemies in citrus orchards. Biocontrol Sci Techn 6:705–709

Zhimo Z, McMurtry JA (1990) Development and reproduction of 3 *Euseius* (Acari: Phytoseiidae) species in the presence and absence of supplementary foods. Exp Appl Acarol 8:233–242

Chapter 11
Past, Present, and Future: Biological Control of Spider Mites on California-Grown Avocados

Jesús R. Lara and Mark S. Hoddle

11.1 Introduction to Spider Mite Biological Control for the California Avocado System

Avocado, *Persea americana* Miller (Laureaceae) originated in southern Mexico from where it was dispersed and domesticated by humans over thousands of years throughout Mesoamerica (Bost et al. 2013). As a result of avocado domestication by humans in its home range, contemporary cultivars of *P. americana* are grouped among three botanical varieties (i.e., races, or ecotypes) commonly known as Mexican (*P. americana* var. *drymifolia* Blake), Guatemalan (*P. americana* var. *guatemalensis* Williams) and West Indian (*P. americana* var. *americana* Miller) (Bost et al. 2013). These three broad varieties differ not only in growth form and in fruit nutrient composition traits, but also in their physiological adaptation to different climates. Avocado cultivars associated with Mexican, Guatemalan and West Indian varieties are best suited to semitropical, subtropical and tropical climates, respectively (Wolstenholme 2013). These inherent climatic adaptations have significant bearing on where cultivars associated with these varieties can be grown successfully for commercial production.

However, cultivars from the three botanical varieties can hybridize naturally through cross-pollination and produce hybrid-cultivars with new combinations of 'variety' traits (Lahav and Lavi 2013). Furthermore, new cultivars can also develop as offspring from avocado seeds because avocado seedlings do not breed true-to-type. With modern advances in agricultural science and growing commerce, this natural process of avocado reproduction created incentives for conducting foreign

J.R. Lara • M.S. Hoddle (✉)
Department of Entomology, University of California, Riverside, Riverside, CA 92521, USA
e-mail: jesus.lara@ucr.edu; mark.hoddle@ucr.edu

D. Carrillo et al. (eds.), *Prospects for Biological Control of Plant Feeding Mites and Other Harmful Organisms*, Progress in Biological Control 19,
DOI 10.1007/978-3-319-15042-0_11

exploration in Mesoamerica and establishing on-going breeding programs for the propagation and distribution of novel marketable cultivars (Crane et al. 2013). In this manner, avocado cultivars have been purposefully introduced and propagated by humans at an accelerated rate in new regions with amenable avocado growing climates where this crop did not previously occur during pre-Columbian times (Bost et al. 2013). For example, during the nineteenth century, avocado was introduced by humans from its native home range to southern California (USA). With growing demand and economic interest for producing exotic avocado fruit, initial commercial avocado production in southern California began in 1908 (Shepherd and Bender 2002). Since then, commercial production in California has expanded to cover approximately 23,406 ha and is currently valued at US$ 435 million (CAC 2014).

The commercial avocado system in California has been established over an extremely brief period, compared to the evolutionary time frame of avocado domestication in its home range, and various aspects of its history in California have been well documented in the literature. The availability of this information provides an opportunity to track the development and progression of past and present avocado pest problems in California. This investigative process can generate valuable insight as to how pest control strategies might be improved. The primary purpose of this chapter is to understand how specific avocado problems related to foliar feeding spider mite (Acari: Tetranychidae) populations in California progressed over time since this host plant was introduced and how these spider mite problems might be resolved as part of future efforts relying on the sustainable use of predators (i.e., biological control).

At least five common foliar-feeding spider mite species occur on avocado in the world. These include *Eotetranychus sexmaculatus* (Riley); *Oligonychus coffeae* Nietner; *O. perseae* Tuttle, Baker and Abattielo; *O. punicae* (Hirst) and *O. yothersi* McGregor (Peña et al. 2013). Presently, California has three of these exotic species, *E. sexmaculatus*, *O. perseae* and *O. punicae*. Since the 1950s, extensive research efforts have been invested in controlling populations of spider mites on California avocados by strategizing the use of candidate beneficial predators under the framework of managed biological control programs. These management strategies stem from three general biological control types known as conservation, augmentation and classical biological control (Van Driesche et al. 2008; Hoy 2011).

Conservation biological control modifies environmental or cultural practices to enhance the performance of resident predator populations. One example of conservation biological control includes the provision of supplemental food resources like pollen to boost predator numbers during key periods of the growing season, when prey and/or plant food resources (e.g., nectar, pollen) are scarce, thereby preventing densities of the target pest from later exceeding economic thresholds and causing crop damage. Pollen can be deployed mechanically or made accessible by planting cover crops adjacent to avocado trees (Maoz et al. 2011a).

Augmentative biological control involves the supplemental release of predators into the intended crop habitat to boost or augment populations of existing predators thereby increasing their numbers in advance of anticipated pest outbreaks. Suppression results from the short term establishment and population buildup of the released natural enemies. Effective predators of the target pest can be mass-reared and released in great numbers (inundative releases) when target pest densities are elevated and require immediate control. Establishment and subsequent reproduction of released predators is not considered a priority in inundative releases. Alternatively, a small number of predators can be released into the crop habitat early in the season (inoculative releases of biological control agents) so that they can establish temporarily and provide seasonal control for target-pest populations. In general, both conservation and augmentative biological control strategies require input of resources by humans to maintain on-going efficacy at times when pest control is needed.

In contrast, the endpoint of classical biological control, or introduction biological control, is to deliberately establish permanent, self-sustaining control by reuniting the target pest species (often an invasive species) with its key co-evolved natural enemies from its home range (Van Driesche et al. 2008). Key natural enemies are identified as those whose populations can maintain, in a density-dependent manner, populations of the target pest in its home range at sub-economic levels. Consequently, the underlying premise of classical biological control is that populations of the target pest have (1) become unmanageable in a new geographic area due to the absence of its co-evolved key natural enemy or enemies and (2) effective natural control provided by the available natural enemy community in the invaded area is lacking.

One of the essential requirements for designing a successful classical biological control program is to determine the home range of the introduced target pest species (Van Driesche et al. 2008). In previous decades obtaining this information was difficult because there were few objective methods and established theoretical principles by which to determine with confidence the precise geographic origin of exotic pest species found in a new area. As will be discussed, this deficiency in technology became a limiting factor while conducting classical biological control efforts against *O. punicae* on California avocados. Molecular DNA-based tools have helped overcome this obstacle and this scientific advancement has had tremendous implications in designing classical biological control programs for economically important pests (Van Driesche et al. 2008), including spider mites such as *O. perseae*.

All three types of biological control have been considered or implemented with varying degrees of success for spider mite control in the California avocado system using natural enemies from the mite family Phytoseiidae. The Phytoseiidae contains approximately 2,709 described species among 91 genera (Demite et al. 2014) and its members display abundant diversity in feeding habits and life history traits (McMurtry and Croft 1997; Gerson et al. 2003; McMurtry et al. 2013). Like spider mites, phytoseiids can inhabit plants, but ecologically phytoseiids function primarily as predators of arthropods, like insects, and some phytoseiid species can feed selectively on spider mite species. For this reason phytoseiids have been used preferentially in spider mite biological control programs. Nevertheless, the compatibility between the life history traits of the target pest and the candidate phytoseiid species on a designated host plant system must be carefully assessed. This implies that the precise application of a phytoseiid species under the framework of biological control will depend on its unique ecology within the intended cropping system at a particular location.

Traditionally, the limited success of phytoseiid-based spider mite control on California avocados has been attributed to the incompatibility between prey-predator life history traits and/or intraguild predation among potential spider mite predators. However, growing evidence suggests that avocado host-plant susceptibility to spider mites like *O. perseae* (Kerguelen and Hoddle 2000) may have also played a significant role in the increased mite problem. This complex dynamic, which is affected by avocado cultivar, is not readily apparent and has to be traced through time. To provide insight into this intricate problem in the context of spider mites on California avocados, this chapter has been divided among six related subsections.

First, historical concerns regarding the introduction of exotic avocado pests in California, including insects, are outlined. Second, the historical economic importance of introduced spider mite species (i.e., *O. punicae* and *E. sexmaculatus*) and the success of their biological control programs are reviewed. Third, evidence regarding the introduction of *O. perseae* into the California avocado system from its home range and its general biology are presented. Fourth, the emergence of *O. perseae* as the key foliar pest of avocados in California is discussed. The development of this phenomenon is discussed from two angles: (1) avocado cultivar susceptibility and (2) the ecological limitations of candidate phytoseiids that are either naturally-occurring or have been introduced into avocado orchards for biological control of *O. perseae*. Factors associated with cultivar susceptibility are attributed to acreage expansion and shifts in cultivar acreage. These components are inter-related and their underlying influence has made attaining effective spider mite biological control difficult. Subsequently, the ecological limitations of *O. perseae* natural enemies are discussed in a fifth section. Finally, the discussion transitions to a sixth section that outlines future prospects for spider mite biological control on California avocados. The previous five sections create a platform on which to discuss future advancements for spider mite control in the California avocado system. Among these is the prospective use of understudied predatory mite families such as the Stigmaeidae. Members of this predator group have not been considered as spider mite biological control agents on California avocados, despite their documented natural presence on avocado in California and other avocado growing areas throughout the world.

11.2 Historical Concerns with Avocado Pests

California is in close geographic proximity to avocado production areas in Mexico. Therefore, the threat of introduction of avocado-feeding arthropods from their home range into California due to growing commerce (legal and illegal) in fresh fruit and plants has existed ever since this crop was introduced to California. In response, quarantine measures were implemented early on to prevent this. The main concern was with organisms that feed directly on avocado fruits. In 1914, the United States Department of Agriculture restricted the entry of fresh avocado fruit from Mexico to protect the growing California avocado industry (and Florida which also grows avocado) from the introduction of the injurious avocado seed weevil, *Heilipus lauri* Boheman (Coleoptera: Curculionidae). Later in 1973, the economic justifications for quarantine restrictions on avocado fruit importation from Mexico were expanded to prevent the introduction of other potentially serious avocado fruit pests such as the avocado seed moth, *Stenoma catenifer* Walsingham (Lepidoptera: Elachistidae), and other weevil species in the genus *Conotrachelus* that can attack avocado (Bellamore 2003). Due to changes in international trade agreements (e.g., NAFTA in 1994), the restrictions on fruit importation were gradually lifted during the late 1990s to allow importation of avocado fruit from approved pest-quarantined avocado production areas in Mexico into the United States (Bellamore 2003). After 2007, the import of avocado fruit was permitted into all U.S. states, including California, and this transition generated concern regarding the potential establishment of exotic species in California such as armored scales (Hemiptera: Diaspididae) which were detected on imported fresh fruit from Mexico (Morse et al. 2009).

Although these quarantine measures may have temporarily reduced the probability for the introduction of avocado feeding arthropods from Mexico, the species diversity of the California avocado arthropod fauna and its geographical distribution within the state grew throughout the twentieth century. Early records of arthropods associated with avocado were documented in the periodicals of the California Avocado Society, established in 1915. This increase in arthropod diversity was most likely the result of three factors. First, new ecological associations were formed between native generalist herbivore species and newly established avocado host plants in California. One example of this is the California-native moth *Amorbia cuneana* (Walsingham) (Lepidoptera: Tortricidae) (Oatman and Platner 1985; Gilligan and Epstein 2009) whose larvae feed on several plant species from several families including *Umbellularia californica* (Hook. & Arn.) Nutt., a native species in California that, like avocado, is a member of the Laureaceae. Secondly, generalist herbivores introduced with other crops from other areas of the world may have formed new associations with avocado in various avocado growing regions. One example of this is the polyphagous scale *Hemiberlesia lataniae* (Signoret) (Hemiptera: Diaspididae) which can feed on host plants representative of 78 families and has been recorded on avocado in several countries (Peña et al. 2013). A more recent example that demonstrates the warranted concern for the development of new-host associations in California between exotic avocado and an exotic pest species is provided by the polyphagous shot hole borer, *Euwallacea* sp. (Coleoptera: Curculionidae: Scolytinae), which was detected in California in 2003. This invasive beetle is native to Asia and has symbiotic relationship with a fungus (*Fusarium* sp.) that is transmitted by the beetle to host plants during the construction of galleries in woody tissue. Fungal infection of vascular tissue in host-plants can produce branch die-back and eventual tree death. Field surveys in California indicate that this beetle can attack 207 species (11 are native to California) that are representative of 58 plant families. About 54 % of host plant species attacked by polyphagous shot hole borer, including avocado, were infected with the fungus carried by this beetle (Eskalen et al. 2013). This pest has also been recorded attacking avocado in Israel. Last but not least, before the establishment of clear organized efforts to prevent the entry of avocado pests and to document their presence, it is possible that more specialized avocado feeding arthropods, including their natural predators, may have been inadvertently introduced from Mesoamerica on avocado breeding material collected from these areas (Moznette 1922). Foliar feeding spider mites and predatory mites associated with avocado in Mexico are some examples of this third factor which will be discussed throughout this chapter.

Most likely, these insect and mites remained undetected or given minor importance on California avocados. This changed as avocado became a popular specialty crop and the expansion of commercial avocado acreage provided a permanent seasonal habitat for populations of native and introduced arthropod species, capable of feeding and reproducing on commercial avocado cultivars in California (McLean 1932; Quayle 1933; McKenzie 1934). At the same time, on-going developments on taxonomic, biological and species-inventory research created opportunities to document the presence and understand the role (e.g., beneficial predators/herbivore pest) of populations of introduced and native species occurring within the California avocado system (Boyce 1947; Ebeling and Pence 1953, 1958). For example, prior to 1990, the two foliar-feeding spider mite species whose presence made commercial California avocado growers wary were *O. punicae* and *E. sexmaculatus*. Populations of *O. punicae* were first noticed infesting California avocados during the 1920s (McGregor 1942) while *E. sexmaculatus* populations were first noticed during the 1950s (Ebeling and Pence 1953). The detection of these exotic spider mite species on avocado occurred roughly 60–90 years after the earliest recorded presence of avocado in California in 1856 (Shepherd and Bender 2002). The precise origin and manner of introduction of these two species into California has not been confirmed.

11.3 Economic Importance of Early Spider Mite Associations with Avocado

As spider mites, motile *O. punicae* and *E. sexmaculatus* individuals use cheliceral stylets to feed on the cell contents of tissue from top surface (adaxial) and undersurface (abaxial) of avocado leaves, respectively (Bailey and Olsen 1990). In general, motile stages of these species can also be distinguished by their characteristic body coloration, location of webbing construction on leaf surfaces and visual feeding damage. Motile *E. sexmaculatus* have a yellow-green appearance and produce complicated webbing, defined as three-dimensional and irregular (Saito 1983). Feeding damage caused by populations of this species is detected as brown-purple discoloration on the leaf undersurface. *O. punicae* individuals have a dark red-brown appearance, produce complicated webbing, and feeding damage caused by populations of this species is visible as bronze discoloration on the top leaf surface (UCANR 2008).

Historically, these spider mites have had different levels of economic importance. Low densities of *E. sexmaculatus*, 5–10 mites per leaf, can induce premature leaf drop (Bailey and Olsen 1990), but California populations of this species are mainly confined to coastal areas, particularly Santa Barbara and San Luis Obispo (UCANR 2008). In these areas avocado acreage is not well represented. In addition, warm weather has been thought to prevent populations from becoming elevated in the more arid inland avocado growing areas of southern California (UCANR 2008). Therefore, *E. sexmaculatus* is not considered a key pest of avocado in California. However, under a different set of avocado growing and climatic conditions like in New Zealand, *E. sexmaculatus* is considered a key pest of avocado (Tomkins 2002; White 2002; Jamieson and Stevens 2007). In contrast, *O. punicae* populations have been recorded throughout the California avocado growing region and densities in this region can naturally build to several hundred mites per leaf (McMurtry and Johnson 1966). On the Hass cultivar, feeding by *O. punicae* populations on foliage was shown to be negatively correlated with stomatal and mesophyll conductance, photosynthesis and transpiration rates on leaves (Sances et al. 1982). Continuous feeding from *O. punicae* populations in the approximate range of 50–70 mites per leaf can also cause premature leaf drop but whether this damage influences yield has not been determined (Sances et al. 1982). Even so, concern in California about foliar damage caused by *O. punicae* and *E. sexmaculatus* populations declined over decades as new information became available from a large body of biological control studies designed to target spider mite populations using candidate species from families of predatory beetles (Coccinelidae) and mites (Phytoseiidae) that are known to be economically important because they can control pest populations (Flint and Dreistadt 1998).

11.3.1 Success of Early Spider Mite Biological Control Efforts

Avocado field studies conducted during the 1950s suggested that populations of *O. punicae* and *E. sexmaculatus* in California were maintained under natural control by a guild of native predators and/or predators that originated from the home range of avocado (Fleschner et al. 1956a; Fleschner 1958; McMurtry and Johnson 1966). Most importantly, these studies revealed that spider mite biological control agents differed in their degree of efficacy. For example, *O. punicae*, which is a pest of avocado in Mexico (Peña et al. 2013), was shown to be under natural control by resident populations of the beetle *Stethorus picipes* Casey (Coleoptera: Coccinellidae) in California. *Stethorus picipes* is a specialized predator of spider mites, as are other *Stethorus* species (Chazeau 1985), and *S. picipes* occurs naturally in western North America (McMurtry 1989). Surprisingly, field studies also revealed that natural populations of predatory mites like *Euseius hibisci* (Chant), a species which is native to Mexico (and presumably California) and occurs naturally on avocado, was not consistent in providing effective control of *O. punicae* populations (McMurtry and Johnson 1963, 1966). Instead, *E. hibisci* populations were considered to be better suited for controlling *E. sexmaculatus* populations as the densities of this spider mite species on California avocados were naturally low in comparison to *O. punicae* (Fleschner et al. 1956a).

With regards to *O. punicae*, which was more abundant and widely distributed, efforts were made to design an augmentative biological control strategy involving field releases of mass-reared *S. picipies* to boost the action of resident conspecific populations. However, this approach was shown to be unsustainable because rearing *S. picipes* beetles in great numbers require a lot of spider mites as prey and this was expensive (Scriven and Fleschner 1960). This observation generated the notion that *S. picipes* was more effective in controlling *O. punicae* populations once densities of this spider mite became elevated (McMurtry and Johnson 1968; McMurtry et al. 1972). Additional research was conducted with various phytoseiid species which could be more easily reared as part of a seasonal inoculative biological control strategy. The justification of this work was that introduced populations of select phytoseiid species could potentially control lower population levels of *O. punicae* much earlier in the season than *S. picipes* because their prey requirements were lower, thus preventing *O. punicae* populations from reaching damaging densities (McMurtry et al. 1984). Several candidate phytoseiid species from multiple countries were considered for this strategy because the origin of *O. punicae* populations in California had not been confirmed (this problem still exists today). This uncertainty over the evolutionary area of origin was complicated by the fact that *O. punicae* has been recorded from other host plants from Latin America and Asia (Jeppson et al. 1975).

Nevertheless, field trials were conducted to assess the efficacy of nine candidate phytoseiid species for control of *O. punicae* populations on avocado (McMurtry et al. 1984). Among them was *Neoseiulus californicus* (McGregor), a species that has historically been used successfully for control of spider mites on other crops (Gerson et al. 2003). Three of these predators, including *N. californicus*, occur naturally in California and the remaining six were introduced from South Africa, the Mediterranean, Brazil, and the Cook Islands (McMurtry et al. 1984). Of the three native species, only *Amblydromalus limonicus* (Garman & McGregor) was naturally found in avocado. Even so, none of the phytoseiid species provided significant control of *O. punicae* populations (McMurtry et al. 1984).

Subsequently, it was presumed that *O. punicae* populations from California originated from Latin America. In this instance, six additional phytoseiid species were collected and introduced into southern California from Mexico and Central America (i.e., Costa Rica, Guatemala and El Salvador). None of these predators established (McMurtry 1989). A possible mechanism that explains these unsuccessful outcomes is that well-established resident populations of *E. hibisci* competed with or interfered with the action of introduced phytoseiid species (McMurtry et al. 1984). Climatic mis-match and insufficient release efforts (i.e., low frequency of predator releases and few individuals released) may have also played a role in failed establishment. It might have been possible to have restricted the search area and devoted more time searching for compatible natural enemies in specific locations had clear information (e.g., available from phylogenetic, taxonomic, or host plant use data) on the geographic origin of California's *O. punicae* populations been available at the time.

Despite the limited success with the introduction of several phytoseiid species, the combination of natural predation provided by resident populations of phytoseiid and insect predator species was deemed to be largely responsible for effective spider mite control in California avocado orchards throughout most of the twentieth century (McMurtry and Johnson 1966; McMurtry 1992). Similar positive results with the biological control programs of other insect herbivores were also documented on California avocados (Fleschner 1954; McMurtry 1992). However, the success of spider mite biological control was not permanent because of the acquisition of an additional invasive mite pest.

11.4 Introduction of *Oligonychus perseae*

Starting in 1990, California avocado production was impacted in an unprecedented manner with an additional series of invasive pests that are economically injurious to avocado and for which effective biological control programs have been difficult to develop. Among the recently introduced pest species is the foliar feeding spider mite *O. perseae* (Fig. 11.1a). This avocado pest was first detected in California in 1990 (Bender 1994), prior to the implemented revisions of trade agreements between Mexico and the United States.

The precise invasion pathway of *O. perseae* into California has not been confirmed but evidence from multiple sources has provided clues as to the likely area of origin. The taxonomic description of *O. perseae* is based on specimens collected on

Fig. 11.1 (**a**) Feeding adult *Oligonychus perseae* (Acari: Tetranychidae) and round-shaped conspecific eggs underneath a webbed-nest (transparent) on avocado leaf undersurface, (**b**) Accumulated necrotic damage on avocado leaf undersurface (concentrated along veins) caused by feeding populations of *O. perseae* and which can lead to pre-mature defoliation on host avocado trees, (**c**) Sunburn (circular black blemish) on maturing avocado fruit due to limited cover protection from foliage

avocado foliage originating from San Luis Potosi, Mexico that was interdicted at a quarantine station in El Paso, Texas in 1975 (Tuttle et al. 1976). Given its feeding habit, *O. perseae* was most likely introduced into the United States on infested avocado plant material originating from Mexico where *O. perseae* has co-evolved with this host plant. However, it is still not clear whether *O. perseae* was introduced into California on infested plant material coming directly from Mexico or indirectly through U.S.-Mexico border states like Texas. Historically, avocado has been grown in Texas (Cooper1949) but climatic conditions do not support a large commercially viable production region as in California. Furthermore, the direct presence of *O. perseae* populations on avocado in Texas has not been recorded in the literature. This may make the Texas-California invasion pathway unlikely and hints at direct introduction into California from Mexico.

Most importantly, the previous lack of updated information regarding species-abundance and detailed life-history information on avocado pests in their native home range such as *O. perseae* made it possible for this spider mite species to remain undetected and/or its potential threat as a serious foliar avocado pest to be given minor consideration until established populations of this spider mite started causing problems in a new avocado production area like California. As a result of these factors, it is possible that *O. perseae* individuals from infested avocado material from Mexico may have been introduced into California at an earlier period (Bender 1994), but this seems unlikely given the rapid spread of this pest after its initial discovery in California. Results from molecular studies (J. Lara, 2013) reveal low mitochondrial genetic diversity in California's *O. perseae* populations compared to populations on avocado from Mexico and this is consistent with the hypothesis that a small founding group of *O. perseae* individuals were likely introduced from its home range in Mexico. Furthermore, field observations by M. Hoddle and J. Lara (2012) indicated that *O. perseae* is distributed naturally over a large geographic region of southern Mexico on backyard avocado trees in various municipalities from states such as Michoacán, México, Puebla, Morelos, and Guanajuato. *Oligonychus perseae* has also been found infesting avocado at the north-eastern (i.e., Baja California) and southernmost (i.e., Chiapas) states of Mexico.

These recent survey results indicate that *O. perseae* is widespread in its home range on avocado, and is often damaging on the Hass cultivar. If this information had been known earlier, it could have prompted earlier concern regarding the potential threat posed by *O. perseae* to commercial avocado production outside its home range. Early efforts to document associated biological control agents in its home range, if any, that could effectively regulate *O. perseae* populations in a density-dependent manner on commercial avocado cultivars could then have been undertaken. Unfortunately, realizing these scenarios is limited by the scarce availability of research resources and the large number of candidate pest species on a given crop, especially co-evolved herbivores in the crop's area of origin. Consequently, invasive species research for agricultural systems is often driven by post-introduction economic losses experienced in other countries rather than interests in pursuing proactive measures to study pest species whose exotic populations may or may not establish and become problematic once introduced into a new geographic area. For this reason, there was little interest in studying *O. perseae* in Mexico, despite documented damage on avocado during the 1970s, and in other minor avocado-growing areas outside of Mexico that are considered to be part of the natural home range of avocado. For example, *O. perseae* was first recorded on avocado in Costa Rica as early as 1974 (Ochoa et al. 1994), but important commercial cultivars like Hass were not introduced into that country until the mid-1980s (Schroeder 1959; MAG 2007). The full extent of the damage that *O. perseae* was capable of producing outside the natural home range of avocado was observed years later, in major commercial avocado production areas of California.

Since its first detection in California, *O. perseae* rapidly established in parts of the southern California avocado growing region, extending from San Luis Obispo County to San Diego County, where outbreaks of this spider mite can be extremely damaging. Subsequently, *O. perseae* has been found infesting avocado in Israel (2001), Spain (2004), Madeira (2005) and the Canary Islands (2005) (González-Fernández et al. 2009; Hernández-Suárez et al. 2010; Maoz et al. 2011a). As a result, *O. perseae* is considered an international avocado pest (Peña et al. 2013). This has fueled substantial interest in understanding the ecology of *O. perseae* on avocado and developing effective pest management strategies for this pest, including the use of natural enemies in biological control programs.

11.4.1 General Biology of **Oligonychus perseae**

The biology of *O. perseae* in California was studied by Aponte and McMurtry (1997a, b) and reviewed by Hoddle and Morse (2013). As with other spider mites, the sequential life stages of *O. perseae* are egg, larva, protonymph, deutonymph and adult. Under 15, 20, 25 and 30 °C, development from egg to adult on foliage of the Hass cultivar can be completed in approximately 35, 17, 14 and 10 days, respectively (Aponte and McMurtry 1997a). Motile stages of *O. perseae* are capable of constructing webbed nests (Fig. 11.1a) which are made on the leaf undersurface, mainly along the midrib and radiating leaf veins (Aponte and McMurtry 1997b). The webbed nests of *O. perseae* demarcate feeding and development arenas and provide mating spaces. Eggs are laid in these nests (Fig. 11.1a), which may provide suitable microclimatic conditions for egg development. In addition, webbing functions as a physical protective cover against potential predators, such as phytoseiids.

Motile life stages of *O. perseae* use cheliceral stylets to feed on the cell contents of parenchyma tissue from the undersurface of avocado leaves (Aponte and McMurtry 1997b). This gives feeding individuals a greenish-yellow appearance. Continual feeding by *O. perseae* produces damaged tissue that is visible as brown necrotic blotches on the leaf undersurface (Fig. 11.1b). Sometimes these blotches are also visible from the top leaf surface. When feeding areas become necrotic, motile *O. perseae* may abandon the nest in search of new colonization areas on the leaf undersurface. Alternatively, under high population densities, *O. perseae* individuals can produce fine silken strands for wind-mediated dispersal. In this manner, *O. perseae* individuals might be carried as founders to other avocado trees which may contain un-infested foliage available for colonization.

This type of passive airborne dispersal using silk, referred to as "ballooning," is not unique to *O. perseae*; *E. sexmaculatus* and *O. punicae* and many other spider mites are capable of doing so (Fleschner et al. 1956b; Kennedy and Smitley 1985). Nevertheless, it represents a natural pathway by which damaging *O. perseae* populations can become distributed within and between avocado orchards in commercial production areas. This dispersal pathway may also explain how, in theory, an initially small population of *O. perseae* was able to increase its distribution once introduced into California. Another contributing factor for the distribution of *O. perseae* within California may have been the human-aided transport of contaminated plant material from one area to another, including infested plants or detached infested leaves inside packing bins moved from orchard to orchard. The underlying factors responsible for the severity of *O. perseae* infestations in the California avocado system are intricate and will be discussed below.

11.5 Potential Factors Explaining the Economic Importance of *Oligonychus perseae*

Like *O. punicae* and *E. sexmaculatus*, the accumulation of feeding damage on avocado foliage caused by *O. perseae* can induce premature leaf drop (Aponte and McMurtry 1997b) and this occurs when 7.5 % of the leaf surface is damaged (Kerguelen and Hoddle 1999). Although *O. perseae*, *O. punicae*, and *E. sexmaculatus* do not attack avocado fruit directly, mite-induce defoliation can thin out the tree canopy and avocado fruit can become sun-burnt (Fig. 11.1c). Typically in California, *O. perseae* densities begin increasing during later spring months (April–May) when new avocado foliage is available for colonization and feeding, followed by a peak in spider mite densities during summer (June–August) and eventual decline during fall (September–November) and winter months (December–February) (Yee et al. 2001a; UCANR 2008; Hoddle and Morse 2013). The main concern with potential damage caused by *O. perseae* populations occurs during the summer, when densities are at peak levels. A custom binomial sampling plan has been developed for *O. perseae* to help avocado pest managers better monitor population densities of this mite on California avocados and to make pest management decisions based on accurate estimates of pest densities (Lara and Hoddle 2013).

The severity of foliar damage resulting from *O. perseae* feeding depends, in part, on the susceptibility of the infested cultivar (Kerguelen and Hoddle 2000). In hindsight, the expansion of avocado acreage and abundance of highly susceptible cultivars to *O. perseae* might explain, in part, why *O. peresea*, in comparison to *O. punicae* and *E. sexmaculatus*, has become a prevalent avocado pest in California. The mechanisms of how acreage expansion and shifts to susceptible cultivars may have enabled and magnified the extent of *O. perseae* problems in California are discussed below. An understanding of these mechanisms is important for overcoming the limitations of current integrated *O. perseae* management programs.

11.5.1 Acreage Expansion of the Hass Cultivar

California avocado bearing acreage has increased from approximately 202 ha in 1925 to 24,131 ha in 2012 (USDA 2014). Until the 1960s, avocado acreage in California was dominated by the Fuerte cultivar (Carman and Craft 1998). Initial Fuerte avocado propagation in California began with budwood collected by Carl Schmidt, during a 1911 avocado expedition, from a backyard avocado tree in the state of Puebla in Mexico. Fuerte was named after the Spanish word for "strong" because trees displayed vigorous growth and survived the severe California winter of 1913 (Rounds 1947). This cultivar produces a green thin-peel fruit and is thought to be a Mexican × Guatemalan hybrid, and molecular data suggests it may have a higher genetic contribution from the Mexican parent (Chen et al. 2009). Due to its frost resistance and good taste compared to other available cultivars during the early 1900s, Fuerte dominated the avocado market in California for several decades and later became one of the dominant commercial cultivars within some areas of its home range in Mexico, where it was introduced with other cultivars developed in California during the 1950s–1970s (Sánchez Colín et al. 2001). However, the economic importance of Fuerte was temporary.

The transitional replacement of Fuerte acreage began with the accidental discovery of the hybrid Hass cultivar in 1926 by Rudolph Hass in La Habra, California (Griswold 1946). Molecular evidence supports previous assumptions that the Hass cultivar carries roughly a similar share of alleles from both Mexican and Guatemalan parents (Chen et al. 2009). This hybridization generated a cultivar with a relatively short fruit maturation period (i.e., faster production compared to Fuerte), some level of cold tolerance, and fruit with a thicker but still easy-to-peel pebbly skin whose color turns from green to dark-purple once the fruit ripens. Most importantly, the Hass fruit is known for its rich nutty and oily taste. This combination of attributes was favorable for commercial production, but during this initial period of discovery the susceptibility of the Hass cultivar to pests was not a research priority to the growing avocado industry. Rudolph Hass patented the Hass cultivar in 1935 and through subsequent marketing and nursery support, Hass avocado gradually gained recognition as the preferred commercial cultivar (Griswold 1946), first in California and subsequently abroad in other avocado producing countries.

During the 1960s, the share of total avocado production in California by cultivar was roughly 60 % Fuerte, 20 % Hass and the remaining 20 % were other cultivars. By the 1990s, the Hass cultivar accounted for roughly 80 % of the total avocado production in California (Carman and Craft 1998). Currently, 96 % of all avocado produced in California is of this cultivar (CAC 2014). Even so, California is currently only the third largest producer of Hass avocado in the world, after Mexico (first) and Chile (second). Mexico has historically been and continues to be the primary producer of avocado in the world with most production occurring in the state of Michoacán. The Hass cultivar was introduced from California into Mexico during the 1960s and since then Hass acreage in Mexico has increased. During the 1970s, 85 % of all commercial avocado trees grown in Michoacán were Hass

(Sánchez Colín et al. 2001). In 2011, Mexico produced approximately 1.26 million metric tons of fresh avocado fruit on 126,598 ha (FAO 2014). The state of Michoacán alone devoted approximately 108,683 ha to the production of Hass avocado (Flores and Berman 2012), more than four times the estimated Hass acreage of California. Examples of other countries which have established Hass cultivar industries include Australia, Colombia, Costa Rica, Dominican Republic, Guatemala, Israel, New Zealand, Peru, South Africa and Spain. Overall, it is estimated that the Hass cultivar accounts for 90 % of global avocado exports (Crane et al. 2013). Despite its economic dominance in the world market, the Hass cultivar is not known for being resistant to pest damage.

11.5.2 Susceptibility of the Hass Cultivar to Spider Mites and Other Arthropods

A decade after the introduction of *O. perseae* in California, a critical study by Kerguelen and Hoddle (2000) revealed that Hass cultivar leaves fed upon by *O. perseae* populations accumulated relatively high leaf area damage (66 %) when compared to five other cultivars (Kerguelen and Hoddle 2000). The most resistant cultivar found in that study was Fuerte, with only 35 % leaf area damage (Kerguelen and Hoddle 2000). Recent work has also demonstrated that the leaf injury caused by *O. perseae* to Hass can translate directly into yield reduction and consequently the recommended action threshold for control of *O. perseae* on this cultivar was determined to be 50–100 mites per leaf (Maoz et al. 2011b).

The combined results from these studies and the gradual increase of Hass acreage lead to the conclusion that the California avocado system had become vulnerable to attack by specialized avocado pest as the industry shifted acreage from pest resistant Fuerte to susceptible Hass avocado production. In addition, the increase in Hass acreage in Mexico over decades, possibly played a role in increasing the risk that herbivores like *O. perseae*, and potentially others (see below), would at some point be introduced from their home range to other expanding and susceptible Hass avocado growing regions. The symptoms of this latent vulnerability became apparent in the 1990s, when the share of California avocado acreage was 80 % Hass and specialist avocado herbivores such as *O. perseae* were introduced. Awareness on this cultivar susceptibility problem has heightened since *O. perseae* was also introduced into the Hass producing regions of Israel and Spain (see Sect. 11.4), where climate permits the successful production of this cultivar.

It is curious as to why *O. punicae* which was already established in California and had a wider distribution than *E. sexmaculatus*, has not become a major spider mite pest of Hass avocado. Similar to *O. perseae*, the reproduction and development of *O. punicae* is comparatively better while feeding on Hass than other avocado varieties such as Fuerte and Criollo (i.e., undomesticated avocado of the Mexican variety grown from seeds) (Cerna et al. 2009). The answer may lie with an assortment of unstudied biotic factors such as the differential use of leaves by populations of these spider mites (*O. perseae* feeds on the undersurface and *O. punicae* feeds on the upper surface of avocado leaves), the outcome of asymmetric competition between populations of these mites on avocado foliage, the differential tolerance of the Hass cultivar to the degree of damage caused by populations of each spider mite species, or the differential efficacy of resident natural enemies against different spider mite species.

Inefficiency of resident natural enemies might be an important factor, as *Stethorus* beetles have not been found in great abundance in California orchards infested with *O. perseae* (Yee et al. 2001a) but they have been shown to be important for controlling *O. punicae*. Also, no recent work has been conducted to determine whether *O. punicae* populations are still under effective natural control by *S. picipes* since the *O. perseae* invasion of California. Future studies will need to confirm the additional influence of these biotic factors as an explanation for the pest status of different spider mite species on avocado in California.

It is also important to recognize that the ecological interactions of spider mites on avocado cultivars like Hass is also mediated by abiotic factors related to climate that operate as part of larger spatial scale processes. As mentioned in Sect. 11.3 of this chapter, *E. sexmaculatus* is not considered a threatening avocado pest in California and previous studies indicate that this spider mite mainly occur in some cool foggy areas near the central California coast, where avocado acreage is not well-represented (Bailey and Olsen 1990; UCANR 2008). In this case, climate may have also restricted the southward spread of *E. sexmaculatus* populations into the expanding Hass cultivar acreage throughout southern California, even though Hass has historically been identified as being susceptible to induced premature defoliation caused by *E. sexmaculatus* feeding (Ebeling and Pence 1953). With regards to *O. perseae* and *O. punicae*, inland areas of southern California (e.g., Riverside and San Bernardino Counties), where avocado acreage is relatively minimal due to high summer heat and winter frost, do not experience elevated densities of these spider mites during the summer as other coastal regions in southern California where avocado acreage is well represented (e.g., San Diego and Ventura Counties). These observations imply that climate influences where cultivars can be grown and also the potential distribution of spider mite infestations within avocado growing regions. For example, within the United States, *O. perseae* has been reported on avocado in Florida (Peña et al. 2013), but the climate in this region is more suitable for the production of West Indian cultivars which are probably not susceptible to feeding by *O. perseae* to the same degree as Hass. *Oligonychus perseae* is not considered a significant pest in Florida and extant populations have not been recently identified despite recent surveys for other invasive pests [e.g., red bay ambrosia beetle (Peña J., 2013, personal communication)].

The abundance and intrinsic susceptibility of the Hass cultivar may also directly explain the prevalence of other spider mites outside of the United States, such as *O. yothersi*. This species is the primary pest of Hass avocado in Chile, where *O. perseae* has not been reported (León Lobos 2003). Also, *O. yothersi* is an occasional avocado pest in Florida (Peña and Johnson 1999), where the Hass cultivar is not widely cultivated. Like *O. punicae*, *O. yothersi* feeds on the upper leaf surface and causes bronzing damage on avocado foliage (Jeppson et al. 1975). Research by Léon Lobos (2003) revealed that life table parameters of *O. yothersi* such as net reproduction rate (R_o), oviposition rate and survivorship of immature stages were lower on Fuerte than on Hass. These results lead to other important conclusions. First, *O. yothersi* has reproductive advantage feeding on Hass foliage that could translate into increased population densities. The extensive cultivation of Hass in Chile may explain why *O. yothersi* became a prevalent foliar avocado pest there. The second conclusion is that the variable degree of susceptibility between avocado cultivars to herbivory is not unique to *O. perseae*.

Like *O. yothersi* on Hass in Chile, a similar prevalence pattern may have developed over time for *E. sexmaculatus* on Hass avocado in New Zealand. Both *O. perseae* and *E. sexmaculatus* damage the leaf undersurface but *O. perseae* does not occur in New Zealand, where *E sexmaculatus* has been recorded since the 1950s (on grapes) and was considered of low economic importance on avocado until the 1990s (Jamieson and Stevens 2007). Similar to California, New Zealand has undergone avocado cultivar shifts and acreage expansion (White 2002) resulting in Hass being the dominant cultivar there. Starting in 1998, severe problems with *E. sexmaculatus* were reported on New Zealand avocados (Tomkins 2002) and this outcome may have been promoted, in part, by increased Hass cultivation.

Overall, these patterns establish the relative susceptibility among avocado cultivars to different species of spider mites under varying climatic conditions. There is also growing evidence for avocado cultivars to vary in susceptibility to arthropods other than spider mites. One example is the avocado thrips, *Scirtothrips perseae* Nakahara (Thysanoptera: Thripidae).

Scirtothrips perseae was new to science when it was discovered in California in 1996 (Nakahara 1997) and, like *O. perseae*, post-introduction research has been necessary to understand the dynamics and severity of this invasive species in the California avocado system. This insect feeds on avocado foliage and small immature fruit (Hoddle 2002). Populations of *S. perseae* have established throughout the California avocado growing region but the severity of these problem in commercial orchards are positively correlated with the increasing proximity of sites to the Pacific coast as a result of cooler temperatures that provide an optimal thermal range (15–20 °C) for this species (Hoddle 2003). The main concern with *S. perseae* is the direct feeding damage it causes to maturing avocado fruit during the spring. Yee et al. (2001b) determined that developing Hass fruits were especially vulnerable to thrips feeding damage in the size range of 0.42–1.42 cm in length. Fruits heavily attacked by *S. perseae* develop scarring and lose economic value as they have to be culled or downgraded. In California, the annual economic losses to commercial avocado production attributed to *S. perseae* populations is approximately $4.45 million (Hoddle et al. 2003). Therefore, *S. perseae* is a key primary pest of Hass fruit in California, even though it is not a specialist on avocado fruit. Molecular studies suggest that California populations of *S. perseae* originated from avocado in Mexico, specifically the municipality of Coatepec-Harinas in the state of México (Rugman-Jones et al. 2008).

Similar to *O. perseae*, *O. punicae*, *E. sexmaculatus* and *O. yothersi*, the varying susceptibility of avocado cultivars to *S. perseae* has also been observed between Hass and Fuerte (UCANR 2008), but quantitative studies are still needed to compare the extent and underlying nature of these differences. Previously documented observations by California avocado researchers also suggested that Fuerte (the Hass cultivar was also present at the time) was one of the least susceptible cultivars to infestations by greenhouse thrips, *Heliothrips haemorrhoidalis* (Bouché), an exotic generalist herbivore species in California known to attack avocado and other crops as well (Ebeling and Pence 1953).

However, Hass is not universally susceptible to all herbivores. For example, Ebeling and Pence (1958) recorded that Fuerte was more susceptible to infestations by insects such as the scale *H. lataniae* and the larval stages of the native herbivore *Sabulodes aegrotata* (Guenée) (Lepidoptera: Geometridae). The precise nature of reduced Hass susceptibility to infestations of both of these herbivores has not been studied. Ebeling and Pence (1958) vaguely indicated that it may have something to do with the "less favorable growing habits" of Hass in comparison to Fuerte. Therefore, to some extent the severity of pest infestations on avocado may be related to the degree of compatibility between the seasonal phenology of cultivars and populations of each herbivore in California. For example, in the case of *S. perseae* it is likely that Fuerte fruit is less susceptible to thrips feeding in the spring due its early fruit development when thrips densities are low. In contrast, development of young Hass fruit occurs later, approximately from March–July (UCANR 2008); this period overlaps with increasing thrips populations during spring and economic damage results.

Ultimately, these observations on varying cultivar susceptibly to pest infestations may be explained by genetic differences. If evolutionary genetic differences between cultivars explain their inherent morphological, phenological, and physiological traits, it would seem possible that these differences could also confer a suite of herbivory resistance traits to avocado cultivars. In general, these resistance traits may be associated with structural/morphological or chemical defenses (Mortensen 2013). For example, the observed resistance of the Fuerte cultivar to *S. perseae* could be conferred by chemical profile differences in foliage. Like *O. perseae*, it is possible that the reproductive potential of *S. perseae* populations is also lower when feeding on Fuerte foliage and this translates into potentially lower populations infesting this cultivar. In turn, the skin of Fuerte fruit is thinner than that of the Hass cultivar, and this morphological difference does not seem compatible with the higher susceptibility of Hass to *S. perseae*. The actual explanation for variable cultivar susceptibility to herbivores requires further study, but the underlying logic would explain why the Fuerte cultivar, which is from southern Mexico and whose genetic makeup is close to the Mexican variety (Chen et al. 2009) would be relatively resistant to populations of *S. perseae* and *O. perseae*, that are also from the same region.

By extension, it is also conceivable that chemical profile differences between cultivars may also have significant ecological implications in explaining the limited efficacy of biological control agents against some pest species.

11.5.3 The Implication of Chemical Profiles of Avocado Cultivars in Relation to Herbivory and Biological Control

Investigations conducted in Mexico suggest that variations in expressed foliar secondary metabolite profiles among and between avocado cultivars may also mediate their susceptibility to avocado herbivores (Bravo-Monzón and Espinosa-García 2008; Rincón-Hernández et al. 2011; Torres-Gurrola et al. 2011). Rincón-Hernández et al. (2011) found that chemical profiles of 250 different trees could be used to distinguish trees of Mexican varieties from three other hybrid crosses, Mexican x Guatemalan (the Hass cultivar belongs to this cross), Mexican x West Indian and Mexican x *P. shiedeana* (this latter species is indigenous to Mexico and Central America and related to *P. americana*). Torres-Gurrola et al. (2011) found that the presence of specific foliar chemicals could explain the incidence of galls produced by *Trioza anceps* Tuthill (Hemiptera: Triozidae) on foliage from trees of the Mexican variety. Furthermore, a study by Bravo-Monzón and Espinosa-García (2008) demonstrated that trees of the Hass cultivar did not produce detectable volatile compounds even when attacked by the specialized avocado stem borer, *Copturus aguacatae* Kissinger (Coleoptera: Curculionidae). Interestingly, macerated leaf extractions showed that the Hass cultivar had the potential to produce volatiles. In contrast, trees of the indigenous Mexican variety produced less volatile compounds when attacked by *C. aguacatae*.

The ecological implications of these results in relation to the trophic interaction between avocado herbivores, potential natural enemies and host avocado plants were not studied, but it is known that plants are capable of emitting volatile compounds (kairomones), elicited by herbivory, that attract natural enemies of herbivores or of producing defensive compounds that interfere directly with herbivory (Gatehouse 2002). A quantitative comparison of the foliar chemical profiles between Fuerte and Hass cultivars damaged by either *O. perseae*, *O. yothersi*, or *S. perseae* has not been conducted. Still, based on the study with *C. aguacatae* on the Hass cultivar and the Mexican variety (associated with Fuerte), there is at least indirect evidence to suggest that foliar chemical differences between Hass and Fuerte may exist, and could, in theory, mediate the observed relative susceptibility of these cultivars to avocado herbivores. It is also important to realize that susceptibility within cultivars may also not be uniform and this may generate differences in pest infestations within avocado growing sites. For example, the study by Torres-Gurrola et al. (2011) suggests that levels of chemical protection may vary between trees even under the same growing conditions. Variation of susceptibility between avocado trees to *O. perseae* was also documented from a research population of Gwen avocado trees at South Coast Research & Extension Center in Irvine, CA (Ashworth V., 2014, personal communication). Similar variation on *O. perseae* infestations between groups of Hass cultivar trees within commercial California orchards has been observed by J. Lara and M. Hoddle (2011). This important variation needs to be accounted for in future studies to elucidate the influence of individual host plant chemical profiles on population dynamics of *O. perseae* within avocado orchards.

One possible resistance mechanism could be that Hass avocado trees either do not initiate or mount effective chemical foliar defenses when attacked by feeding *O. perseae* in comparison to Fuerte. A second important mechanism could be that the lack of volatile production on Hass avocado trees attacked by *O. perseae* limits the recruitment of natural enemies to infested foliage that are normally capable of using herbivore induced plant volatiles (HIPVs) to find potential prey. Olfactometer studies demonstrated that phytoseiids and *Stethorus* beetles can sense and respond behaviorally to blends of plant volatiles emitted from spider mite infested foliage from other crops (Sabelis and Dicke 1985; Nimet et al. 2009). In addition, McMurtry et al. (1992) demonstrated, under laboratory conditions, that *E. hibsici* responded positively to volatiles from extracts of ground *O. punicae* individuals on avocado leaves deposited on filter paper.

In theory, the lack of response of resident predators in commercial avocado orchards, such as *E. hibisci* and *S. picipes*, to building *O. perseae* infestations could be due to the inability of Hass trees to produce volatiles that could stimulate those predators to actively search for spider mite patches. For example, *O. perseae* populations display aggregated distributions on avocado leaves (J. Lara, personal observation); early in the growing season (spring) some leaves will contain a few to several hundred *O. perseae* individuals, while others remain spider-mite-free. Depending on cultural pruning practices, cultivar growth habits, host plant size and age, an avocado tree canopy can in some cases consist of several hundred leaves and when this number is summed among other trees present in an orchard, the sampling universe of leaves becomes very large. At low *O. perseae* densities, a condition is created within each tree such that few infested leaves are "hidden" in the tree canopy. Consequently, the inability of resident predators to effectively find scarce local patches of *O. perseae* among several thousand leaves during early infestation stages could cause populations of resident predators to fail to respond numerically to increasing levels of that pest. Ultimately, *O. perseae* populations would escape effective control by the natural predator community and surpass action thresholds, because predators were not attracted to infestation sites.

At the moment, there is no clear field evidence to suggest that HIPVs are used by resident predators in the California avocado system to locate *O. perseae* infestations. Results from other spider mite systems (De Boer and Dicke 2004; Woods et al. 2011; Gadino et al. 2012) suggest that there is potential merit in using synthetic HIPVs (e.g., methyl salicylate) for this purpose. Furthermore, results from a pilot field study by M. Hoddle (2007) showed that counts of *O. perseae* were significantly higher on sticky cards in close proximity to avocado trees treated with methyl salicylate dispensers in comparison to control plots that lacked this semiochemical. This finding suggested that methyl salicylate may have been repellant and caused *O. perseae* to prematurely disperse from infested leaves. It is also possible that *O. perseae* dispersal was further instigated by higher rates of predation on Hass trees treated with methyl salicylate because of recruitment. However, predator populations were not monitored. Further studies will need to elucidate and quantify the beneficial mechanisms of HIPV deployment (e.g., pest repellency, natural enemy recruitment) on avocado and establish optimal lure release rates for attraction of predators (Woods et al. 2011).

However, the possible lack of kairomone production by susceptible cultivars like Hass is one of several potential factors that may explain the high prevalence of *O. perseae* in California and the limitations of resident biological control agents to suppress populations of this pest. Additional studies have revealed that life history traits of candidate phytoseiid predators may have also played a direct role in *O. perseae* populations performing well on Hass.

11.6 Present Biological Control Program of *Oligonychus perseae* Using Phytoseiids

Phytoseiidae has a strong historical legacy as being a source of effective spider mite biological control agents in agricultural systems (Gerson et al. 2003; Hoy 2011). Biological control of *O. perseae* using phytoseiids has been studied extensively in California and has either relied on natural control provided by resident populations of *E. hibisci* and seasonal inoculative releases of a commercial strain of *N. californicus*. These same species were previously used with limited success for the biological control of *O. punicae* on avocado (McMurtry and Johnson 1966; McMurtry et al. 1984). Their ability to reduce *O. perseae* populations has also been limited. This outcome stresses that the degree of compatibility of life histories between the candidate natural enemy and the target pest in relation to the intended release habitat is important. For more than two decades the main problem in implementing a successful phytoseiid-based biological control program for reduction of *O. perseae* populations during the summer, when pest densities increase, has been dealing with the ecological and biological limitations of natural enemies that are easily mass-reared (McMurtry and Scriven 1965a) and readily available in California (McMurtry and Johnson 1966), but either fail to establish and/or do not respond numerically to mite outbreaks and/or require supplementary food resources to promote persistence. A brief review of the ecology of these phytoseiid species in relation to *O. perseae* on California Hass avocado is provided next.

11.6.1 Natural Control Using Euseius hibisci

Euseius hibisci was originally described from *Hibiscus* sp. leaves from the state of Sonora in Mexico (1959) and this species was re-described by Congdon and McMurtry (1985) from California populations occurring on avocado. Initial monitoring data from 1992 (McMurtry 1993) showed that populations of *E. hibisci*, an abundant and widely distributed native species in California avocado orchards

(McMurtry and Johnson 1966), failed to control populations of *O. perseae* which reached densities as high as 1,600 spider mites per leaf during that season. Furthermore, during 1998–2000 Yee et al. (2001a) monitored populations of *O. perseae* and indigenous phytoseiid populations (mainly *E. hibisci*) at three Hass orchards that were representative of coastal, intermediate and warm-inland climates of the avocado growing region in southern California. Similarly, the results of that study indicated that there were no consistent trends between *O. perseae* and resident phytoseiid densities despite the fact that phytoseiids were the most abundant and continually present group of natural enemies on avocado foliage at all three sites. In these studies, *E. hibisci* failed to show any type of density dependent or numerical response to increasing densities of pest mites.

More recent studies have provided supporting evidence for the limited role of *Euseius* species as effective biological control agents of *O. perseae*. During 2009–2010, J. Lara (unpublished data) studied the cross-sectional spatial dynamics of resident *Euseius* species populations (most likely *E. hibisci*) from nine commercial Hass orchards infested by *O. perseae*. In four orchards, resident phytoseiids displayed a statistically significant association with infested *O. perseae* leaves but overall populations of these predators failed to display a biologically significant numerical response to high densities of *O. perseae*, which in most cases exceeded the action threshold boundary of 100 mites per leaf suggested by Moaz et al. (2011b). This cross-sectional study revealed that *O. perseae* populations readily escaped effective natural control from resident phytoseiids. Furthermore, as part of a long-term study, J. Lara and M. Hoddle (2014) monitored densities of resident *Euseius* species and *O. perseae* during 2002–2013 on a group of untreated Hass trees. Cross-correlation analyses revealed a lack of synchrony between *O. perseae* and resident phytoseiids and the inability of predator populations to respond numerically to peak *O. perseae* densities observed during summer, over the 11 year study period.

There are several potential reasons for this outcome. Members of *Euseius* are regarded as pollen specialists (McMurtry and Croft 1997). Furthermore, laboratory studies confirmed that *E. hibici* develops faster on a pollen diet than on spider mites alone (McMurtry and Scriven 1964; Zhimo and McMurtry 1990) and imbibing avocado leaf cell content may provide nutrients required for optimal reproduction (McMurtry and Scriven 1965a; Porres et al. 1975). Second, behavioral observations revealed that *Euseius* species could not penetrate the webbed nests of *O. perseae* and could only feed on exposed spider mites outside of the nest (Fig. 11.2a) (McMurtry 1993). Together, these results suggested *E. hibisci* is a generalist facultative predator primarily dependent on obtaining plant-based nutrition rather than feeding exclusively on spider mites such as *O. perseae*. Additionally, the leaf feeding habit might explain why populations of *E. hibisci* can subsist year round in avocado orchards, even in the absence of pollen and at times when levels of *O. perseae* and other potential prey species are low.

The ecological limitations of *E. hibisci* as an effective predator of *O. perseae* are also compounded when the potential influence of cultivar-susceptibility (see Sect. 11.5.3) is taken into account, as it allows *O. perseae* populations to (1) have an

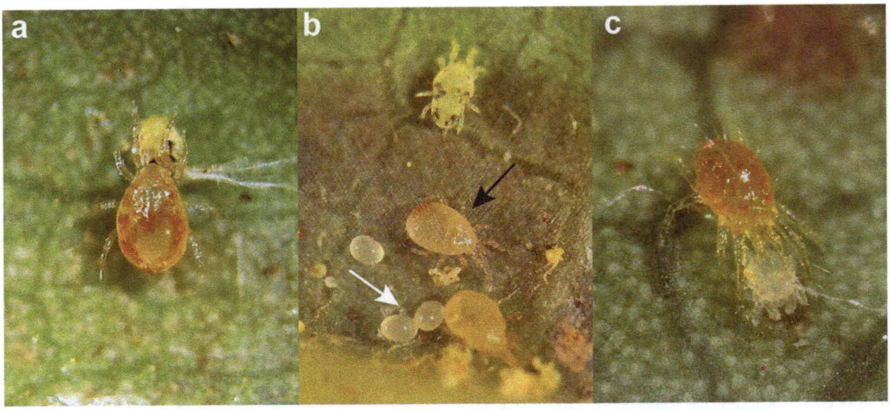

Fig. 11.2 (**a**) Adult *Euseius* sp. (Acari: Phytoseiidae) feeding on a motile *Oligonychus perseae* (Acari: Tetranychidae) individual on avocado leaf undersurface, (**b**) Adult (*black arrow*) *Neoseiulus californicus* (Acari: Phytoseiidae) resting after feeding on spider mites inside transparent *O. perseae* webbed-nest (*white arrow* points to oval-shaped phytoseiid eggs outside of the nest), (**c**) Adult predatory mite (Acari: Stigmaeidae) feeding on a motile *O. perseae* individual (Color figure online)

increased reproductive potential, and (2) avoid plant-mediated detection by predators early in the season when spider mite populations are increasing and effective predation is needed (this hypothesis requires further study).

11.6.2 Augmentative Biological Control Using Neoseiulus californicus

Searches for alternative biological control agents of *O. perseae* in California have focused on screening commercially available phytoseiid species for deployment in commercial avocado orchards (Hoddle et al. 1999, 2000). The current recommended *O. perseae* biological control strategy depends on inoculative releases of *N. californicus*, a selective spider mite predator that can also reproduce on non-spider mite food sources (Croft et al. 1998). This species was initially described from California lemons (McGregor 1954). On avocado, *N. californicus* can penetrate the webbed nest of *O. perseae* (Fig. 11.2b) and feed on all life stages of this pest (Takano-Lee and Hoddle 2002). However, making the recommended seasonal releases of 2,000 *N. californicus* predators per tree during the growing season to obtain *O. perseae* control is not practical or cost-effective for commercial orchards when large areas need treatment.

Also, despite their generalist feeding habits, *N. californicus* populations cannot remain at detectable numbers year-round in California avocado orchards, probably as the result of two non-mutually exclusive mechanisms. The first is the inability of this predator to utilize the varying types of food sources available on the avocado plants during the year. Unlike *E. hibisci*, mass reared *N. californicus* is unable to subsist year-round on the specific food resources available within the avocado system (e.g., avocado pollen, supplemental feeding on avocado leaf cell content, *O. perseae*, other insect and mite species). Consequently, *N. californicus* may be competitively excluded by populations of *Euseius* species which may not only be better adapted for obtaining nutrition from the same available resources in avocado orchards, but may also have a behavioral advantage in being more effective in detecting and finding these resources on avocado (see McMurtry and Johnson 1966). The reproductive potential of *N. californicus* and *Euseius* species reared on *O. perseae* has not been evaluated and results from this type of fundamental study might offer supporting evidence on the overall limited performance of both of these predators in the field when feeding exclusively on *O. perseae* (see Escudero and Ferragut 2005), compared with a diet consisting of a combination of *O. perseae* and avocado pollen or some other readily obtainable plant resource.

The second potential mechanism explaining the lack of long-term establishment of *N. calfornicus* on avocado could be asymmetric intraguild predation by *Euseius* species on certain *N. californicus* life stages (e.g., larva). This process could lead to a reduction of *N. californicus* populations and ultimately translate to reduced *O. perseae* suppression. This type of intraguild interaction has also been shown to occur between *Euseius stipulatus* (Athias-Henriot) and *N. californicus* for control of *Tetranychus urticae* Koch on clementine plants (Abad-Moyano et al. 2010a, b) and could occur in avocado.

11.6.3 Conservation Biological Control Using Euseius *Species*

Due to the practical and economic limitations of making *N. californicus* releases, rescuing indigenous phytoseiid populations which can subsist year-round in avocado orchards so that they can provide biological control services has been the focus of several studies outside California, in places where *O. perseae* is also a pest. Research in Israel (Maoz et al. 2011a) and Spain (González-Fernández et al. 2009) demonstrated that supplemental pollen provisioning can boost levels of indigenous *Euseius* species. However, in a more recent study (Montserrat et al. 2013), even with pollen provisioning, field releases of *E. stipulatus* and *N. californicus* were insufficient to control *O. perseae*. Monserrat et al. (2013) posit that this result was most likely due to a disruption of biological control when populations of phytoseiids experienced average temperatures above an estimated optimal value of 24.7 °C. This potential influence of abiotic components points to a third mechanism that in some cases might explain why the biological control efficacy of these natural enemies on avocado is limited. However, Monserrat et al. (2013) did not rule out the possibility that intraguild predation between *E. stipulatus* and *N. californicus* may have also been a factor in explaining the lack of *O. perseae* suppression.

11.7 Future Prospects for Improving Biological Control of *Oligonychus perseae* in California

The information reviewed in Sects. 11.1, 11.2, 11.3, 11.4, 11.5, and 11.6 lays the foundation on which to discuss future recommendations for improving the spider mite biological control program of the California avocado system. Although this information focuses on *O. perseae*, the underlying principles could be used to study (i.e., identify key crop production vulnerabilities that are conducive to pest problems) and improve biological control programs of spider mites on other perennial crops.

Although avocado was introduced from Mesoamerica to California in the mid-nineteenth century (see Sect. 11.1), problems with avocado-feeding arthropods (i.e., spider mites and insects) developed in the early twentieth century, as production of commercial avocado cultivars expanded and the avocado system became available as a perennial habitat to populations of avocado feeding pests (see Sect. 11.2). Initial concern with the increased presence of these herbivores prompted investigators, starting in the 1950s, to design and improve natural pest control strategies under the theoretical framework of biological control, a top-down ecological approach for pest management (see Sect. 11.3). Therefore, spider mite control research on commercial avocado host-plants (first trophic level) focused on exploiting the regulation of leaf-feeding populations of *O. punicae* and *E. sexmaculatus* (second trophic level) using natural enemies (third trophic level). Results from field studies indicated that, for the most part, populations of resident predator species such as the coccinellid *S. picipes* and the phytoseiid *E. hibisci* were effective natural enemies of spider mites in California avocado orchards for at least 40 years (1950s–1990) (see Sect. 11.3.1).

Interestingly, the biological resistance provided by resident predator populations against established tetranychid species in the California avocado system since the 1990s has been absent (e.g., *S. picipes*) or ineffective (e.g., phytoseiids) for control of *O. perseae* (see Sect. 11.4). Research has emphasized top-down control of *O. perseae*, but success with phytoseiid-based biological control strategies has been limited (see Sect. 11.6). Most likely the limited biological resistance of the California avocado system against *O. perseae* has been mediated by the growing availability of susceptible cultivars, as Hass, that overlap with the phenology of feeding *O. perseae* populations, allowing pest populations to escape effective top-down control. Other reasons may include the limited recruitment and searching ability of predators on *O. perseae* infested Hass foliage and other problems having to do with the incompatibility of life-history traits between *O. perseae* and extant candidate phytoseiid species (see Sects. 11.5.3 and 11.6). This process may have also occurred with other spider mite species outside of California, such as *O. yothersi* in Chile, where Hass avocado is also commercially produced (see Sects. 11.5.1 and 11.5.2). Better understanding of these factors may help restore the biological resistance of the California avocado system against key pests and improve the current avocado pest management program (UCANR 2008).

Hass is still the major commercial cultivar in California, so populations of *O. perseae* are expected to continue to be problematic in the future. In fact, economically damaging densities of *O. perseae* were observed region-wide in commercial avocado orchards during 2013. This outbreak of *O. perseae* populations resulted in the dissemination of an official area-wide pest alert from the California Avocado Commission (CAC 2013) to warn growers. Due to the lack of effective biological control, the integrated pest management program of *O. perseae* has shifted its reliance almost exclusively to chemical control and this less than ideal situation can lead to other unintended problems such as the development of pesticide resistance among populations (Hoddle and Morse 2013). The increased reliance on chemical control has also been observed with other invasive arthropods that have been introduced into California (Hoddle 2005), such as *S. perseae* (see Sect. 11.5.2), furthering less grower reliance on biological control and moving IPM programs from being biologically-based to being pesticide-focused.

To promote judicious pesticide use, a binomial sampling plan for *O. perseae* was developed by Lara and Hoddle (2013). This sampling plan can be used by pest managers to monitor densities of *O. perseae* populations in commercial orchards and is compatible with the long-standing concept of controlling pest populations when densities exceed research-based action thresholds that once crossed will result in economic losses (Stern et al. 1959; Maoz et al. 2011b). Sampling plans are also being developed (J. Lara and M. Hoddle, 2014) that can be used to monitor densities of resident phytoseiids in commercial orchards. These sampling plans have potential to allow avocado pest managers to concurrently maintain accurate chronological and quantitative records on *O. perseae* and phytoseiid populations. Subsequently, these assessments can be used to make informed decisions on pesticide use to target *O. perseae* populations and to determine whether predatory mites are likely to provide some level of control.

Compared to *O. perseae*, these types of custom sampling strategies were not developed for *O. punicae* and *E. sexmaculatus* on California avocados. This reflects the success of natural biological control for these latter species which obviated the pressing need for implementing these types of statistically-based sampling tools. Several alternatives exist for improving the biological control program of *O. perseae* in California. These alternatives are discussed below and include (1) the development of a classical biological control program, (2) the development of augmentation or conservation biological control strategies using largely unstudied resident populations of stigmaeid species present in commercial avocado orchards from California, and (3) addressing the issue of cultivar susceptibility to *O. perseae*. The first two research avenues emphasize top-down control while the third option stresses the idea of exploiting bottom-up (host plant based) strategies for the management of *O. perseae* populations.

11.7.1 Classical Biological Control Using Phytoseiids

Classical biological control is the deliberate practice of prospecting for co-evolved natural enemies in the evolutionary center of origin of the pest. Areas within the native range may be selected preferentially for searching based on climate similarities between the donor (i.e., native range) and receiving (i.e., invaded area) regions. Biological control theory suggests that natural enemies pre-adapted to the prevailing climate in the receiving range are more likely to establish than natural enemies from areas where there is climatic mis-match. Additionally, molecular-based studies may pinpoint areas within a large native range from within which the invading pest population might have originated. Searching in these areas where there is a close genetic match between the source and invasive populations may result in the discovery of natural enemies adapted to the dominant pest genotype in the invaded range. These two factors, climatic and genetic matching, may improve the chances of finding natural enemies that could be effective classical biological control agents. Natural enemies released as part of a classical biological control program are expected to establish, spread, and provide permanent suppression of pest populations without ongoing human management. Perennial agro-ecosystems like avocado orchards provide long-term habitat stability that increases the likelihood of natural enemy establishment.

Surveys for mites associated with avocado in Michoacán (Mexico), the largest Hass producing region in the world, resulted in the documentation of 18 species in nine families, of which five in three families were phytophagous and 12 species in four families likely exhibit some level of predatory behavior (see Table 1 in Estrada-Venegas et al. 2002). *Oligonychus perseae* and *O. punicae* were commonly encountered during surveys and it was noted that they were of high economic importance in Hass orchards. Phytoseiids previously considered in biocontrol programs for spider mites on California avocado that were found in Michoacán include *A. limonicus*, *E. hibisci*, and an undetermined species of *Neoseiulus*. *Amblydromalus limonicus* and *E. hibisci* are residents in California avocado orchards but have provided limited control of spider mites (McMurtry and Scriven 1965b; McMurtry et al. 1984; Estrada-Venegas et al. 2002). *Neoseiulus californicus* is a representative of the genus mentioned by Estrada-Venegas et al. (2002), but as previously discussed (see Sect. 11.6.2) populations of that genus have not established on California avocados for permanent spider mite control (McMurtry et al. 1984; Hoddle et al. 2000).

To complement the work of Estrada-Venegas et al. (2002), additional surveys were conducted by M. Hoddle and J. Lara (unpublished data) in 2012 for predator mites associated with *O. perseae* on Hass and non-Hass cultivars in five Mexican states, Guanajuato, México, Michoacán, Morelos and Puebla. A total of 21 sites were sampled and a total of 585 natural enemy specimens were collected, of which, 92 % were members of the Phytoseiidae and Stigmaeidae. Further taxonomic and molecular studies are needed to generate a list of reliably identified predator species present in Mexico that are associated with *O. perseae*. From this species list, novel *O. perseae* predators (not present in California) can be selected for deliberate introduction into California avocado orchards as part of a classical biological control program. However, these selected agents, if found, would have to demonstrate superior prey searching, prey consumption and numerical response abilities for controlling *O. perseae* when compared to natural enemies already present in California. In addition to predatory mites, 2,268 *O. perseae* were collected. These specimens are being used in molecular studies in an attempt to identify the possible source area in Mexico from which the invasive California *O. perseae* population originated. This area of origin, should it be identified, will further refine our interpretation of the importance of natural enemy species associated with *O. perseae* in its native home range.

However, recurring outbreaks of *O. perseae* and *O. punicae* on Hass in the presence of a diverse resident predatory guild that is unable to reliably keep these two pest mites below economically damaging densities, indicate that the susceptibility of the Hass cultivar to spider mites could be a key factor driving the pest status of these two species, especially *O. perseae*. This observation was striking at times in Mexico when Hass cultivar trees were inter-planted with non-Hass trees; the Hass cultivar was always noticeably more damaged by *O. perseae* than unidentified non-Hass varieties.

11.7.2 Biological Control Using Stigmaeids

Compared to the phytoseiids, the use of stigmaeids as biological control agents for spider mites has not been well studied and this deficiency has been noted previously (Huffaker et al. 1969; Gerson et al. 2003, Chap. 7 of this book). Nevertheless, there are some examples of stigmaeids that under certain circumstances have been shown to provide effective biological control of spider mites. Two well-documented examples include *Zetzellia malli* (Ewing) on apple for control of *Panonychus ulmi* (Koch) and *Agistemus longisetus* González-Rodríguez on citrus for control of *Panonychus citri* (McGregor) (Croft and MacRae 1993; Jamieson et al. 2005). The use of stigmaeids as effective biological control agents of *O. perseae* in commercial avocado orchards remains to be demonstrated.

11.7.2.1 Field Studies in Hass Orchards

The work by Yee et al. (2001a) is the only published avocado field study including seasonal monitoring of *O. perseae*, stigmaeid (identified only to family) and phytoseiid (mainly *E. hibisci*) populations in commercial avocado orchards in California. Results from that work indicated that stigmaeids were not as abundant as phytoseiids and overall populations of both of these predator groups did not respond numerically to increases in *O. perseae* densities. More recently, naturally-occurring stigmaeid and phytoseiid populations were monitored monthly during 2012–2013 (J. Lara, unpublished data) on an untreated research plot of Hass cultivar trees in Irvine, California. Again, densities of stigmaeids and phytoseiids (most likely *E. hibisci*) did not reflect a strong numerical response to increases in *O. perseae* during the summer, when average densities peaked at more than 50 (2012) and 126 (2013) *O. perseae* per leaf. Still, under field conditions, stigmaeids and phytoseiids were found naturally at other times of the year, when *O. perseae* levels were low, suggesting that they likely have a generalist feeding habit, are capable to co-exist to some extent and are permanently present in the avocado system.

Progress in stigmaeid research on avocado has benefited from the revision of mites of this family by González-Rodríguez (1965). Details about the taxonomy and other aspects of the stigmaeids are presented in Chap. 7 of this book. There are at least nine stigmaeid species historically recorded on avocado throughout the world, including *Agistemus arcypaurus* González-Rodíguez, *A. ecuadoriensis* González-Rodríguez, *A. fleschneri* Summers, *A. longisetus* González-Rodríguez, *A. simplex* González-Rodríguez, *A. striolatus* González-Rodríguez, *A. terminalis* (Quayle), *A. tucumanensis* González-Rodríguez and *Z. yusti* Summers (Gonzalez-Rodriguez 1965). Among these, *A. arcypaurus*, *A. fleschneri*, *A. longisetus* and *A. terminalis* have been documented on California avocados (Gonzalez-Rodriguez 1965) but adequate knowledge on relevant aspects of their ecology within this crop system is unknown (e.g., geographic distribution, phenology, interactions with other mite species, etc.). This information is essential to design experiments and investigate their role as spider mite natural enemies in avocado orchards. Interestingly, *A. longisetus* and *A. terminalis* have been collected on avocado from Mexico and Central America (Gonzalez-Rodriguez 1965), which include the geographic home range of *O. perseae*. This potentially significant ecological connection between *O. perseae* and stigmaeids has received little attention. This observation may reflect the bias in spider mite biological control research that leads to primary experimentation with candidate phytoseiid species because of their (1) documented historical and economic importance and (2) commercial availability.

11.7.2.2 Additional Observations in Hass Orchards

A survey of the California avocado mite fauna was conducted in 2011 by J. Lara (unpublished data). After Phytoseiidae, Stigmaeidae was the second most abundant group of predatory mites recorded. A total of 172 phytoseiid females and 102 motile stigmaeid specimens were collected from 700 leaves. The presence of stigmaeids was detected in several Hass orchards from southern California, including San Diego, Ventura, Santa Barbara and San Luis Obispo Counties. These are the principal commercial avocado production counties in California. Stigmaeids and phytoseiids were found predominantly on the underside of avocado leaves, mainly by natural shelters (i.e., domatia) at leaf vein junctions. The ecological interaction between host plant structures and predators has been previously observed for both stigmaeids (*A. longisetus*) and phytoseiids [*E. elinae* (Schicha)] (Walter and O'Dowd 1992).

During the course of this survey, an abundant population of stigmaeids was found in an organic Hass orchard in Santa Paula, California that was also infested with *O. perseae*. An additional larger sample of avocado leaves was subsequently collected from this orchard. Of 144 randomly selected mature leaves (one leaf per tree), 43 % contained at least one motile stigmaeid. Phytoseiids (*Euseius* spp.) were also present, but visual inspection of leaves in the field showed them to be less abundant than stigmaeids.

Stigmaeids from the Santa Paula orchard were tentatively identified as *A. longisetus* (Fig. 11.2c) using the taxonomic key of González-Rodríguez (1965), which includes descriptions of adults of both sexes. Another species of the same genus, *A. tucumanensis*, was recorded by González-Rodríguez (1965) on avocado from Argentina, along with a phytoseiid species identified as *E. hibisci* (probably a different but similar species, as *E. hibisci* has not been reported from that country in subsequent samplings; see Guanilo et al. 2008). *Agistemus tucumanensis* is morphologically similar to *A. longisetus*, but it has not been reported from California. Therefore, the sampled stigmaeid from the orchard in Santa Paula is most likely *A. longisetus*. Lines of this stigmaeid were reared in the laboratory on excised avocado leaves on a diet of either ice plant pollen (*Mesembryanthemum* sp.) or *O. punicae* to conduct behavioral studies similar to those conducted with candidate phytoseiids for the control of *O. perseae* (Takano-Lee and Hoddle 2002).

In general, *A. longisetus* is slow-moving in contrast to *Euseius* species. Even so, similar to *E. hibisci*, *A. longisetus* cannot penetrate the webbed nest of *O. perseae* and both species are only capable of attacking *O. perseae* individuals wandering outside of the nests (Fig. 11.2c) (J. Lara, 2011). For this reason, *A. longisetus* was reared on *O. punicae* instead of *O. perseae* on the adaxial leaf surface of avocado leaves where the former spider mite is found (see Sect. 11.3). *Oligonychus punicae* populations consisting of several hundred mites can be easily reared on excised non-Hass cultivar leaves and these can sustain feeding *O. punicae* populations for approximately 1–2 weeks. Interestingly, mass-rearing *O. perseae* on a Hass leaf diet was not possible in the laboratory, as leaves would deteriorate quickly and populations would not increase as quickly as *O. punicae*.

Under laboratory rearing conditions, specimens of *A. longisetus* have access to all stages of *O. punicae* and can lay eggs on the webbing of this spider mite species, but these predators have some difficulty moving easily on webbing produced by that prey. On a pollen diet, *A. longisetus* can move easily over the leaf surface, laying eggs on cotton strands. Still, *A. longisetus* reproduction on ice plant pollen or *O. punicae* alone was observably lower than *E. hibisci* and *N. californicus* on ice plant pollen and *O. punicae*, respectively. Occasionally, adults of *A. longisetus* were engaged in intraspecific predation of eggs, but cannibalism did not adversely affect the continuous production of these predators over a 2 year period in the laboratory when pollen supply was plentiful.

11.7.2.3 Potential Application of Stigmaeids

The laboratory observations presented in the previous section revealed potential issues for using candidate stigmaeids such as *A. longisetus* for the biological control of *O. perseae*. However, drawing conclusions from these observations about their efficacy for biological control of pest mites in California avocado orchards is not advisable until quantitative laboratory and field studies similar to those conducted for *Euseius* spp. (McMurtry and Scriven 1964; McMurtry and Johnson 1966) are conducted. Nevertheless, the inability of *A. longisetus* to enter the webbed nest of *O. perseae* indicates that the ecological potential of this predator to control *O. perseae* is likely constrained when pest populations build during spring and are physically protected from predators by webbed nests. Part of the biological control success for controlling *P. ulmi* and *P. citri* with stigmaeids (Croft and MacRae 1993; Jamieson et al. 2005) may be due to the fact that these *Panonychus* species lack complex webbing structures (Saito 1983). Their webbing life type perhaps facilitates access not only to stigmaeids but also other potential spider mite predators to all *Panonychus* life stages (Jamieson and Stevens 2009). The little webbing life style of *O. punicae* contrasts significantly with the complicated and webbed-nest structures of *O. perseae*, which has been demonstrated to physically restrict access to predators (McMurtry and Johnson 1966; Takano-Lee and Hoddle 2002).

Ideally, it would be expected that the presence of less threatening *O. punicae* populations in avocado orchards might serve as an alternative natural food source for released or natural populations of stigmaeids. Predator populations could build upon this resource and "spill over" onto *O. perseae*. Whether or not this type of apparent competition exists in avocado orchards and would translate into effective *O. perseae* control is unknown. The prospect of promoting this dynamic in avocado is unlikely for *A. longisetus* because this predator has trouble accessing *O. perseae* nests and will not readily reproduce on spider mites like *O. punicae*. These same mobility and reproduction problems were observed with *E. hibisci* in the context of *O. punicae* biological control (McMurtry and Johnson 1966).

The relatively low reproductive potential of *A. longisetus* on diets consisting of *O. punicae* and pollen requires that other suitable diets be evaluated for its eventual mass rearing, such as pollen of other plants, other factitious spider mites or other arthropod species, etc. A natural and alternative rearing prey for stigmaeids might be tydeoid mites (Tydeoidea) which were also frequently encountered in commercial avocado orchards sampled in 2011 and were found at leaf junctions where stigmaeids and phytoseiids are likely to be found. Again, this potential component of the predator-prey system on avocado has not been studied.

11.7.3 Cultivar Improvements for Pest Control

During the past decades Hass fruit has become the benchmark of fruit excellence in most subtropical commercial avocado growing areas and this has resulted in this cultivar dominating global avocado fruit production (see Sect. 11.5.1). However, optimal production of Hass avocado can be limited by several biological factors such as susceptibility of trees to cold temperatures, inconsistent fruit yield between years (i.e., alternate-bearing), and tree susceptibility to other potential stress factors (e.g., soil salinity, diseases, and herbivore feeding) (Bergh 1985). In the context of foliar-feeding spider mite pests present in the California avocado system, the Hass cultivar is relatively more susceptible to attack by *O. perseae*. Furthermore, the impact of *O. perseae* on California avocado production may have been exacerbated by the region-wide cultivation of Hass trees which produced a monoculture of susceptible host plants. A similar intensified dynamic between foliar-feeding spider mite populations and avocado host plants may be in effect in other areas where Hass is grown (see Sects. 11.5.2 and 11.5.3).

This emerging pattern implies that integrated pest management of spider mites, and potentially other arthropods pests (e.g. *S. perseae*), on avocado in areas like California could be improved naturally (without the use of pesticides) by implementing bottom-up (i.e., host plant based) strategies that complement biological control efforts. One potential strategy involves the identification and selective use of novel, marketable cultivars that bolster the biological resistance of the California avocado system against populations of key arthropod pests. The development of herbivore-resistant cultivars to target pest populations could be facilitated by identifying and exploiting key host plant mechanism(s) (e.g., physiological, biochemical characteristics) responsible for the reproductive advantage or disadvantage of pests. It is likely that these mechanisms are influenced by the interaction of cultivar genetics and input from the environment (climate) since these components are known to influence cultivar phenotypes and define optimal avocado growing requirements (see Sects 11.1 and 11.5.1).

Furthermore, diversifying avocado cultivar acreage is essential to minimize crop losses due to pests. Avocado cultivars with improved levels of herbivore-resistance could be planted selectively by growers in locations with specific climatic conditions that are amenable to producing high quality avocado fruit that is competitive with Hass but that offer reduced reproductive advantages to key avocado pests. For example, the Hass cultivar grows favorably in coastal areas but *O. perseae* and *S. perseae* populations are also capable of doing well in this environment. Therefore, a variety of commercially-acceptable cultivars that are resistant to these pests could be grown in these areas in addition to Hass.

With regards to *O. perseae*, the use of herbivore-resistant cultivars may promote the following ecological mechanisms: (1) resistant cultivars will offer a natural buffer against elevated *O. perseae* densities during spring and summer in areas with climates permissive to spider mite outbreaks and (2) a lower baseline of *O. perseae* infestations could allow resident or introduced species of predatory mite populations (see Sects. 11.7.1 and 11.7.2) to display improved efficacy because lower densities of pests need control. In addition, these beneficial mechanisms could lower grower concerns over spider mite outbreaks, promote reduced pesticide use and ultimately restore the natural spider mite balance of the California avocado system as it was known prior to the *O. perseae* invasion (Fleschner et al. 1956a). Reaching this end-point will require a long-term investment on interdisciplinary research that could be adopted as part of future research efforts of the California avocado breeding program.

The long-term goal of the California avocado breeding program is to screen and select new marketable cultivars (scions) and compatible rootstocks. Fuerte is known to be resistant to *O. perseae* populations but this cultivar displays erratic yield and is unlikely to regain acreage dominance in California and replace Hass avocado. Some improved cultivars related to Hass (e.g., GEM and Lamb-Hass) have been identified in California. GEM is a promising cultivar (Crane et al. 2013) that has potential to replace Hass acreage in California, but its susceptibility to *O. perseae* has not been evaluated (Arpaia M.L., 2014, personal communication). Lamb-Hass is known to be resistant to *O. perseae*, but its susceptibility to other herbivores in California has also not been evaluated. In any case, the replacement of Hass acreage with herbivore-resistant cultivars is uncertain and will depend on market factors such as consumer preference. In addition, the California industry will need time to study and optimize the benefits of potential acreage transitions to remain competitive with other major Hass-producing countries that export avocado to the United States.

A more acceptable host plant based approach could be the exploration of rootstock induced herbivore-resistance for Hass scions. Rootstocks could be evaluated and selected based on the ability to tolerate challenging edaphic conditions (e.g., high salinity), or their tolerance to *Phytophthora cinnamomi* Rands, which causes avocado root rot. Aside from these selection factors, it is also known that rootstocks can influence the phenotype of commercial cultivars. Clear evidence of this was documented for Hass cultivar and the rootstock Martin Grande (G755). Martin Grande confers root rot resistance to Hass scions but it also induces the expression of negative traits like lower tree yield and alternate bearing on Hass (Crane et al. 2013). Since rootstocks can influence cultivar phenotypes then it might be possible to screen for a Hass-compatible rootstock that (1) increases or at least does not hinder consistent yield, (2) does not negatively alter favorable Hass fruit characteristics, (3) confers tolerance to root rot, and (4) mediates scion-resistance to key pests of avocado in California, as *O. perseae*. The search for rootstock-scion combinations that meet these requirements has not been conducted in California but this approach also has potential to restore the biological resistance of the California avocado system against exotic pests and promote competitive Hass cultivar production with reduced reliance on pesticides.

References

Abad-Moyano R, Urbaneja A, Schausberger P (2010a) Intraguild interactions between *Euseius stipulatus* and the candidate biocontrol agents of *Tetranychus urticae* in Spanish clementine orchards: *Phytoseiulus persimilis* and *Neoseiulus californicus*. Exp Appl Acarol 50:23–34

Abad-Moyano R, Urbaneja A, Hoffmann D, Schausberger P (2010b) Effects of *Euseius stipulatus* on establishment and efficacy in spider mite suppression of *Neoseiulus californicus* and *Phytoseiulus persimilis* in clementine. Exp Appl Acarol 50:329–341

Aponte O, McMurtry JA (1997a) Biology, life table and mating behavior of *Oligonychus perseae* (Acari: Tetranychidae). Int J Acarol 23:199–207

Aponte O, McMurtry JA (1997b) Damage on 'Hass' avocado leaves, webbing and nesting behaviour of *Oligonychus perseae* (Acari: Tetranychidae). Exp Appl Acarol 21:265–272

Bailey JR, Olsen KN (1990) Control of two avocado mite pests. Calif Agric 44:31–32
Bellamore T (2003) Mexican avocados: history...the full story. Calif Avocado Soc Yearb 86:51–71
Bender GS (1994) A new mite problem in avocados. Calif Avocado Soc Yearb 77:73–77
Bergh B (1985) Avocado varieties for California. Calif Avocado Soc Yearb 68:75–93
Bost JB, Smith NJH, Crane JH (2013) History, distribution and uses. In: Schaffer B, Wolstenholme BN, Whiley AW (eds) The avocado: botany, production and uses, 2nd edn. CABI, Boston, pp 10–30
Boyce AM (1947) Pests of the avocado. Calif Avocado Soc Yearb 30:84–85
Bravo-Monzón AE, Espinosa-García FJ (2008) Volatile emissions in *Persea americana* in response to the stem borer *Copturus aguacatae* attack. Allelopath J 21:165–173
CAC (2013) Pest alert: persea mite, September 12, 2013. http://californiaavocadogrowers.com/articles/pest-alert-persea-mite-september-12-2013. Accessed 10 Apr 2014
CAC (2014) California avocado commission annual report 2012–2013. http://www.californiaavocadogrowers.com/sites/default/files/documents/2012-13%20CAC%20Annual%20Report%20-%20FINAL.pdf. Accessed 1 Mar 2014
Carman HF, Craft RK (1998) An economic evaluation of California avocado industry marketing programs, 1961–1995. Giannini foundation research report number 345, University of California Agriculture and Natural Resources (UCANR), California
Cerna E, Badii MH, Ochoa Y, U Aguirre LA, Landeros J (2009) Tabla de vida de *Oligonychus punicae* Hirst (Acari: Tetranychidae) en hojas de aguacate (Persea americana Mill) variedad hass, fuerte y criollo. Universidad y Ciencia 25:133–140
Chazeau J (1985) Predaceous insects. In: Helle W, Sabelis MW (eds) Spider mites: their biology, natural enemies and control, vol 1B, World crop pests. Elsevier, Amsterdam, pp 211–246
Chen H, Morrell PL, Ashworth VETM, De La Cruz M, Clegg MT (2009) Tracing the geographic origins of major avocado cultivars. J Hered 100:56–65
Congdon BD, McMurtry JA (1985) Biosystematics of *Euseius* on California citrus and avocado with the description of a new species (Acari: Phytoseiidae). Int J Acarol 11:23–30
Cooper WC (1949) The Texas avocado society. Calif Avocado Soc Yearb 33:137–138
Crane JH, Douhan G, Faber BA, Arpaia ML, Bender GS, Balerdi CF, Barrientos-Priego AF (2013) Cultivars and rootstocks. In: Schaffer B, Wolstenholme BN, Whiley AW (eds) The avocado: botany, production and uses, 2nd edn. CABI, Boston, pp 201–233
Croft BA, MacRae IV (1993) Biological control of apple mites: impact of *Zetzellia mali* (Acari: Stigmaeidae) on *Typhlodromus pyri* and *Metaseiulus occidentalis* (Acari: Phytoseiidae). Environ Entomol 22:865–873
Croft BA, Monetti LN, Pratt PD (1998) Comparative life histories and predation types: are *Neoseiulus californicus* and *N. fallacis* (Acari: Phytoseiidae) similar type II selective predators of spider mites? Environ Entomol 27:531–538
De Boer JG, Dicke M (2004) The role of methyl salicylate in prey searching behavior of the predatory mite *Phytoseiulus persimilis*. J Chem Ecol 30:255–271
Demite PR, McMurtry JA, de Moraes GJ (2014) Phytoseiidae database: a website for taxonomic and distributional information on phytoseiid mites (Acari). Zootaxa 5:571–577
Ebeling W, Pence RJ (1953) Pests of the avocado. Calif Avocado Soc Yearb 37:113–133
Ebeling W, Pence RJ (1958) Avocado pests. Calif Avocado Soc Yearb 41:25–33
Escudero LA, Ferragut F (2005) Life-history of predatory mites *Neoseiulus californicus* and *Phytoseiulus persimilis* (Acari: Phytoseiidae) on four spider mite species as prey, with special reference to *Tetranychus evansi* (Acari: Tetranychidae). Biol Control 32:378–384
Eskalen A, Stouthamer R, Lynch SC, Rugman-Jones PF, Twizeyimana M, Gonzalez A, Thibault T (2013) Host range of *Fusarium* dieback and its ambrosia beetle (Coleoptera: Scolytinae) vector in southern California. Plant Dis 97:938–951
Estrada-Venegas EG, Rodríguez-Navarro S, McMurtry JA (2002) Some avocado mites from Michoacan Mexico. Int J Acarol 28:387–393
FAO (2014) FAOSTAT. http://faostat3.fao.org/faostat-gateway/go/to/download/Q/QC/E. Accessed 1 Mar 2014

Fleschner CA (1954) Biological control of avocado pests. Calif Avocado Soc Yearb 38:125–129
Fleschner CA (1958) Field approach to population studies of tetranychid mites on citrus and avocado in California. In: Baker EC (ed) Proceedings of the tenth international congress of entomology, Montreal, 17–25 Aug 1956
Fleschner CA, Hall JC, Ricker DW (1956a) Natural balance of mite pests in an avocado grove. Calif Avocado Soc Yearb 39:155–162
Fleschner CA, Badgley ME, Ricker DW, Hall JC (1956b) Air drift of spider mites. J Econ Entomol 49:624–627
Flint ML, Dreistadt SH (1998) Natural enemies handbook: the illustrated guide to biological pest control. University of California Press, Berkeley
Flores D, Berman D (2012) Mexico: avocado annual production and exports forecast higher. In: USDA foreign agricultural service report number MX2084. http://gain.fas.usda.gov/Recent%20GAIN%20Publications/Avocado%20Annual_Mexico%20City_Mexico_11-26-2012.pdf. Accessed 2 Mar 2014
Gadino AN, Walton VM, Lee JC (2012) Evaluation of methyl salicylate lures on populations of *Typhlodromus pyri* (Acari: Phytoseiidae) and other natural enemies in western Oregon vineyards. Biol Control 63:48–55
Gatehouse JA (2002) Plant resistance towards insect herbivores: a dynamic interaction. New Phytol 156:145–169
Gerson U, Smiley RL, Ochoa R (2003) Mites (Acari) for pest control. Blackwell, Malden
Gilligan TM, Epstein ME (2009) *Amorbia cuneana*. In: LBAM ID: tools for diagnosing light brown apple moth and related western U.S. leafrollers (Tortricidae: Archipini). Colorado State University, California Department of Food and Agriculture, and Center for Plant Health Science and Technology, USDA, APHIS, PPQ. http://itp.lucidcentral.org/id/lep/lbam/Amorbia_cuneana.htm. Accessed 20 Mar 2014
González-Fernández JJ, de la Peña F, Hormaza JI, Boyero JR, Vela JM, Wong E, Trigo MM, Montserrat M (2009) Alternative food improves the combined effect of an omnivore and a predator on biological pest control. A case study in avocado orchards. Bull Entomol Res 99:433–444
Gonzalez-Rodriguez RH (1965) A taxonomic study of the genera *Mediolata*, *Zetzellia* and *Agistemus* (Acarina: Stigmaeidae). Univ Calif Publ Entomol 41:1–64
Griswold HB (1946) The Hass avocado. Calif Avocado Soc Yearb 30:27–31
Guanilo AD, de Moraes GJ, Toledo S, Knapp M (2008) Phytoseiid mites (Acari: Phytoseiidae) from Argentina, with description of a new species. Zootaxa 1884:1–35
Hernández-Suárez E, Torres-Luis E, López-González AA, Perera-González S, Saavedra-Oliva OL (2010) Evaluación de eficacia de acaricidas y sueltas inundativas de *Neoseiulus californicus* para el control de la araña cristalina *Oligonychus persea*e en aguacate. http://www.agrocabildo.org/publica/Publicaciones/frut_299_ara%C3%B1a_cristalina_aguacate.pdf. Accessed 20 Sept 2014
Hoddle MS (2002) Development and reproductive biology of *Scirtothrips perseae* Nakahara (Thysanoptera: Thripidae): a new avocado pest in California. Bull Entomol Res 92:279–285
Hoddle MS (2003) Phenology of *Scirtothrips perseae* Nakahara (Thysanoptera: Thripidae) and associated natural enemies in southern California avocado orchards. Calif Avocado Soc Yearb 86:127–145
Hoddle MS (2005) Invasions of leaf feeding arthropods: why are so many new pests attacking California-grown avocados? Calif Avocado Soc Yearb 87:65–81
Hoddle MS, Morse JG (2013) The persea mite invasion into California: history, biology, management and current status. Calif Avocado Soc Yearb 95:106–136
Hoddle MS, Aponte O, Kerguelen V, Heraty J (1999) Biological control of *Oligonychus perseae* (Acari: Tetranychidae) on avocado: I. evaluating release timings, recovery, and efficacy of six commercially available phytoseiids. Int J Acarol 25:211–219
Hoddle MS, Robinson L, Virzi J (2000) Biological control of *Oligonychus perseae* (Acari: Tetranychidae) on avocado: III. evaluating the efficacy of varying release rates and release frequency of *Neoseiulus californicus* (Acari: Phytoseiidae). Int J Acarol 26:203–214

Hoddle MS, Jetter KM, Morse JG (2003) The economic impact of *Scirtothrips perseae* Nakahara (Thysanoptera: Thripidae) on California avocado production. Crop Prot 22:485–493

Hoy MA (2011) Agricultural acarology: introduction to integrated mite management. CRC Press, Boca Raton

Huffaker CB, van de Vrie M, McMurtry JA (1969) The ecology of tetranychid mites and their natural control. Annu Rev Entomol 14:125–174

Jamieson LE, Stevens PS (2007) Development rates, longevity and fecundity of six-spotted mite (*Eotetranychus sexmaculatus*) at constant temperatures. N Z Plant Prot 60:72–77

Jamieson LE, Stevens PS (2009) Miticides against citrus red mites (*Panonychus citri*). N Z Plant Prot 62:302–309

Jamieson LE, Charles JG, Stevens PS, McKenna CE, Bawden R (2005) Natural enemies of citrus red mite (*Panonychus citri*) in citrus orchards. N Z Plant Prot 58:299–305

Jeppson LR, Keifer HH, Baker EW (1975) Mites injurious to economic plants. University of California Press, Berkeley

Kennedy GG, Smitley DR (1985) Dispersal. In: Helle W, Sabelis MW (eds) Spider mites: their biology, natural enemies and control, vol 1A, World crop pests. Elsevier, Amsterdam, pp 233–242

Kerguelen V, Hoddle MS (1999) Biological control of *Oligonychus perseae* (Acari: Tetranychidae) on avocado: II. Evaluating the efficacy of *Galendromus helveolus* and *Neoseiulus californicus* (Acari: Phytoseiidae). Int J Acarol 25:221–229

Kerguelen V, Hoddle MS (2000) Comparison of the susceptibility of several cultivars of avocado to the persea mite, *Oligonychus perseae* (Acari: Tetranychidae). Sci Hortic 84:101–114

Lahav E, Lavi U (2013) Genetics and breeding. In: Schaffer B, Wolstenholme BN, Whiley AW (eds) The avocado: botany, production and uses, 2nd edn. CABI, Boston, pp 51–85

Lara JR, Hoddle MS (2013) Sampling guidelines for persea mite in California avocado orchards. Calif Avocado Soc Yearb 95:151–173

León Lobos OA (2003) Estudio de los parámetros de vida de *Oligonychus yothersi* McGregor (Acarina: Tetranychidae) en dos cultivares de palto (*Persea americana* Mill.), Hass y Fuerte. Dissertation, Austral University of Chile

MAG (2007) Región central oriental: caracterización de la agrocadena de aguacate, zona de los Santos. Ministerio de Agricultura y Ganadería. http://www.mag.go.cr/bibliotecavirtual/a00076.pdf. Accessed 25 Apr 2014

Maoz Y, Gal S, Argov Y, Coll M, Palevsky E (2011a) Biocontrol of persea mite, *Oligonychus perseae*, with an exotic spider mite predator and an indigenous pollen feeder. Biol Control 59:147–157

Maoz Y, Gal S, Zilberstein M, Izhar Y, Alchanatis V, Coll M, Palevsky E (2011b) Determining an economic injury level for the persea mite, *Oligonychus perseae*, a new pest of avocado in Israel. Entomol Exp Appl 138:110–116

McGregor EA (1942) The avocado mite of California, a new species. Calif Avocado Soc Yearb 27:68

McGregor EA (1954) Two new mites in the genus *Typhlodromus* (Acarina: Phytoseiidae). Bull South Calif Acad Sci 53:89–92

McKenzie HL (1934) Latest developments in avocado pest control. Calif Avocado Soc Yearb 18:33–36

McLean RR (1932) Pest control – the growing problem of the industry. Calif Avocado Soc Yearb 16:69–80

McMurtry JA (1989) Utilizing natural enemies to control pest mites on citrus and avocado in California, U.S.A. In: Channabasavanna GP, Viraktamath CA (eds) Progress in acarology. VII international congress of acarology, Bangalore, Aug 1986, vol 2. Oxford and IBH Publishing Co. Pvt. Ltd., New Delhi, pp 325–336

McMurtry JA (1992) The role of exotic natural enemies in the biological control of insect and mite pest of avocado in California. In: Lovatt CJ (ed) The shape of things to come. Proceedings of the second world avocado congress, Orange, Apr 1991, vol 1. University of California, Riverside, pp 247–252

McMurtry JA (1993) Biological control of mites and greenhouse thrips on avocado. In: California avocado research symposium. California Avocado Society and University of California, Riverside. http://www.avocadosource.com/ARAC/SUM_1993/SYMP_1993_PG_76.pdf. Accessed 2 Mar 2014

McMurtry JA, Croft BA (1997) Life-styles of phytoseiid mites and their roles in biological control. Annu Rev Entomol 42:291–321

McMurtry JA, Johnson HG (1963) The avocado brown mite in relation to its natural enemies. Calif Avocado Soc Yearb 46:76–81

McMurtry JA, Johnson HG (1966) An ecological study of the spider mite *Oligonychus punicae* (Hirst) and its natural enemies. Hilgardia 37:363–402

McMurtry JA, Johnson HG (1968) Preliminary studies on releasing *Stethorus* beetles for control of the avocado brown mite. Calif Avocado Soc Yearb 51:173–176

McMurtry JA, Scriven GT (1964) Studies on the feeding, reproduction and development of *Amblyseius hibisci* (Acarina: Phytoseiidae) on various food substances. Ann Entomol Soc Am 57:649–655

McMurtry JA, Scriven GT (1965a) Insectary production of phytoseiid mites. J Econ Entomol 58:282–284

McMurtry JA, Scriven GT (1965b) Life-history studies of *Amblyseius limonicus*, comparative observations on *Amblyseius hibisci* (Acarina: Phytoseiidae). Ann Entomol Soc Am 58:106–111

McMurtry JA, Johnson HC, Gustafson CD (1972) Further studies on control of avocado brown mite with releases of *Stethorus picipes*. Calif Avocado Soc Yearb 55:128–134

McMurtry JA, Johnson HC, Badii MH (1984) Experiments to determine effects of predator releases on populations of *Oligonychus punicae* (Acarina: Tetranychidae) on avocado in California. Entomophaga 29:11–19

McMurtry JA, de Moraes GJ, Johnson HG (1992) Arrestment response of some phytoseiid mites to extracts of *Oligonychus punicae*, *Tetranychus urticae* and pollen. Isr J Entomol 25–26:29–34

McMurtry JA, de Moraes GJ, Sourassou NF (2013) Revision of the lifestyles of phytoseiid mites (Acari: Phytoseiidae) and implications for biological control strategies. Syst Appl Acarol 18:297–320

Montserrat M, Guzmán C, Sahún RM, Belda JE, Hormaza JI (2013) Pollen supply promotes, but high temperatures demote, predatory mite abundance in avocado orchards. Agric Ecosyst Environ 164:155–161

Morse JG, Rugman-Jones PF, Watson GW, Robinson LJ, Bi JL, Stouthamer R (2009) High levels of exotic armored scales on imported avocados raise concerns regarding USDA-APHIS' phytosanitary risk assessment. J Econ Entomol 102:855–867

Mortensen B (2013) Plant resistance against herbivory. Nat Educ Knowl 4:5

Moznette GF (1922) The possibility of transmitting insects on budwood with special reference to the avocado. Calif Avocado Soc Yearb 8:56–57

Nakahara S (1997) *Scirtothrips perseae* (Thysanoptera: Thripidae), a new species infesting avocado in southern California. Insecta Mundi 11:189–192

Nimet SG, Kumral NA, Sivritepe HO, Seidi M, Susurluk H, Senturk B (2009) Olfactory response of the ladybird beetle *Stethorus gilvifrons* to two preys and herbivore-induced plant volatiles. Phytoparasitica 37:217–224

Oatman ER, Platner GR (1985) Biological control of two avocado pests. Calif Agric 39:21–23

Ochoa R, Aguilar H, Vargas C (1994) Phytophagous mites of Central America: an illustrated guide. CATIE, Turrialba

Peña JE, Johnson F (1999) Insect management in avocados. http://manatee.ifas.ufl.edu/lawn_and_garden/master-gardener/gardening-manatee-style/a/avocado-insect-management.pdf. Accessed 2 Mar 2014

Peña JE, Hoddle MS, Aluja M, Palevsky E, Ripa R, Wysoki M (2013) Insect and mites pests. In: Schaffer B, Wolstenholme BN, Whiley AW (eds) The avocado: botany, production and uses, 2nd edn. CABI, Boston, pp 423–488

Porres MA, McMurtry JA, March RB (1975) Investigations of leaf sap feeding by three species of phytoseiid mites by labeling with radioactive phosphoric acid ($H_3{}^{32}PO_4$). Ann Entomol Soc Am 68:871–872

Quayle HJ (1933) The trend of avocado pests and their control. Calif Avocado Soc Yearb 17:38–39

Rincón-Hernández CA, Sánchez-Pérez JDLL, Espinosa-García FJ (2011) Caracterización química foliar de los árboles de aguacate criollo (*Persea americana* var. *drymifolia*) en los bancos de germoplasma de Michoacán, México. Rev Mex Biodivers 82:395–412

Rounds MB (1947) The Fuerte avocado. Calif Avocado Soc Yearb 30:54–56

Rugman-Jones PF, Hoddle MS, Stouthamer R (2008) Population genetics of *Scirtothrips perseae*: tracing the origin of a recently introduced exotic pest of Californian avocado orchards, using mitochondrial and microsatellite DNA markers. Entomol Exp Appl 124:101–115

Sabelis MW, Dicke M (1985) Long-range dispersal and searching behaviour. In: Helle W, Sabelis MW (eds) Spider mites: their biology, natural enemies and control, vol 1B, World crop pests. Elsevier, Amsterdam, pp 141–157

Saito Y (1983) The concept of "life types" in Tetranychidae. An attempt to classify the spinning behaviour of Tetranychidae. Acarologia 24:377–391

Sances FV, Toscano NC, Hoffman MP, Lapré LF, Johnson MW, Bailey JB (1982) Physiological responses of avocado leaves to avocado brown mite feeding injury. Environ Entomol 11:516–518

Sánchez Colín S, Mijares Oviedo P, López-López L, Barrientos-Priego AF (2001) Historia del aguacate en México. http://www.avocadosource.com/journals/cictamex/cictamex_1998-2001/CICTAMEX_1998-2001_PG_171-187.pdf. Accessed 1 Mar 2014

Schroeder CA (1959) Avocado materials and observations in Costa Rica, Ecuador and Peru. Calif Avocado Soc Yearb 42:72–76

Scriven GT, Fleschner CA (1960) Insectary production of *Stethorus* species. J Econ Entomol 53:982–985

Shepherd J, Bender G (2002) A history of the avocado industry in California. Calif Avocado Soc Yearb 85:29–50

Stern VM, Smith RF, van den Bosch R, Hagen KS (1959) The integrated control concept. Hilgardia 29:81–101

Takano-Lee M, Hoddle M (2002) Predatory behaviors of *Neoseiulus californicus* and *Galendromus helveolus* (Acari: Phytoseiidae) attacking *Oligonychus perseae* (Acari: Tetranychidae). Exp Appl Acarol 26:13–26

Tomkins AR (2002) Sustainable management of sixspotted spider mite (*Eotetranychus sexmaculatus* (Riley)) on Avocados. N Z Avocado Grow Assoc Annu Res Rep 2:1–9

Torres-Gurrola G, Delgado-Lamas G, Espinosa-García FJ (2011) The foliar chemical profile of criollo avocado, *Persea americana* var. *drymifolia* (Lauraceae), and its relationship with the incidence of a gall-forming insect, *Trioza anceps* (Triozidae). Biochem Syst Ecol 39:102–111

Tuttle DM, Baker EW, Abbatiello MJ (1976) Spider mites of Mexico (Acari: Tetranychidae). Intl J Acarol 2:1–102

UCANR (2008) Integrated pest management for avocados, publication 3503. University of California Agriculture and Natural Resources, Oakland

USDA (2014) California historical commodity data. Available via NASS. http://www.nass.usda.gov/Statistics_by_State/California/Historical_Data/Avocados.pdf. Accessed 1 Mar 2014

Van Driesche R, Hoddle M, Center T (2008) Control of pests and weeds by natural enemies: an introduction to biological control. Blackwell, Malden

Walter DE, O'Dowd DJ (1992) Leaves with domatia have more mites. Ecology 73:1514–1518

White J (2002) The avocado industry in New Zealand. Calif Avocado Soc Yearb 85:77–91

Wolstenholme BN (2013) Ecology: climate and soils. In: Schaffer B, Wolstenholme BN, Whiley AW (eds) The avocado: botany, production and uses, 2nd edn. CABI, Boston, pp 86–117

Woods JL, James DG, Lee JC, Gent DH (2011) Evaluation of airborne methyl salicylate for improved conservation biological control of two-spotted spider mite and hop aphid in Oregon hop yards. Exp Appl Acarol 55:401–416

Yee WL, Phillips PA, Rodgers JL, Faber BA (2001a) Phenology of arthropod pests and associated natural predators on avocado leaves, fruit, and leaf litter in southern California. Environ Entomol 30:892–898

Yee WL, Phillips PA, Faber BA, Rodgers JL (2001b) Relationships between *Scirtothrips perseae* (Thysanoptera: Thripidae) populations on avocado leaves, fruit, and scarring damage on fruit. Environ Entomol 30:932–938

Zhimo Z, McMurtry JA (1990) Development and reproduction of three *Euseius* (Acari: Phytoseiidae) species in the presence and absence of supplementary foods. Exp Appl Acarol 8:233–242

Chapter 12
Mite Pathogens and Their Use in Biological Control

Vitalis W. Wekesa, Fabien C.C. Hountondji, and Surendra K. Dara

12.1 Introduction

The expansion of cultivated areas, monocropping and the irrational use of acaricides has led many mites to reach pest status. Among these, spider mites (Tetranychidae) were unknown as pests before intensive use of synthetic pesticides became widespread (Penman and Chapman 1988). The concept of secondary pest outbreaks was first developed when differential mortality caused by pesticides significantly reduced populations of biological control agents of spider mites, such as coccinellid beetles (Coccinellidae) and phytoseiid mites (Phytoseiidae), leading to increased levels of spider mite populations (McMurtry et al. 1970). In other cases, pesticides such as imidacloprid, have been observed to lead to increase in spider mite populations by increasing their fecundity (James and Price 2002; Raupp et al. 2004).

The onset of pesticide resistance of some key pests mites (Messing and Croft 1996; Hoy 1998) and public concerns about environmental impact and safety of

V.W. Wekesa (✉)
Department of Applied and Technical Biology, Technical University of Kenya, 52428-00200, Nairobi, Kenya
e-mail: vwwekesa@gmail.com

F.C.C. Hountondji
Université de Parakou, Ecole Nationale Supérieure des Sciences et Techniques Agronomiques, 73, Djougou, Republic of Benin
e-mail: fabienho@yahoo.com

S.K. Dara
Strawberry and Vegetable Crops Advisor and Affiliated IPM Advisor, San Luis Obispo, Santa Barbara, CA, USA

Ventura Counties, UC Cooperative Extension, 2156 Sierra Way, Ste. C, San Luis Obispo, CA 93401, USA
e-mail: skdara@ucdavis.edu

D. Carrillo et al. (eds.), *Prospects for Biological Control of Plant Feeding Mites and Other Harmful Organisms*, Progress in Biological Control 19,
DOI 10.1007/978-3-319-15042-0_12

chemical applications have stimulated agricultural practitioners to change their pest management approach. In fact, pesticide resistance is an issue not just for mite pests, but also for mite parasites of humans and domesticated animals (Van Leeuwen et al. 2010).

Microbial pesticides may be a solution to the concerns raised by the excessive use of chemical pesticides and could play an important role in IPM. Practical use of microbial pesticides is derived from the study of biological control agents such as fungi, bacteria, protozoa or viruses that attack phytophagous invertebrates. Apart from their natural occurrence, some pathogens have been manipulated for managing pests, especially in high value crops. In recent times, some pathogens have become important tools in pest management (Hajek 2004).

Many pathogens have been investigated as potential agents for mite control, including fungi, viruses, bacteria, rickettsiae, protozoa and nematodes. However, most of these agents were shown to be of little commercial use, largely because of difficulties in their mass production or poor performance under field conditions, often due to their inefficiency in infecting mites (Poinar and Poinar 1998). Two species of fungal pathogens known to be specific to some mite hosts, *Hirsutella thompsonii* Fisher and *Neozygites floridana* Weiser and Muma, are important natural regulators of mite populations (Chandler et al. 2000). The production of these fungal pathogens *in vitro* remains a challenge and this has hampered their adoption in commercial scale. However, they remain the only pathogens with great known potential to regulate mites of agricultural importance in natural environments when conditions are favorable.

Viral diseases have long been known to affect mites, namely the citrus red mite, *Panonychus citri* (McGregor), and the European red mite, *Panonychus ulmi* (Koch) (Van Der Geest et al. 2000). Bacterial pathogens seem less common as natural enemies of phytophagous mites. Because of the sucking habit of phytophagous mites, contamination through ingestion, necessary for bacteria to affect the host organism is less probable, although toxins of *Bacillus thuringiensis* Berliner have been shown to be toxic to the two spotted spider mite, *Tetranychus urticae* Koch (Royalty et al. 1990; Guo et al. 1993).

Although mite-specific pathogens have not been commercialized, the number of biocontrol agents being screened for their efficacy and impact to the environment has increased. Important groups that have been screened for their efficacy against spider mites are mainly fungi. This is because fungi are the most common microorganisms with the potential to regulate pest mite populations through the occurrence of epizootics (Van Der Geest et al. 2000). Studies on their nutritional requirements are therefore necessary to understand growth dynamics and to facilitate their mass production. Also, documentation of epizootics can provide impetus for further studies of these pathogens and for development of products with low undesirable environmental effects.

12.2 Natural Incidence of Fungal Pathogens

Pathogenic fungi play a major role in the natural regulation of spider mite populations. Species of the genus *Neozygites* Witlaczil have been observed to infect spider mites on various crops (Keller 1997; Smitley et al. 2011). Weiser and Muma (1966) described one of the most common species of this genus, *Neozygites floridana*, as a pathogen of the Texas citrus mite, *Eutetranychus banksi* McGregor, on citrus in Lake Alfred, Florida.

Neozygites floridana has subsequently been observed to infect several other mites in the family Tetranychidae. In Florida, it was found infecting *Tetranychus tumidus* Banks on Cotton (Saba 1971) while in India it was found on *Tetranychus ludeni* Zacher infesting bean plants (Ramaseshiah 1971). A decade later, *N. floridana* was found causing epizootics in North Eastern Brazil on the red spider mite, *Tetranychus evansi* Baker and Pritchard on tomato (Humber et al. 1981). Other reports on the occurrence of *N. floridana* in spider mites included *Oligonychus hondoensis* Ehara on cedar in Japan (Brandenburg and Kennedy 1982) and *Mononychellus tanajoa* (Bondar) on cassava in Venezuela (Agudelo-Silva 1986), Kenya (Bartkowski et al. 1988) and Brazil (Delalibera et al. 1992).

In general, epizootics of Entomophthorales commonly develop when environmental conditions such as relative humidity and temperature favor their development, and when host mites are in sufficient numbers to sustain an epizootic (Smitley et al. 1986). In the case of the genus Neozygites, two important reasons are associated with the occurrence of epizootics in mites. First, their life cycles are rapid and are regulated by abiotic factors. The time from the initial host contact with a capilliconidium to the death of the host and fungal sporulation can be as short as 3 days (Oduor et al. 1995). Second, a large number of primary conidia are produced per host cadaver which enables contamination of healthy hosts resulting in rapid proliferation of infections in the entire population.

Besides Entomophthorales, deuteromycetous fungi (Deuteromycetes) of the genus *Hirsutella* also cause epizootics on mites of different families. *Hirsutella thompsonii* Fisher, the most common species of this genus, has been extensively reported in literature on eriophyid mites [e.g. *Phyllocoptruta oleivora* (Ashmead), McCoy and Kanavel (1969); *Calacarus flagelliseta* Flechtmann, Moraes & Barbosa, Hountondji et al. (2010); *Aceria guerreronis* Keifer, Al-Shanfari et al. (2013)] and on tetranychid mites [*M. tanajoa* and *Oligonychus gossypii* Zacher, Yaninek et al. (1996)]. It was recently found to be causing severe epizootics on *C. flagelliseta* on papaya in Oman (Hountondji et al. 2010). Although extensive studies have been conducted about the possible use of this pathogen (Van der Geest et al. 2000; Aghajanzadeh et al. 2006; Kumar and Singh 2008), much still remains to be known about their potential use. Wide scale investigations are still necessary to assess the biodiversity and potential of this group.

Hirsutella thompsonii, considered an important biological control agent of the citrus rust mite, *P. oleivora*, contains toxic metabolites that kill the host (McCoy et al. 1992). Vey et al. (1993) demonstrated that broth cultures of *H. thompsonii* contain insecticidal activity. Two protein toxins, hirsutellin A and B were later detected in the filtrates of the cultures. Omoto and McCoy (1998) isolated the toxic protein, hirsutellin A, from the filtrate during liquid fermentation of *H. thompsonii*. The hirsutellin A toxins were tested against citrus rust mite where mite mortality was positively correlated with an increase in concentration. These toxins were also reported to affect fecundity of the mites.

12.3 Potential of Other Control Agents

Apart from natural epizootics caused by *Neozygites* species, few studies have been conducted to evaluate other fungal acaropathogens against spider mites. Yet, more than 60 years ago conidia of the fungus *Beauveria bassiana* (Balsamo) Vuillemin formulated in dust form was reported to produce 71 % mortality of two spotted spider mite (Dresner 1949). The red palm mite, *Raoiella indica* Hirst (Tenuipalpidae) was reported to have been infected by a fungus, possibly *Hirsutella* sp., in Florida on palms (Peña et al. 2006). So far, *Lecanicillium psalliotae* Treschew has been the only other fungus reported in association with *R. indica*, in Saint Lucia (ARSEF 2009).

Cladosporium is one of the largest genera of hyphomycetes (Crous et al. 2007) that has been isolated from insects and mites. An unidentified species of this genus has been isolated from the two spotted spider mite (ARSEF 2009). On the other hand, a formulated *Fusarium semitectum* Berkeley & Ravenel was reported to suppress the population of *Polyphagotarsonemus latus* (Banks) on pepper (Mikunthan and Manjunatha 2006a). The same species has been reported to be selective for the control of sucking pests, a characteristic that could make it very suitable for incorporation in integrated pest management programs (Mikunthan and Manjunatha 2006b).

Beauveria, *Metarhizium*, *Isaria* and *Verticillium* have not been found infecting spider mites under natural conditions. Several isolates of *B. bassiana* and *Metarhizium anisopliae* (Metschnikoff) Sorokin have been reported as pathogenic to various mite groups (Mwangi et al. 1995; Alves et al. 2002). They have been considered to have potential for practical use in inundative or inoculative approaches in agriculture (Wekesa et al. 2005, 2006; Maniania et al. 2008). Deuteromycetes also produce toxins that cause death of the host before the fungus completes colonization of the latter (Delalibera et al. 1992).

Not much has been done in relation to possible use of pathogens for the control of parasitic mites such as those associated with birds. Most efforts in this sense have been dedicated to the study of fungi for tick control. While a systematic commercial approach to tick control with fungi is still unpopular, several studies show that acaropathogenic fungi are very effective against these parasites (Kaaya and Hassan 2001). In Brazil, *B. bassiana* and *M. anisopliae* var. *anisopliae* were isolated from the cattle tick *Rhipicephalus* (*Boophilus*) *microplus* (Canestrini) (da Costa et al. 2002). Polar et al. (2005) also found a species of Cordycipitaceae, *Simplicillium lamellicola* Gams & Zare (ARSEF1367) to be pathogenic to ticks. In certain countries, products based on *B. bassiana* and *M. anisopliae* have been reported as effective in controlling ticks on cattle (Kirby et al. 2010).

12.4 Host Specificity and Interactions with Other Control Agents (Predators)

The use of acaropathogens is usually considered safe for humans and other non-target organisms (Lacey et al. 2001; Shah and Pell 2003). They are mostly specific to their hosts or have a narrow host range (Van Der Geest et al. 2000). Few species of the main groups of mite pathogens have been reported to infect more than one host species of the same family to which their main hosts belong (e.g. *N. floridana*) and sometimes of different families (e.g. *H. thompsonii* and the protozoa *Pleistophora oribatei* Purrini & Weiser). Recent findings with *N. floridana* illustrate the need for confirmation of the true identity of the pathogen. Infection by populations of this pathogen has been reported on several tetranychid species, including *M. tanajoa*. Isolates of these populations showed different infectivity potential, suggesting varying levels of specificity (Yaninek et al. 1996; Hountondji et al. 2002a; Delalibera et al. 2004). Molecular and pathogenicity tests showed that the isolate infecting *M. tanajoa* constituted a new species, then named *Neozygites tanajoae* Delalibera, Hajek & Humber (Delalibera et al. 2004). Further studies using molecular tools will therefore shed light on the specificity of acaropathogens and may enhance their potential for commercialization. However, pathogenicity to multiple species appears to be more common in deuteromycetous fungi, particularly within the genera *Hirsutella*, *Isaria* and *Verticillium*. Pathogens with broader host ranges have been considered for use in cases of mixed mite infestations (Chandler et al. 2000; Jacobson et al. 2001).

The bacteria *Cardinium*, *Wolbachia* and some non-occluded viruses have been reported in association with several mite species (Nakamura et al 2009; Ros et al 2012). However, identification of these groups to species level is still necessary to confirm their specificity. Identification of pathogens has been largely based on morphological description and their classification still needs to be supported by molecular tools.

Particularly important is the interaction of biocontrol pathogens with other control agents (predators, parasites or other pathogens), especially with predatory mites. The latter, especially those of the family Phytoseiidae, are important natural enemies of mite pests, as discussed in the other chapters of this book. These predatory mites are commonly found associated with phytophagous mites infected by pathogens, but a few pathogens of phytophagous mites have been reported to infect predatory mites (Table 12.1). Wekesa et al. (2007) studied the interaction between *N. floridana* and the phytoseiid *Phytoseiulus longipes* Evans, both of which are natural enemies of *T. evansi* on tomato. They found that although not pathogenic to

Table 12.1 Infection of predatory mites by pathogens

Pathogen	Mite host	Reference
Entomophthorales		
Neozygites acaridis	*Euseius citrifolius* (Phytoseiidae)	Furtado et al. (1996), Keller (1997)
Unidentified species	*Macrocheles peregrinus* (Macrochelidae)	Milner (1985)
Tarichium subglobosum, T. acaricolum, T. pusillum, T. svalbardense, T. tenuisculpturatum	*Pergamasus* sp. (Parasitidae)	Bałazy and Wiśniewski (1984)
Erynia phalangicidae		
Deutoromycetes		
Hirsutella rostrata	*Dendrolaelaps tetraspinosus* (Digamasellidae) *Proctolaelaps* sp. (Melicharidae)	Bałazy and Wiśniewski (1989)
Protozoa		
Microsporidium sp.	*Neoseiulus barkeri* *Neoseiulus cucumeris* (Phytoseiidae)	Beerling and Van Der Geest (1991a, b)
Bacteria		
Rickettsiella phytoseiuli	*Phytoseiulus persimilis* (Phytoseiidae)	Šutáková (1988)
Rickettsia sp.	*Galendromus occidentalis* (Phytoseiidae)	Hess and Hoy (1982)
Serratia marcescens		Lighthart et al. (1988)
Wolbachia sp.	*N. barkeri* *N. bibens* *G. occidentalis* *P. persimilis* (Phytoseiidae)	Steiner (1993), Johanowicz and Hoy (1998), Breeuwer and Jacobs (1996)
Virus		
Non-occluded virus	*N. cucumeris, P. persimilis* (Phytoseiidae)	Steiner (1993), Bjørnson et al. (1997)

P. longipes, the fungus interfered with the behavior of the predator, reducing its predation capacity. This effect resulted from the increased time spent on grooming off capilliconidia sticking to the body of the predator. However, lack of lethal effect of *N. floridana* on *P. longipes* is important because both have been considered for use as biological control agents of *T. evansi*. On the other hand, another phytoseiid, *Typhlodromalus aripo* De Leon showed no preference for healthy or *N. floridana-infected M. tanajoa* and consumed equal numbers of healthy and infected mites in laboratory studies (Ariori and Dara 2007). This ability to feed on infected phytophagous mites may ensure the survival of *T. aripo* in field conditions, but how this interferes with the microbial control is not clear.

In rare instances, phytoseiids have been reported to be infected by *Neozygites*, *Hirsutella*, bacteria, protozoa and non-occluded viruses. However, infections are often low or erratic under natural conditions (Van Der Geest et al. 2000) and the pathogens are considered less detrimental than insecticides in relation to the effect on predatory mites (Goettel and Hajek 2000). Nevertheless, careful selection of microbial control candidates should include the evaluation of their specificity in relation to predatory mites.

12.5 Effect of Environmental Factors on Pathogens and Their Mite Hosts

Some of the main factors affecting the activity of pathogens and their hosts are environmental temperature (Oduor et al. 1996a, b; Fargues et al. 1997; Filotas and Hajek 2007) and relative humidity (Delalibera et al. 2006). The effects of these factors are intimately related because some fungi may tolerate temperature extremes at higher air relative humidity. These factors affect the infective ability of pathogens, their growth rate within the hosts (Feng et al. 1999) as well as the susceptibility of arthropods (Stacey and Fellowes 2002) and their overall mortality (Wekesa et al. 2010).

Sporulation of entomophthoralean species has been investigated over a range of temperatures (Wekesa et al. 2010). In many studies, sporulation has been observed to occur between 5 and 25 °C. Temperatures outside of this range are normally inhibitive to conidial release (Oduor et al. 1995; Wekesa et al. 2010). Overall, the wide range of temperature allowing sporulation also allows the formation of secondary conidia, although temperatures close to 25 °C may enable maximum formation of conidia and facilitate infection and general mycosis. Major differences between fungus performance at optimal and other temperature levels may explain why in spite of the broad range of thermal tolerance, each species of entomophthoralean fungi causes epizootics during specific seasons and not throughout the year (Inglis et al. 2001).

The ability to germinate over a wide range of temperatures may offer adaptive advantages because germinated spores can guarantee infection and increase the chances of establishment by the fungus when introduced to new areas (Yeo et al. 2003). When fungal pathogens are considered for classical biological control, their adaptations to varying climatic conditions in their native region can be an important factor in determining their suitability for release in a new area (Papierok et al. 1993).

Although high levels of humidity are known to be required for conidial germination, humidity thresholds have not been determined for most species. Moisture approaching saturation is usually necessary for sporulation, germination and host invasion (Oduor et al 1996b; Delalibera et al. 2006). However, this factor is not always limiting because sporulation and germination of some fungi, particularly of entomophthoralean species, occur at night near the boundary layer of the leaf surface, where relative humidity is higher than in the environment (Delalibera et al. 2006).

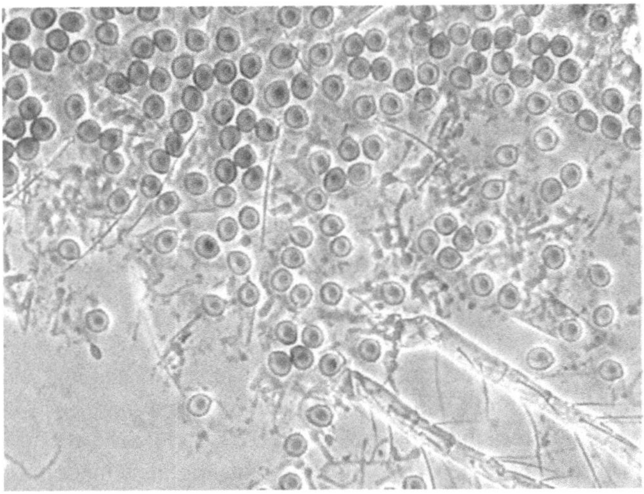

Fig. 12.1 Resting spores from the cadaver of the tomato red spider mite, *Tetranychus evansi*, collected from a greenhouse at Piracicaba, Sao Paulo, Brazil. Photo by Vitalis Wafula Wekesa

Resting spores are resistance structures produced by some pathogens when conditions are unfavorable (Fig. 12.1). These are capable of germinating when favorable conditions are again met. *Neozygites fresenii* frequently produce resting spores on aphids attacking cowpea, whereas *N. tanajoae* produce them erratically on *M. tanajoa* attacking cassava. The factors triggering their formation are not well understood.

12.6 Potential for Exploitation of Pathogens in Mite Control

Novel mass-production methods and improved growth media for most fungal acaropathogens, especially the Deuteromycetes, have been developed allowing the production of large amounts of relatively inexpensive fungal inocula to be used for the control of several agricultural pests (Kassa et al. 2008). Members of Neozygitaceae are among the most advanced of the Entomophthorales due to their high degree of host specialization and growth requirements (Agboton et al. 2009). Numerous attempts to grow *Neozygites in vitro* have failed (Mietkiewski et al. 1993). Some researchers, as Leite et al. (2000) and Delalibera et al. (2003), reported *in vitro* culture of *Neozygites* isolates, but that was limited to the hyphal body stage, which cannot be used for field release. Lack of advancement in the production of Entomophthorales on artificial media (Soper and Ward 1981; Trinci et al. 1990; Leite et al. 2000; Delalibera et al. 2003) limited their commercialization potential.

The production and deployment of fungal cultures in the form of dried cadavers has long been considered the best option for the utilization of these fungal pathogens.

Most small-scale production methods that have been developed frequently depend on laboratory-infection of healthy insects or mites through exposure to sporulating cadavers. Laboratory production is considered suitable for strain maintenance, permitting a continuous supply for use in basic research. Alternatively, cadavers can be field collected during epizootics and stored for future use (Steinkraus and Boys 2005). This procedure can be an option to acquire large quantities of fungal inocula, but their viability may vary considerably (Wekesa and Delalibera 2008). Also, these methods are time-consuming and so far inefficient for the production of adequate quantities for commercial field applications.

12.7 Release Methods of Acaropathogens

The application of acaropathogens formulated in powder or liquid form can be made using similar equipment as used for conventional chemical pesticides. Microbial pest control agents are used in two main strategies, namely classical and augmentative biological control.

Classical biological control is premised on the idea that pests that are introduced and get established in a new area are usually without the natural enemies that control them in their native range. It is considered that the most effective natural enemies of many pests are those that have coevolved in the same native habitat and therefore classical biological control aims at finding useful natural enemies in the native range of the pest, introducing them into the area invaded by the target pest, and allowing them to permanently establish so that they can provide uninterrupted pest control with little or no additional human intervention. For example, *H. thompsonii* var. *synnematosa* Samson, McCoy & O'Donnell was introduced from Zimbabwe to Argentina in 1985 for the control of *Eriophyes sheldoni* (Ewing) and *P. oleivora* and had about 50 % infection for both mite species after release but persistence is unknown (Sosa Gomez and Moscardi 1991; Sosa Gomez 1987). *Neozygites tanajoae* Delalibera, Hajek & Humber was introduced from Brazil to Benin between 1998 and 2000 for the control of *M. tanajoa* (Yaninek and Hanna 2003); it was proven to have been established and remained active as a natural regulator of that spider mite (Hountondji et al. 2002b).

Augmentative biological control refers to the release of additional numbers of a natural enemy when too few are present to effectively control a pest. This strategy relies on the ability to mass-produce large numbers of the natural enemy *in vitro* (Hajek 2004). For instance, there are native species of acaropathogens which attack and kill pest mites on several crops. However, they do not occur in sufficient quantities to cause epizootics all the time, therefore, growers can buy large numbers of these pathogens to augment those already on their crops so as to achieve the desired levels of biological control.

12.8 Compatibility with Chemical Control

Compatibility of acaropathogenic fungi with chemicals or other management options is important in integrated pest management. Many species show great potential for inclusion in such management programs because of their compatibility with chemical pesticides (Samson and Rombach 1985). The use of fungal acaropathogens as alternatives to acaricides, or the combined application of acaricides with fungal acaropathogens, is considered important for pesticide resistance management (Hajek and St. Leger 1994).

Farmers tend to apply multiple pesticides in a tank mix to minimize the cost of application as well as in an attempt to increase control efficacy. Tank mixes can include insecticides, acaricides and fungicides, among other materials that can potentially influence the efficacy of acaropathogens. Triflumuron, a chitin synthesis inhibitor sometimes used as an acaricide, negatively impacted the efficacy of *B. bassiana* against *T. urticae* (Irigaray et al. 2003) in laboratory studies. In a greenhouse study where the insect growth regulator diflubenzuron and *B. bassiana* were applied to control green peach aphid, *Myzus persicae* Sulzer, a similar antagonistic effect of the chemical on the fungus was observed (Olson and Oetting 1999). The use of the neonicotinoid insecticide imidacloprid also had a significant negative impact on the infection level of *M. tanajoa* by *N. floridana* (Dara and Hountondji 2001).

Neem oil (seed extract of Azadirachta indica A. Juss.) and sulfur are used as acaricides and fungicides in multiple cropping systems. These could interfere with the efficacy of acaropathogenic fungi used as insecticides or acaricides. For example, neem oil was reported to have inhibited the germination and growth of *B. bassiana* and *M. anisopliae* (Bajan et al. 1998; Hirose et al. 2001; Islam et al. 2010). Additionally, frequent use of fungicides for disease management may interfere with the use of acaropathogenic fungi even when they are not applied together. Tedders (1981) evaluated the inhibitory effect of triphenyltin hydroxide, benomyl, zineb, dodine, sulfur, and dinocap on *B. bassiana* and *M. anisopliae*. Dara and Dara (2013) also reported that some of the fungicides commonly used for disease management in strawberries inhibited the infectivity of *B. bassiana*. Sometimes, not the active ingredient, but the inert ingredients used in pesticide formulations can be detrimental to acaropathogenic fungi. Such an impact was reported for different formulations of *B. bassiana*, making some ingredients better than others when it comes to use with acaropathogenic fungi (Anderson and Roberts 1983; Anderson et al. 1989).

Considering how agricultural pesticides can be detrimental to acaropathogenic fungi, it is necessary to find out ways to improve their compatibility by choosing less toxic products or by applying them at safe time intervals. Preliminary studies have indicated that proper time interval between applications of certain fungicides and *B. bassiana* can improve the result of their use in pest management (Dara et al. 2014b). Increasing time intervals between *B. bassiana* and some fungicides improved their compatibility. Another study showed that except for captan and thiram, broad spectrum fungicides which have been used in agriculture for a long time, showed zero to negligible impact on *B. bassiana* at different time intervals (Dara et al. 2014c). These results address the concern for using entomopathogenic fungi in cropping systems where fungicides are frequently used and thus promote microbial control.

Conversely, several pesticides have been shown to be either compatible with acaropathogenic fungi or to have a synergistic effect on conidial germination and infection process of *B. bassiana*, *H. thompsonii* and *M. anisopliae* (Anderson et al. 1989; Hassan and Charnley 1989; Boucias et al. 1996; Kaakeh et al. 1997; Quintela and McCoy 1997; Delgado et al. 1999; Dara and Hountondji 2001; Shi and Feng 2006). In China, better control of the citrus red mite, *P. citri*, was achieved when *B. bassiana* was combined with pyridaben (Shi and Feng 2006). Similarly, a field study conducted in the United States showed that a combination of lower rates of *B. bassiana* and bifenazate was equally effective as a higher rate of bifenazate alone against *T. urticae* in California strawberries (Dara unpublished). In a laboratory study, imidacloprid was found to increase conidial germination of *H. thompsonii* (Dara and Hountondji 2001).

Rather than using acaropathogens as stand-alone products, combining and rotating them with chemical pesticides can be recommended in certain cases. Recent studies in California strawberries demonstrated that acaropathogenic fungi, such as *B. bassiana*, can play a significant role in the integrated management of various pests, including *T. urticae*, when used in a rotation program with chemical pesticides or applied together in a tank-mix with chemical or botanical pesticides (Dara 2013, 2014a). Substituting one or more chemical applications with acaropathogens will reduce chemical pesticide use and contribute to sustainable pest management practices.

Some studies show that using sub-lethal or low rates of chemical pesticides along with acaropathogenic fungi improved the efficacy of the latter (Shi et al. 2005; Ye et al. 2005; Dara et al. 2014a). However, the use of low rates of pesticides is not recommended because of the risk of pesticide resistance development. Bartlett (1968) assessed many of the direct and indirect effects of pesticides on spider mites where the need for careful selection of the pesticides in management of mites was suggested as being among the most important factors to be considered. Roberts and Campbell (1977) reported on possible disruption of pathogenic fungi and subsequent host responses following the application of selected fungicides. For instance, field studies have frequently shown that *N. floridana* is associated with dramatic declines in the abundance of spider mites. However, the reduction in incidence of infection by *N. floridana* has also been associated with the application of benomyl in fields infested with spider mites (Brandenburg and Kennedy 1983). These findings demonstrate the importance of understanding pesticide interactions because of their potential in inducing secondary pest outbreaks.

12.9 Main Constraints to the Practical Use of Acaropathogens in Mite Control

Virulence determination is a great challenge in the selection of promising candidates for microbial control. Laboratory, greenhouse and field performance of acaropathogens are not always concordant. Repeated evaluation of virulence of *N. tanajoae* failed to reveal significant differences between the indigenous Beninese isolate with low field epizootic potential and exotic Brazilian isolates with high epizootic potential (Hountondji et al. 2007). Interactions within the acaropathogen-mite-host plant system may well play a role in the expression of virulence of acaropathogens (Hountondji et al. 2005, 2006). Virulence evaluation in the laboratory may therefore not be sufficient for selection of good candidates for microbial control. Greenhouse and/or experimental releases may be essential in determining the real pathogenicity and ability to control the pest.

Mass-production and formulation of acaropathogens are key steps in developing microbial control agents for field use. Their development depends on the type of pathogen involved. Protozoa, virus, rickettsia-like bacteria and some fungi have to be produced on the host mite. Therefore, lack of reliable substrates for mass production can be a major constraint for utilization of these acaropathogens.

Farmers and researchers often consider biopesticides as less effective and more inconsistent in their efficacy than chemical pesticides because of their mode of action, sensitivity to harsh environmental conditions and slow activity. Despite the many encouraging results obtained under laboratory and greenhouse conditions, field studies demonstrating the true potential of fungal pathogens are few (Shi and Feng 2006). This is one of the reasons why uptake for the technology is slow because most growers consider biopesticides to be less competitive.

In a review on the future of insect pathogens, Lacey et al. (2001) highlighted the challenges to the successful use of acaropathogens. Even for acaropathogens that have been proven to have good field performance, practical use is not as simple as the use of chemical pesticides. As live organisms, their efficiency is dependent upon special care to assure their purity during the production process, adequate formulation and storage and proper timing for releases, taking into account the suitable climatic conditions for each pathogen species (Lacey et al. 2001; Shah and Pell 2003).

Lastly, the nature of biopesticide production and formulation often make them expensive and steer farmers to cheaper alternatives. Chemical pesticides are generally less expensive than microbial pesticides. Unless formulation technology makes significant progress and microbial pesticides become more affordable, it will be difficult for them to compete with chemical pesticides. As such, the use of pathogens for mite control is at present mostly limited to organic agriculture or greenhouse conditions, although farmers consider using microbial pesticides when managing certain pests for which chemical pesticides are ineffective due to resistance or when the nature of the pest requires the use of biopesticides.

12.10 Future Prospects for the Use of Acaropathogens in Mite Control

Natural epizootics are very common especially in spider mite populations during hot humid conditions (Humber et al. 1981; Brandenburg and Kennedy 1982). Dependence on the natural epizootics of entomophthoralean fungi and difficulties to predict their occurrences makes them less reliable management strategy, however, in some systems (or countries) where natural epizootics are common, they still play a significant role. The intentional induction of *N. floridana* epizootics in spider mites could enhance the possibility of timing and harvesting the infested cadavers to store for future use, especially through manipulation of conditions in plastic houses (Wekesa and Delalibera 2008). Production of resting spores by some of the acaropathogens during harsh climatic conditions has not been explored. These resting spores have the potential to be used as commercial inocula for pathogen inoculation in fields that are frequently attacked by mites. However, no studies have been commissioned to harvest, induce their germination and apply these spores in biological control of mites, despite their common occurrence during dry or cold conditions.

The acaropathogenic fungi *B. bassiana* and *M. anisopliae* have the potential to engage in fungus-plant interactions and this has generated the interest to deploy them as endophytes. This means that they can be used as dual-function systemic biopesticides. When applied for seed treatment, crops colonized endophytically by these fungi can be protected against plant feeding mites (Bacon and White 2000). However, the ability of an endophytically colonized plant to facilitate death or reduction on foliage-feeding by mites need to be investigated as an alternative method for deployment of these mite pathogens.

Preliminary greenhouse studies in the United States suggest a positive impact of entomopathogenic fungi through endophytic interactions with strawberries (Dara 2014a, b). Soil treatment of strawberry plants with *M. anisopliae* helped the plants withstand *T. urticae* better than untreated plants. Similarly soil treatment of *B. bassiana* appeared to have a negative impact on the green peach aphid, *Myzus persicae* Sulzer. These studies seem to show direct involvement of the pathogens and their effect on pests attacking the crops and are indicative of the possibility of developing endophytism as a management strategy that can reduce operational costs for managing mite pests.

Generalist predators are likely to survive in case of an epizootic than specialized predators because there are other sources of food they depend on after their primary prey is killed by a disease. These aspects have not been conclusively investigated and should be considered for effective use of predators in the management of mites. In most cases, it has been feared that pathogens would hinder predator performance just like predator foraging would interfere with pathogen performance. For instance, the interest of most mite control specialists is to obtain pathogen-free phytoseiid mites to ensure high efficacy in biological pest control as pathogens may affect the performance of predators or alter their reproduction and behavior (Schütte and Dicke 2008).

There is need for a more thorough evaluation of the interaction between pathogens and predatory mites as infections are not always readily visible and may not present obvious gross symptoms. These infections can present a lot of challenges in commercial mass-production units. Because of the interactions of pathogens with the predators, those collected from the field should be subjected to pathogen tests before introduction into the laboratory or mass rearing units.

References

Agboton BV, Hanna R, Hountondji FCC, Von Tiedemann A (2009) Pathogenicity and host specificity of Brazilian and African isolates of the acaropathogenic fungus *Neozygites tanajoae* to mite species associated with cassava. J Appl Entomol 133:651–658

Aghajanzadeh S, Mallik B, Chandrashekar SC (2006) Bioefficacy of six isolates of *Hirsutella thompsonii* Fisher against citrus rust mite, *Phyllocoptruta oleivora* Ashmead (Acari: Eriophyidae) and twospotted spider mite *Tetranychus urticae* Koch (Acari: Tetranychidae). Pak J Biol Sci 9:871–875

Agudelo-Silva P (1986) A species of *Triplosporium* (Zygomycetes: Entomophthorales) infecting *Mononychellus progressivus* (Acari: Tetranychidae) in Venezuela. Fla Entomol 69:444–446

Al-Shanfari A, Hountondji FCC, Al-Zawamri H, Rawas H, Al-Mashiki Y, De Moraes GJ, Moore D, Gowen RS (2013) Occurrence and seasonal prevalence of the coconut mite, *Aceria guerreronis* (Eriophyidae), and associated arthropods in Oman. Exp Appl Acarol 60:139–151

Alves SB, Rossi LS, Lopes RB, Tamai MA, Perera RM (2002) *Beauveria bassiana* yeast on agar medium and its pathogenicity against *Diatraea saccharalis* (Lepidoptera: Crambidae) and *Tetranychus urticae* (Acari: Tetranychidae). J Invertebr Pathol 81:70–77

Anderson TE, Roberts DW (1983) Compatibility of *Beauveria bassiana* isolates with insecticides formulation used in Colorado potato beetle (Coleoptera: Chrysomelidae) control. J Econ Entomol 76:1437–1441

Anderson TE, Hajek AE, Roberts DW, Preisler HK, Robertson JL (1989) Colorado potato beetle (Coleoptera: Chrysomelidae): effects of combinations of *Beauveria bassiana* with insecticides. J Econ Entomol 82:83–89

Ariori S, Dara SK (2007) Predation of *Neozygites tanajoae*-infected cassava green mites by *Typhlodromalus aripo* (Acari: Phytoseiidae). Agric Conspec Sci 72:169–172

ARSEF (ARS Collection of Entomopathogenic Fungal Cultures) (2009) Catalog of species. USDA ARS Biological Integrated Pest Management Research – US Department of Agriculture, Agricultural Research Service, Ithaca, p 435

Bacon CW, White J (2000) Microbial endophytes. CRC Press, Roca Baton, Florida, p 500

Bajan C, Kmitowa K, Popowska EN (1998) Reaction of various ecotypes of entomopathogenic fungus *Beauveria bassiana* to the botanical preparation NEEMTM and pyrethroid Fastak. Arch Phytopathol Plant Protect 31:369–375

Bałazy S, Wiśniewski J (1984) Records on some lower fungi occurring in mites (Acarina) from Poland. Acta Mycol 20:159–172

Bałazy S, Wiśniewski J (1989) Pathogene Pilze bei Milben. Mikrokosmos 78:299–304

Bartkowski J, Odindo MO, Otieno WA (1988) Some fungal pathogens of the cassava green spider mite *Mononychellus* spp. (Tetranychidae) in Kenya. Insect Sci Appl 9:457–459

Bartlett BR (1968) Outbreaks of twospotted spider mites and cotton aphids following pesticide treatment. Pest stimulation vs natural enemy destruction as the cause of outbreaks. J Econ Entomol 61:297–303

Beerling EAM, Van Der Geest LPS (1991a) A *microsporidium* (Microspora: Pleistophoridae) in mass-rearings of the predatory mites *Amblyseius cucumeris* and *A. barkeri* (Acarina: Phytoseiidae): analysis of a problem. IOBC/WPRS Bull 14:5–8

Beerling EAM, Van Der Geest LPS (1991b) Microsporidiosis in mass-rearings of the predatory mites *Amblyseius cucumeris* and *A. barkeri* (Acarina: Phytoseiidae). Proc Exp Appl Entomol NEV Amsterdam 2:157–162

Bjørnson S, Steiner MY, Keddie BA (1997) Birefringent crystals and abdominal discoloration of the predatory mite *Phytoseiulus persimilis* (Acari: Phytoseiidae). J Invertebr Pathol 69:85–91

Boucias DG, Stokes C, Storey G, Pendland JC (1996) The effects of imidacloprid on the termite *Reticulitermes flavipes* and its interaction with the mycopathogen *Beauveria bassiana*. Pflanzenschutz-Nachr Bayer 49:105–151

Brandenburg RL, Kennedy GG (1982) Relationship of *Neozygites floridana* (Entomophthorales: Entomophthoraceae) to two-spotted spider mite (Acari: Tetranychidae) populations in field corn. J Econ Entomol 75:691–694

Brandenburg RL, Kennedy GG (1983) Interactive effects of selected pesticides on the 2-spotted spider-mite and its fungal pathogen *Neozygites floridana*. Entomol Exp Appl 34:240–244

Breeuwer JAJ, Jacobs G (1996) *Wolbachia*: intracellular manipulators of mite reproduction. Exp Appl Acarol 20:421–434

Chandler D, Davidson G, Pell JG, Ball BV, Shaw K, Sunderland KD (2000) Fungal biocontrol of Acari. Biocontrol Sci Tech 10:357–384

Crous PW, Braun U, Schubert K, Groenewald JZ (2007) Delimiting *Cladosporium* from 356 morphologically similar genera. Stud Mycol 58:33–56

da Costa GL, Sarquis MI, de Moraes AM, Bittencourt VR (2002) Isolation of *Beauveria bassiana* and *Metarhizium anisopliae* var. *anisopliae* from *Boophilus microplus* tick (Canestrini, 1887), in Rio de Janeiro State, Brazil. Mycopathologia 154:207–209

Dara S (2013) Microbial control as an important component of strawberry IPM. CAPCA Adviser 16:29–32

Dara SK (2014a) New strawberry IPM studies with chemical, botanical, and microbial solutions. CAPCA Adviser 17:34–37

Dara SK (2014b) *Beauveria bassiana* and California strawberries: endophytic, mycorrhizal, and entomopathogenic interactions. Annual meetings of the Society for Invertebrate Pathology, Mainz, 3–7 Aug, p 148

Dara SK, Dara SS (2013) Compatibility of the entomopathogenic fungus, *Beauveria bassiana* with some fungicides commonly used in strawberries. Univ Calif Coop Ext Strawberries and Vegetables eNewsletter. http://ucanr.edu/blogs/blogcore/postdetail.cfm?postnum=9626

Dara SK, Hountondji FCC (2001) Effects of formulated imidacloprid on two mite pathogens, *Neozygites floridana* (Zygomycotina: Zygomycetes) and *Hirsutella thompsonii* (Deuteromycotina: Hyphomycetes). Insect Sci Appl 21:133–138

Dara SK, Dara SR, Dara SS (2014a) Endophytic colonization and pest management potential of *Beauveria bassiana* in strawberries. J Berry Res 3:203–211

Dara SS, Dara SSR, Dara SK (2014b) Optimal time intervals for using insect pathogenic *Beauveria bassiana* with fungicides. University of California Cooperative Extension Central Coast Agriculture Highlights Newsletter, Feb 2014, pp 4–5. http://cesantabarbara.ucanr.edu/newsletters/Central_Coast_Agriculture_Highlights50500.pdf

Dara SSR, Dara SS, Sahoo A, Bellam H, Dara SK (2014c) Can entomopathogenic fungus, *Beauveria bassiana* be used for pest management when fungicides are used for disease management? University of California Cooperative Extension Strawberries and Vegetables e-Newsletter. http://ucanr.edu/blogs/blogcore/postdetail.cfm?postnum=15671

Delalibera I Jr, Sosa-Gomez DR, de Moraes GJ, de Alencar JA, Farias-Araújo W (1992) Infection of *Mononychellus tanajoa* (Acari: Tetranychidae) by the fungus *Neozygites* sp. in Northeastern Brazil. Fla Entomol 75:145–147

Delalibera I Jr, Hajek AE, Humber RA (2003) Use of cell culture media for cultivation of the mite pathogenic fungi *Neozygites tanajoae* and *Neozygites floridana*. J Invertebr Pathol 84:119–127

Delalibera I Jr, Hajek AE, Humber RA (2004) *Neozygites tanajoae* sp. nov., a pathogen of the cassava green mite. Mycologia 96:1002–1009

Delalibera I Jr, Demetrio CGB, Manly BFJ, Hajek A (2006) Effect of relative humidity and origin of isolates of *Neozygites tanajoae* (Zygomycetes: Entomophthorales) on production of conidia from cassava green mite, *Mononychellus tanajoa* (Acari: Tetranychidae) cadavers. Biol Control 39:489–496

Delgado FX, Britton JH, Onsager JA, Swearingen W (1999) Field assessment of *Beauveria bassiana* (Balsamo) Vuillemin and potential synergism with diflubenzuron for control of Savanna grasshopper complex (Orthoptera) in Mali. J Invertebr Pathol 73:34–39

Dresner E (1949) Culture and use of entomogenous fungi for the control of insects. Contrib Boyce Thompson Inst 15:319–335

Fargues J, Goettel MS, Smits N, Ouedraogo A, Rougier M (1997) Effect of temperature on vegetative growth of *Beauveria bassiana* isolates from different origins. Mycologia 89:383–392

Feng M-G, Poprawski TJ, Nowierski RM, Zeng Z (1999) Infectivity of *Pandora neoaphidis* (Zygomycetes: Entomophthorales) to *Acyrthosiphon pisum* (Homoptera: Aphididae) in response to varying temperature and photoperiod regimes. J Appl Entomol 123:29–35

Filotas MJ, Hajek AE (2007) Variability in thermal responses among *Furia gastropachae* isolates from different geographic origins. J Invertebr Pathol 96:109–117

Furtado IP, De Moraes GJ, Keller S (1996) Infection of *Euseius citrifolius* (Acari: Phytoseiidae) by an entomophthoralean fungus in Brazil. Rev Ecossistema 21:85–86

Goettel MS, Hajek AE (2000) Evaluation of non-target effects of pathogens used for management of arthropods. In: Wajnberg E, Scott JK, Quimby PC (eds) Evaluating indirect ecological effects of biological control. CABI Publishing, Wallingford, pp 81–95

Guo YL, Zuo G, Zhao JH, Wang NY, Jiang JW (1993) A laboratory test of thuringiensin to *Tetranychus urticae* (Acari: Tetranychidae) and *Phytoseiulus persimilis* (Acari: Phytoseiidae). Chin J Biol Control 9:151–155

Hajek A (2004) Natural enemies; an introduction to biological control. Cambridge University Press, Cambridge

Hajek AE, St. Leger R (1994) Interactions between fungal pathogens and insect hosts. Annu Rev Entomol 39:293–322

Hassan AEM, Charnley AK (1989) Ultrastructural study of the penetration by *Metarhizium anisopliae* through Dimilin-affected cuticle of *Manduca sexta*. J Invertebr Pathol 54:117–124

Hess RT, Hoy MA (1982) Microorganisms associated with the spider mite predator *Metaseiulus* (*Typhlodromus*) *occidentalis*: electron microscope observations. J Invertebr Pathol 40:98–106

Hirose E, Neves PMOJ, Zequi JAC, Martins LH, Peralta CH, Moino A Jr (2001) Effect of biofertilizers and neem oil on the entomopathogenic fungi *Beauveria bassiana* (Bals.) Vuill. and *Metarhizium anisopliae* (Metsch.) Sorok. Braz Arch Biol Technol 44:419–423

Hountondji FCC, Yaninek JS, de Moraes GJ, Oduor GI (2002a) Host specificity of the cassava green mite pathogen *Neozygites floridana*. BioControl 47:61–66

Hountondji FCC, Lomer CJ, Hanna R, Cherry AJ, Dara SK (2002b) Field evaluation of Brazilian isolates of *Neozygites floridana* (Entomophthorales: Neozygitaceae) for the microbial control of cassava green mite in Benin, West Africa. Biocontrol Sci Tech 12:361–370

Hountondji FCC, Sabelis M, Hanna R, Janssen A (2005) Herbivore-induced plant volatiles trigger spoprulation in entomopathogenic fungi: the case of *Neozygites tanajaoe* infecting the cassava green mite. J Chem Ecol 31:1003–1021

Hountondji FCC, Hanna R, Sabelis MW (2006) Does methyl salicylate, a component of herbivore-induced plant odour, promote sporulation of the mite-pathogenic fungus *Neozygites tanajoae*? Exp Appl Acarol 39:63–74

Hountondji FCC, Hanna R, Cherry AJ, Sabelis MW, Agboton B, Korie S (2007) Scaling-up tests on virulence of the cassava green mite fungal pathogen *Neozygites tanajoae* (Entomophthorales: Neozygitaceae) under controlled conditions: first observations at the population-level. Exp Appl Acarol 41:153–168

Hountondji FCC, De Moraes J, Al-Zawamri H (2010) Mites (Acari) on coconut, date palm and associated plants in Oman. Syst Appl Acarol 15:228–234

Hoy MA (1998) Myths, modes and mitigations of resistance to pesticides. Philos Trans R Soc Lond B 353:1787–1795

Humber RA, De Moraes GJ, Dos Santos JM (1981) Natural infection of *Tetranychus evansi* (Acarina: Tetranychidae) by a *Triplosporium* sp. (Zygomycetes: Entomophthorales) in northeastern Brazil. Entomophaga 26:421–425

Inglis GD, Goettel MS, Butt TM, Strasser H (2001) Use of hyphomycetes fungi for managing insect pests. In: Butt TM, Jackson C, Magan N (eds) Fungi as biological control agents. CAB International, Wallingford, pp 23–27

Irigaray FJS, Marco-Mancebón V, Pérez-Moreno I (2003) The entomopathogenic fungus *Beauveria bassiana* and its compatibility with triflumuron: effects on the twospotted spider mite *Tetranychus urticae*. Biol Control 26:168–173

Islam MT, Olleka A, Shunxiang R (2010) Influence of neem on susceptibility of *Beauveria bassiana* and investigation of their combined efficacy against sweetpotato whitefly, *Bemisia tabaci* on eggplant. Pestic Biochem Physiol 98:45–49

Jacobson RJ, Chandler D, Fenlon J, Russel KM (2001) Compatibility of *Beuveria bassiana* (Balsamo) Vuillemin with *Amblyseius cucumeris* Oudemans (Acarina: Phytoseiidae) to control *Frankliniella occidentalis* Pergande (Thysanoptera: Thripidae) on cucumber plants. Biocontrol Sci Technol 11:391–400

James DG, Price TS (2002) Fecundity in twospotted spider mite (Acari: Tetranychidae) is increased by direct and systemic exposure to imidacloprid. J Econ Entomol 95:729–732

Johanowicz DL, Hoy MA (1998) *Wolbachia* in a predator–prey system: 16S ribosomal DNA analysis of two phytoseiids (Acari: Phytoseiidae) and their prey (Acari: Tetranychidae). Ann Entomol Soc Am 89:435–441

Kaakeh W, Reid BL, Bohnert TJ, Bennett GW (1997) Toxicity of imidacloprid in the German cockroach (Dictyoptera: Blattellidae), and the synergism between imidacloprid and *Metarhizium anisopliae* (Imperfect fungi: Hyphomycetes). J Econ Entomol 90:473–482

Kaaya GP, Hassan S (2001) Entomogenous fungi as promising biopesticides for tick control. Exp Appl Acarol 24:913–926

Kassa A, Brownbridge M, Parker BL, Skinner M, Gouli V, Gouli S, Guo M, Lee F, Hata T (2008) Whey for mass production of *Beauveria bassiana* and *Metarhizium anisopliae*. Mycol Res 112:583–591

Keller S (1997) The genus *Neozygites* (Zygomycetes: Entomophthorales) with special reference to species found in tropical regions. Sydowia 49:118–146

Kirby C, Safford III, Allan SA (2010) Field applications of entomopathogenic fungi *Beauveria bassiana* and *Metarhizium anisopliae* F52 (Hypocreales: Clavicipitaceae) for the control of *Ixodes scapularis* (Acari: Ixodidae). J Med Entomol 47:1107–1115

Kumar SP, Singh L (2008) Enabling mycelial application of *Hirsutella thompsonii* for managing the coconut mite. Exp Appl Acarol 46:169–182

Lacey LA, Frutos R, Kaya HK, Vail P (2001) Insect pathogens as biological control agents: do they have a future? Biol Control 21:230–248

Leite LG, Smith L, Moraes GJ, Roberts DW (2000) *In vitro* production of hyphal bodies of the mite pathogenic fungus *Neozygites floridana*. Mycologia 92:201–207

Lighthart B, Sewall D, Thomas DR (1988) Effect of several stress factors on the susceptibility of the predatory mite, *Metaseiulus occidentalis* (Acari: Phytoseiidae), to the weak bacterial pathogen *Serratia marcescens*. J Invertebr Pathol 52:33–42

Maniania NK, Bugeme DM, Wekesa VW, Delalibera I Jr, Knapp M (2008) Role of entomopathogenic fungi in the control of *Tetranychus evansi* and *Tetranychus urticae* (Acari: Tetranychidae), pests of horticultural crops. Exp Appl Acarol 46:259–274

McCoy CW, Kanavel RF (1969) Isolation of *Hirsutella thompsonii* from the citrus rust mite, *Phyllocoptruta oleivora* and its cultivation on various synthetic media. J Invertebr Pathol 14:386–390

McCoy CW, Omoto C, Mazet I, Vey A (1992) Biological activity of crude filtrates and hirsutellin A to mosquitoes and phytophagous arthropods. Abstract 120 XXV annual meeting. Society of Invertebrate Pathology, Heidelberg

McMurtry JA, Huffaker CB, Van de Vrie M (1970) Ecology of tetranychid mites and their natural enemies. A review on tetranychid enemies; their biological characters and the impact of spray practices. Hilgardia 40:331–370

Messing RH, Croft BA (1996) Pesticide resistance in Eriophyoid mites, their competitors and predators. In: Lindquist EE, Sabelis MW, Bruin J (eds) Eriophyoid mites, their biology, natural enemies and control. Elsevier, Amsterdam, pp 689–694

Mietkiewski RT, Balazy S, Van Deer Geest LPS (1993) Observations on a mycosis of spider mites (Acari: Tetranychidae) caused by *Neozygites floridana* in Poland. J Invertebr Pathol 61:317–319

Mikunthan G, Manjunatha M (2006a) *Fusarium semitectum*, a potential mycopathogen against thrips and mites in chilli, *Capsicum annuum*. Commun Agric Appl Biol Sci 71:449–463

Mikunthan G, Manjunatha M (2006b) Pathogenicity of *Fusarium semitectum* against crop pests and its biosafety to non-target organisms. Commun Agric Appl Biol Sci 71:465–473

Milner RJ (1985) *Neozygites acaridis* (Petch) comb. nov. An entomophthoralean pathogen of the mite, *Macrocheles peregrinus*, in Australia. Trans Br Mycol Soc 85:641–647

Mwangi EN, Kaaya GP, Essuman S (1995) Experimental infections of the tick *Riphicephalus appenticulatus* with pathogenic fungi, *Beauveria bassiana* and *Metarhizium anisopliae*, and natural infections of some ticks with bacteria and fungi. J Afr Zool 109:151–160

Nakamura Y, Kawai S, Yukuhiro F, Ito S, Gotoh T, Kisimoto R, Yanase T, Matsumoto Y, Kageyama D, Noda H (2009) Prevalence of *Cardinium* bacteria in planthoppers and spider mites and taxonomic revision of "*Candidatus* Cardinium hertigii" based on detection of a new *Cardinium* group from biting midges. Appl Environ Microbiol 75:6757–6763

Oduor GI, Moraes GJ, Yaninek JS, Van Der Geest LPS (1995) Effect of temperature, humidity and photoperiod on mortality of *Mononychellus tanajoa* (Acari: Tetranychidae) infected by *Neozygites* cf. *floridana* (Zygomycetes: Entomophthorales). Exp Appl Acarol 19:571–579

Oduor GI, Yaninek JS, Van Der Geest LPS, De Moraes GJ (1996a) Germination and viability of capilliconidia of *Neozygites floridana* (Zygomycetes: Entomophthorales) under constant temperature, humidity and light conditions. J Invertebr Pathol 67:267–278

Oduor GI, De Moraes GJ, Van der Geest LPS, Yaninek JS (1996b) Production and germination of primary conidia of *Neozygites floridana* (Zygomycetes: Entmophthorales) under constant temperatures, relative humidities and photoperiods. J Invertebr Pathol 68:213–222

Olson DL, Oetting RD (1999) Compatibility of insect growth regulators and *Beauveria bassiana* (Balsamo) Vuillemin in controlling green peach aphids (Homoptera: Aphididae) on greenhouse chrysanthemums. J Entomol Sci 34:286–294

Omoto C, McCoy CW (1998) Toxicity of purified fungal toxin hirsutellin A to the citrus rust mite, *Phyllocoptruta oleivora* (Ash.). J Invertebr Pathol 72:319–322

Papierok B, Rafanomezantsoa-Randriambololona BN, Ziat N (1993) Nouvelles donne'es sur l'ecologie et le comportement entomopathoge'ne experimental de l'entomophthorale *Conidiobolus coronatus* (Zygomycetes). Entomophaga 38:299–312

Peña JE, Mannion CM, Howard FW, Hoy MA (2006) *Raoiella indica* (Prostigmata: Tenuipalpidae): the red palm mite: a potential invasive pest of palms and bananas and other tropical crops in Florida. University of Florida IFAS Extension, ENY-837, 414. http://edis.ifas.ufl.edu/IN681

Penman DR, Chapman RB (1988) Pesticide induced mite outbreaks: pyrethroids and spider mites. Exp Appl Acarol 4:265–276

Poinar G Jr, Poinar R (1998) Parasites and pathogens of mites. Annu Rev Entomol 43:449–469

Polar P, Kairo MTK, Moore D, Pegram R, John S, Peterkin D (2005) Assessment of fungal isolates for the development of a myco-acaricide for cattle tick control. Vector Borne Zoonotic Dis 5:276–284

Quintela ED, McCoy CW (1997) Pathogenicity enhancement of *Metarhizium anisopliae* and *Beauveria bassiana* to first instars of *Diaprepes abbreviatus* (Coleoptera: Curculionidae) with sublethal doses of imidacloprid. Environ Entomol 26:1173–1182

Ramaseshiah G (1971) Occurrence of an Entomophthora on tetranychid mites in India. J Invertebr Pathol 24:218–223

Raupp MJ, Webb R, Szczepaniec A, Booth D, Ahern R (2004) Incidence, abundance, and severity of mites on hemlocks following applications of imidacloprid. J Arboric 30:108–113

Roberts DW, Campbell AS (1977) Stability of entomopathogenic fungi. Misc Publ Entomol Soc Am 10:19–76

Ros, VID, Fleming VM, Feil EJ, Breeuwer JAJ (2012) Diversity and recombination in *Wolbachia* and *Cardinium* from *Bryobia* spider mites. BMC Microbiol. http://www.biomedcentral.com/1471-2180/12/S1/S13

Royalty RN, Hall FR, Taylor RAJ (1990) Effects of *thuringiensin* on *Tetranychus urticae* (Acari: Tetranychidae) mortality, fecundity and feeding. J Econ Entomol 83:792–798

Saba F (1971) Population dynamics of tetranychids in subtropical Florida. In: Proceeding of the 3rd international congress of Acarology, Prague. Junk, The Hague, pp 237–240

Samson RA, Rombach MC (1985) Biology of the fungi *Verticilium* and *Aschersonia*. In: Hussey NW, Scopes N (eds) Biological pest control, the glasshouse experience. Cornell University Press, Ithaca, pp 34–42

Schütte C, Dicke M (2008) Verified and potential pathogens of predatory mites (Acari: Phytoseiidae). Exp Appl Acarol 46(1–4):307–328

Shah PA, Pell JK (2003) Entomopathogenic fungi as biological control agents. Appl Microbiol Biotechnol 61:413–423

Shi W-B, Feng M-G (2006) Field efficacy of application of *Beauveria bassiana* formulation and low rate pyridaben for sustainable control of citrus red mite *Panonychus citri* (Acari: Tetranychidae) in orchards. Biol Control 39:210–217

Shi W-B, Jian Y, Feng M-G (2005) Compatibility of ten acaricides with *Beauveria bassiana* and enhancement of fungal infection to *Tetranychus cinnabarinus* (Acari: Tetranychidae) eggs by sublethal application rates of pyridaben. Appl Entomol Zool 40:659–666

Smitley DR, Brooks W, Kennedy GG (1986) Environmental effects on production of primary and secondary conidia, infection, and pathogenesis of *Neozygites floridana*, a pathogen of the two-spotted spider mite, Tetranychus urticae. J Invertebr Pathol 47:325–332

Smitley DR, Kennedy GG, Brooks WM (2011) Role of the entomogenous fungus, *Neozygites floridana*, in population declines of the twospotted spider mite, *Tetranychus urticae*, on field corn. Entomol Exp Appl 41:255–264

Soper RS, Ward MG (1981) Production, formulation and application of fungi for insect control. In: Papavizas CG (ed) Biological control in crop production. Osmoun, Montclair, pp 161–180

Sosa Gomez DR (1987) Control microbiano de *Phyllocoptruta oleivora* (Ashm). y *Eriophyes sheldoni* Ewing mediante pulverizaciones de conídios de tres variedades de *Hirsutella thompsonii* Fisher. XI Congr Bras Entomol Res 1:167

Sosa Gomez DR, Moscardi F (1991) Microbial control and insect pathology in Argentina. Ciencia e Cultura 43(5):375–379

Stacey DA, Fellowes MDE (2002) Influence of temperature on pea aphid (*Acyrthosiphon pisum*) resistance to natural enemy attack. Bull Entomol Res 92:351–357

Steiner M (1993) Quality control requirements for pest biological control agents. Alberta Government Publication AECV93-R6, Alberta Environmental Centre, Vegrevile

Steinkraus DC, Boys GO (2005) Mass harvesting of the entomopathogenic fungus, *Neozygites fresenii*, from natural field epizootics in the cotton aphid, *Aphis gossypii*. J Invertebr Pathol 88:212–217

Šutáková G (1988) Electron microscopic study of developmental stages of *Rickettsiella phytoseiuli* in *Phytoseiulus persimilis* Athias-Henriot (Gamasoidea: Phytoseiidae) mites. Acta Virol 32:50–54

Tedders WL (1981) *In vitro* inhibition of the entomopathogenic fungi *Beauveria bassiana* and *Metarhizium anisopliae* by six fungicides used in pecan culture. Environ Entomol 10:346–349

Trinci APJ, Lane BS, Humphreys AM (1990) Optimization of cultural conditions for production and longevity of entomopathogenic fungi. Proceeding of the Vth international colloquium on invertebrate pathology and microbial control, Adelaide, 20–24 Aug 1990, pp 121–125

Van Der Geest LPS, Elliot SL, Breeuwer JAJ, Beerling EAM (2000) Diseases of mites. Exp Appl Acarol 24:497–560

Van Leeuwen T, Vontas J, Tsagkarakou A, Dermauw W, Tirry L (2010) Acaricide resistance mechanisms in the twospotted spider mite *Tetranychus urticae* and other important Acari: a review. Insect Biochem Mol Biol 40:563–572

Vey A, Quiot JM, Mazet I, McCoy CW (1993) Toxicity and pathology of crude broth filtrate produced by *Hirsutella thompsonii* var. *thompsonii* in shake culture. J Invertebr Pathol 61:131–137

Weiser J, Muma MH (1966) *Entomophthora floridana* n.s. (Phycomycetes: Entomophthoraceae) a parasite of the Texas citrus mite *Tetranychus banksi*. Fla Entomol 49:155–159

Wekesa VW, Delalibera I Jr (2008) Long-term preservation of *Neozygites tanajoae* (Entomophthorales: Neozygitaceae) in cadavers of *Mononychellus tanajoa* (Acari: Tetranychidae). Biocontrol Sci Technol 18:621–627

Wekesa VW, Maniania NK, Knapp M, Boga HI (2005) Pathogenicity of *Beauveria bassiana* and *Metarhizium anisopliae* to the tobacco spider mite *Tetranychus evansi*. Exp Appl Acarol 36:41–50

Wekesa VW, Knapp M, Maniania NK, Boga HI (2006) Effects of *Beauveria bassiana* and *Metarhizium anisopliae* on mortality, fecundity and egg fertility of *Tetranychus evansi*. J Appl Entomol 130:155–159

Wekesa VW, Moraes GJ, Knapp M, Delalibera I Jr (2007) Interactions of two natural enemies of *Tetranychus evansi*, the fungal pathogen *Neozygites floridana* (Zygomycetes: Entomophthorales) and the predatory mite, *Phytoseiulus longipes* (Acari: Phytoseiidae). Biol Control 41:408–414

Wekesa VW, Moraes GJ, Ortega EMM, Delalibera I Jr (2010) Effect of temperature on sporulation of *Neozygites floridana* isolates from different climates and their virulence against the tomato red spider mite, *Tetranychus evansi*. J Invertebr Pathol 103:36–42

Yaninek S, Hanna R (2003) Cassava green mite in Africa—a unique example of successful classical biological control of a mite pest on a continental scale. In: Neuenschwander P, Borgemeister C, Langewald J (eds) Biological control in IPM systems in Africa. CAB International, Wallingford, pp 61–75

Yaninek JS, Saizonou S, Onzo A, Zannou I, Gnanvossou D (1996) Seasonal and habitat variability in the fungal pathogens: *Neozygites* cf. *floridana* and *Hirsutella thompsonii*, associated with cassava mites in Benin. Biocontrol Sci Technol 6:23–33

Ye SD, Dun YH, Feng M-G (2005) Time and concentration dependent interactions of *Beauveria bassiana* with sublethal rates of imidacloprid against the aphid pests *Macrosiphoniella sanborni* and *Myzus persicae*. Ann Appl Biol 146:459–468

Yeo H, Pell JK, Alderson PG, Clark SJ, Pye BJ (2003) Laboratory evaluation of temperature effects on the germination and growth of entomopathogenic fungi and on their pathogenicity to two aphid species. Pest Manag Sci 59:156–165

MIX
Papier aus verantwortungsvollen Quellen
Paper from responsible sources
FSC® C105338

If you have any concerns about our products, you can contact us on
ProductSafety@springernature.com

In case Publisher is established outside the EU, the EU authorized representative is:
Springer Nature Customer Service Center GmbH
Europaplatz 3, 69115 Heidelberg, Germany

Printed by Libri Plureos GmbH
in Hamburg, Germany